International Federation of Automatic Control

DISTRIBUTED COMPUTER CONTROL SYSTEMS 1982

Other Titles in the IFAC Proceedings Series

AKASHI: Control Science and Technology for the Progress of Society, 7 Volumes
ATHERTON: Multivariable Technological Systems
BANKS & PRITCHARD: Control of Distributed Parameter Systems
CAMPBELL: Control Aspects of Prosthetics and Orthotics
Van CAUWENBERGHE: Instrumentation and Automation in the Paper, Rubber, Plastics and Polymerisation Industries
CICHOCKI & STRASZAK: Systems Analysis Applications to Complex Programs
CRONHJORT: Real Time Programming 1978
CUENOD: Computer Aided Design of Control Systems
De GIORGO & ROVEDA: Criteria for Selecting Appropriate Technologies under Different Cultural, Technical and Social Conditions
DUBUISSON: Information and Systems
ELLIS: Control Problems and Devices in Manufacturing Technology 1980
GHONAIMY: Systems Approach for Development (1977)
HAASE: Real Time Programming 1980
HAIMES & KINDLER: Water and Related Land Resource Systems
HARRISON: Distributed Computer Control Systems
HASEGAWA: Real Time Programming 1981
HASEGAWA & INOUE: Urban, Regional and National Planning — Environmental Aspects
HERBST: Automatic Control in Power Generation Distribution and Protection
ISERMANN: Identification and System Parameter Estimation
ISERMANN & KALTENECKER: Digital Computer Applications to Process Control
JANSSEN, PAU & STRASZAK: Dynamic Modelling and Control of National Economics
LAUBER: Safety of Computer Control Systems
LEONHARD: Control in Power Electronics and Electrical Drives
LESKIEWICZ & ZAREMBA: Pneumatic and Hydraulic Components and Instruments in Automatic Control
MAHALANABIS: Theory and Application of Digital Control
MILLER: Distributed Computer Control Systems 1981
MUNDAY: Automatic Control in Space
NAJIM & ABDEL-FATTAH: Systems Approach for Development 1980
NIEMI: A Link Between Science and Applications of Automatic Control
NOVAK: Software for Computer Control
O'SHEA & POLIS: Automation in Mining, Mineral and Metal Processing
OSHIMA: Information Control Problems in Manufacturing Technology (1977)
RAUCH: Control Applications of Nonlinear Programming
REMBOLD: Information Control Problems in Manufacturing Technology (1979)
RIJNSDORP: Case Studies in Automation related to Humanization of Work
SAWARAGI & AKASHI: Environmental Systems Planning, Design and Control
SINGH & TITLI: Control and Management of Integrated Industrial Complexes
SMEDEMA: Real Time Programming 1977
SUBRAMANYAM: Computer Applications in Large Scale Power Systems
TITLI & SINGH: Large Scale Systems: Theory and Applications
Van WOERKOM: Automatic Control in Space 1982

NOTICE TO READERS

Dear Reader

If your library is not already a standing/continuation order customer to this series, may we recommend that you place a standing/continuation order to receive immediately upon publication all new volumes. Should you find that these volumes no longer serve your needs, your order can be cancelled at any time without notice.

ROBERT MAXWELL
Publisher at Pergamon Press

DISTRIBUTED COMPUTER CONTROL SYSTEMS 1982

*Proceedings of the Fourth IFAC Workshop
Tallinn, U.S.S.R., 24-26 May 1982*

Edited by

R. W. GELLIE

*Commonwealth Scientific & Industrial Research Organization
Division of Manufacturing Technology
Fitzroy, Australia*

and

R.-R. TAVAST

*Academy of Sciences of Estonian S.S.R.
Institute of Cybernetics
Tallinn, U.S.S.R.*

Published for the

INTERNATIONAL FEDERATION OF AUTOMATIC CONTROL

by

PERGAMON PRESS

OXFORD · NEW YORK · TORONTO · SYDNEY · PARIS · FRANKFURT

U.K.	Pergamon Press Ltd., Headington Hill Hall, Oxford OX3 0BW, England
U.S.A.	Pergamon Press Inc., Maxwell House, Fairview Park, Elmsford, New York 10523, U.S.A.
CANADA	Pergamon Press Canada Ltd., Suite 104, 150 Consumers Road, Willowdale, Ontario M2J 1P9, Canada
AUSTRALIA	Pergamon Press (Aust.) Pty. Ltd., P.O. Box 544, Potts Point, N.S.W. 2011, Australia
FRANCE	Pergamon Press SARL, 24 rue des Ecoles, 75240 Paris, Cedex 05, France
FEDERAL REPUBLIC OF GERMANY	Pergamon Press GmbH, Hammerweg 6, D-6242 Kronberg-Taunus, Federal Republic of Germany

Copyright © 1983 IFAC

All Rights Reserved. No part of this publication may be reproduced, stored in a retrieval system or transmitted in any form or by any means: electronic, electrostatic, magnetic tape, mechanical, photocopying, recording or otherwise, without permission in writing from the copyright holders.

First edition 1983

Library of Congress Cataloging in Publication Data

IFAC Workshop on Distributed Computer Control
Systems (4th: 1982: Tallinn, Estonia)
Distributed computer control systems 1982.
(IFAC proceedings)
1. Automatic control—Data processing—Congresses.
2. Electronic data processing—Distributed processing—
Congresses. I. Gellie, R.W. II. Tavast, R.-R.
(Raul-R.) III. International Federation of Automatic
Control.
TJ212.2.I34 1982 629.8'95 83-2388

British Library Cataloguing in Publication Data

IFAC Workshop DCCS: (*4th 1982: Tallinn*)
Distributed computer control systems 1982. —
(IFAC Proceedings)
1. Automatic control—Data processing—
Congresses 2. Electronic data processing—
Distribution processing—Congresses
I. Title. II. International Federation of
Automatic Control. III. Gellie, R.W.
IV. Tavast, R.R. V. Series
629.8'95 TJ212
ISBN 0-08-028675-5

In order to make this volume available as economically and as rapidly as possible the authors' typescripts have been reproduced in their original forms. This method unfortunately has its typographical limitations but it is hoped that they in no way distract the reader.

Printed in Great Britain by A. Wheaton & Co. Ltd., Exeter

FOURTH IFAC WORKSHOP ON DISTRIBUTED COMPUTER CONTROL SYSTEMS

Organized by
U.S.S.R. National Committee on Automatic Control
Institute of Cybernetics, Academy of Sciences of the Estonian S.S.R.
Tallinn Technical University

Sponsored by
IFAC Technical Committee on Computers

Co-sponsored by
IFAC Technical Committee on Education

Workshop Chairman
Academician B. Tamm

International Program Committee
R.-R. Tavast, U.S.S.R. (Chairman)
V. G. Bochmann, Canada
R. W. Gellie, Australia
J. Gertler, Hungary
T. J. Harrison, U.S.A.
Th. Lalive d'Epinary, Switzerland
L. Motus, U.S.S.R.
K. D. Müller, F.R.G.
S. Narita, Japan
E. A. Trachtengerts, U.S.S.R.

PREFACE

IFAC Workshops have proven to be a popular and effective forum for presentation and in-depth discussion of ideas by competent experts in emerging areas of automatic control. Indeed, the Workshops on Real Time Programming, which date back to 1971 and have been held annually since that time, have proved so successful that IFAC has published a special booklet "Guidelines for Organizers of IFAC Sponsored Workshops" to assist and encourage more events of this type.

During his term as Chairman of the IFAC Computers Committee, Mr. Charles Doolittle established the Workshops on Distributed Computer Control Systems in recognition of the great interest and activity in this area. The first event in this series was held in Tampa, Florida (1979) with subsequent events in Ste. Adele, Quebec (1980), Beijing, China (1981) and now Tallinn, Estonia in 1982. Every event in this series has been very successful in terms of quality of papers, numbers of participants, and the degree to which the attendees have contributed to lively discusson and exchange of ideas.

As can be noted from the list of participants the Tallinn Workshop was attended by 75 experts from 14 countries. The quality of the papers presented and the discussions which followed may be judged by the reader.

In these proceedings the papers are published in the order in which they were presented. The discussions which took place were recorded and subsequently transcribed. Some editing was done to improve clarity and avoid repetition but it is hoped that the text retains a sense of spontaneity. The Workshop program concluded with a panel session. The initial presentations by the panel members and the ensuing discussion are also included.

The two papers by H.G. Mendelbaum, G. de Sablet., and Wu Zhimei, Zhang Wenkuan, Zhang Yingzhong, Cheng Yunyi were not presented at the Workshop because the authors were unable to attend. However we felt that their inclusion would add to the value of this volume.

I wish to gratefully acknowledge the part played by my co-editor, Raul Tavast, who performed the difficult task of transcribing and editing the discussions.

R.W. Gellie.
November 23, 1982.
Fitzroy, Victoria, Australia.

LIST OF PARTICIPANTS

IFAC 4th WORKSHOP ON DISTRIBUTED COMPUTER CONTROL SYSTEMS DCCS-82

Olympic Yachting Centre, Tallinn, Estonian S.S.R., U.S.S.R., 24-26 May 1982

O. Aarna
Tallinn Technical University
Ehitajate tee 5
Tallinn   200026   USSR

H. Aben
Institute of Cybernetics
Akadeemia tee 21
Tallinn   200026   USSR

A. Abreu
Central Institute of Digital Research
198 No. 1713 Cubanacan
Havana   CUBA

Z. Apostolova
State Committee of Science and Technology,
ul. Slavjanska 8
Sofia   BULGARIA

A. Ariste
Institute of Cybernetics
Akadeemia tee 21
Tallinn   200026   USSR

S.I. Baranov
Institute for Analytical Instrumentation
Sci. & Techn. Corp.,
USSR Academy of Sciences
Prospect Ogorodnikova 26
198103   Leningrad   USSR

B. Becski
Technical University of Budapest
Visepradi u.30
1132   Budapest   HUNGARY

G. Bingzhen
Research Institute of Electronical Technical Applications
Beijing   CHINA

J. Davidson
Tecsult International Limité
85, rue Ste-Catherine Ouest,
Montreal, Quebec H2X 3T4   CANADA

C. Dimitrov
Institute for Scientific Research in Telecommunications
Haidushka Poljana str. 8
Sofia   1612   BULGARIA

D.G. Dimmler
Senior Scientist
Brookhaven National Laboratory
Upton.
New York   11973   USA

A. Divitakov
ZNIKA
str. I. Vishovgradsky 46,
Room 608   Sofia   BULGARIA

V.N. Dragunova
INEUM
Moscow V-334   117812   GSP   USSR

T.O. Dzjubek
Institute of Cybernetics
Academy of Sciences
Kiev   USSR

J. Ehrlich
Technische Hochschule Leipzig
Karl-Liebknecht-Strasse 132
DDR-703   LEIPZIG

W. Engmann
Technische Hochschule Ilmenau
DDR-63   Ulmenau-Ehrenbert

W. Enkelmann
Zentralinstitut für Hernforschung, Rossendorf
8051   Dresden, P.O. 19
DDR

G. Evstratov
Politechnical Institute Kharkov
Frunze str. 21.
310002   Kharkov   USSR

A. Gościński
Institute of Computer Science
Stanislaw Staszik
University of Mining & Metallurgy
al. Mickiewicza 30
30-059   Krakow   POLAND

List of Participants

*,** T.J. Harrison
    IBM Corporation
    P.O. Box 1328 Boca Raton
    Florida 33432 USA

    H. Hetzheim
    Academy of Sciences
    GDR

    I.G. Ilzinya
    Institute of Electronics &
    Computer Systems
    Akademijas 14
    Riga 6   226006   USSR

*   A. Inamoto
    Computer Systems Works
    325 Kamimachiya
    Kamakura City
    Kanagawa Prefecture
    JAPAN 247

    U. Jaaksoo
    Institute of Cybernetics
    Akadeemia tee 21
    Tallinn   200026   USSR

    H. Jinwei
    System Development Division of China
    Computer Technical Service Corporation
    Beijing   CHINA

    K.A. Joudu
    Moscow Institute of Avionics
    Volokolamskoe 4
    Moscow GSP   125871   USSR

** A. Keevallik
    Tallinn Technical University
    Ehitajate tee 5
    Tallinn   200026   USSR

    B. Kovacs
    Computer & Automation Institute
    Hungarian Academy of Sciences
    Kende u. 13/17
    1111   Budapest   HUNGARY

    K. Kralev
    State Committee of Science &
    Technology
    ul. Slavjanska 8
    Sofia   BULGARIA

    V. Krüger
    Academy of Sciences
    GDR

*,** K. Kääramess
    Institute of Cybernetics
    Akadeemia tee 21
    Tallinn   200026   USSR

*   J. Lan
    Department of Computer Engineering &
    Science
    Qinghua University
    Beijing   CHINA

    J. Lukács
    Central Research Institute for Physics
    P.O. Box 49,
    H-1525 Budapest   HUNGARY

    U. Luoto
    Ekono Oy
    P.O. Box 27
    SF-00131   Helsinki   FINLAND

    G.G. Mask
    Central Bureau of Statistics
    Endla 15
    Tallinn   200105   USSR

    M. Mantseva
    State Committee of Science and
    Technology
    ul. Slavjanska 8
    Sofia   BULGARIA

    M. Martin
    Zentralinstitut für Kybernetik
    und Informationsprozesse
    Dresden   8027 Dresden
    Maeskelstr. 20   DDR

    M. Maxwell
    Manager of Control Systems
    Colgata-Palmolive Co
    105 Hudson St.
    Jersey City
    New Jersey   USA

    I. Meiszterics
    Technical University of Budapest
    Vérhalom u.29
    1025   Budapest   HUNGARY

*,** L. Motus
    Institute of Cybernetics
    Akadeemia tee 21
    Tallinn   200026   USSR

*,** S. Narita
    Electrical Engineering Dept.
    Waseda University
    3-4-1 Okubo, Shinjuku-ku
    Tokyo 160   JAPAN

    L.F. Natiello
    Exxon Research & Engineering Company
    P.O. Box 101
    Florham Park
    New Jersey   07932   USA

    D. Nedo
    Central Institute of Cybernetics &
    Information Processing
    Kurstr. 33
    1086   Berlin   GDR

*   M. Ollus
    Technical Research Centre
    Electrical Engineering Lab.
    VTT/SAH
    SF-02150 Espoo 15   FINLAND

Z. Pengzu
Lab. North-China Institute of
Computing Technology
Beijing   CHINA

D. Penkin
Soviet-Bulgar Institute
Stambulinskaja 62-64
Sofia   BULGARIA

V.L. Pertchuk
Institute of Automation and
Control Processes
Suhhanova 5A
Vladivostok   690600   USSR

B. Petkov
Ministry of Communication
Computing Centre
Sofia   BULGARIA

K. Petrov
State Committee of Science &
Technology
ul. Slavjanska 8
Sofia   BULGARIA

J. Pino
Central Institute of Digital Research
198 No. 1713 Cubanacan
Havana   CUBA

W. Qinsheng
Computer Industry of the Ministry of
Electronical Industry
Beijing   CHINA

* M.G. Rodd
University of the Witwatersrand
1 Jan Smuts Avenue
Johannesburg   2000   SOUTH AFRICA

L. Rozsa
Computer & Automation Institute
Hungarian Academy of Science
Kendue u. 13
Budapest   1111   HUNGARY

* S.M. Rotanov
Institute of Electronics & Computer
Systems
Akademijas 14
Riga   6   226006   USSR

I.M. Shenbrot
Central Institute of Complex
Automation
Olhovskaya 25
Moscow B-66   107816   USSR

A.V. Shileiko
Moscow Institute of Railway Engineers
F. Engels 36-25
Moscow B-5   107005   USSR

B.T. Shreiber
Central Institute of Complex
Automation
Olhovskaya 25
Moscow B-66   107816   USSR

V. Slivinskas
Institute of Mathematics & Cybernetics
K. Pozelos 54
Vilnius 232600   USSR

A.A. Sternberg
INEUM
Moscow V-334   117812   GSP   USSR

* G.J. Suski
Lawrence Livermore National Laboratory
P.O. Box 808,
Livermore
California   94550   USA

* K. Takezawa
Toshiba Fuchu Works - Toshiba Corp.
1, Toshiba-cho, Fuchu
Tokyo   183   TOKYO

** B. Tamm
Tallinn Technical University
Ehitajate tee 5
Tallinn   200026   USSR

H. Tani
Institute of Cybernetics
Akadeemia tee 21
Tallinn   200026   USSR

** R. Tavast
Institute of Cybernetics
Akadeemia tee 21
Tallinn   200026   USSR

I. Tepandi
Tallinn Technical University
Ehitajate tee 5
Tallinn   200026   USSR

P. Tianchian
Research Institute of Metallurgical
Automation
Beijing   CHINA

*,** E.A. Trakhtengerts
Institute of Control Problems
Profsoyuznaya 81
Moscow V-279   117806   USSR

I.S. Ukolov
Scientific Council of Cybernetics USSR
Academy of Sciences
40, Vavilov str.
117333   Moscow   USSR

V.I. Utkin
Institute of Control Problems
Profsoyuznaya 81
Moscow V-279   117806   USSR

T. Vamos
Computer & Automation Institute
Hungarian of Academy of Sciencies
P.O. Box 63
Budapest 112   HUNGARY

L. Vohandu
Tallinn Technical University
Ehitajate tee 5
Tallinn   200026   USSR

B. Werner
VEB PCK Schwedt
DDR-1330   Schwedt   DDR

*   G.G. Wood
Foxboro Yoxall
Redhill
Surrey RH1 2HL   UK

**   A. Work
Institute of Cybernetics
Akadeemia tee 21
Tallinn   200026   USSR

K. Zieliński
Academy of Mining & Metallurgy
Institute of Computer Science
al. Mickiewicza 30
30-059   Krakow   POLAND

---

*   Author

**   Program/Organizing Committee

CONTENTS

Welcome address - T. Vamos, President of IFAC — xvii

Welcome address - B. Tamm, Workshop Chairman — xix

Session 1. R.-R. Tavast, Chairman

Software Design Specifics for Multiprocessor Control Systems
E.A. Trakhtengerts, Yu.M. Shuraits — 1

Discussion — 11

Session 2. L.F. Natiello, Chairman

IEEE Project 802: Local Area Network Standard
A March 1982 status report
T.J. Harrison — 13

Discussion — 25

Standardization Work for Communication Among Distributed
Computer Control Systems
G.G. Wood — 27

Discussion — 31

Session 3. L. Rosza, Chairman

Development and Quantitative Evaluation of Distributed
Sensor Based Management System
T. Muto, C. Inamochi, A. Inamoto, S. Kato — 33

Discussion — 43

Nova Control System - Goals, Architecture and System Design
G.J. Suski, J.M. Duffy, D.G. Britton, F.W. Halloway,
I.R. Krammen, R.G. Ozarski, J.R. Sevevyn, P.J. Van Arsdall — 45

Discussion — 57

Session 4. A. Work, Chairman

A Memory Intensive Functional Architecture for Distributed
Computer Control Systems
D.G. Dimmler — 59

Discussion — 73

Implementation and Performance Evaluation of a Distributed
Splitted-Bus Multiple Computer System
Lan Jin, Wei-min Zheng, Ding-xing Wang, Mei-ming Sheng — 75

Discussion — 81

Session 5.   M. Maxwell, Chairman

On the Design of Hierarchical Process Control Computer Systems
J. Davidson, J.L. Houle …83

Discussion …91

A Model Based Design of Distributed Control Systems Software
L. Motus, K. Kääramees …93

Discussion …101

Session 6.   M.G. Rodd, Chairman

Parallel Processing for Real-Time Control and Simulation of DCCS
H. Kasahara, S. Narita …103

Discussion …115

On the Distribution of Tasks in Automation Systems
M. Ollus, B. Wahlström …117

Discussion …127

Session 7.   Lan Jin, Chairman

Software Test Facilities with Distributed Architecture
K. Takezawa …129

Discussion …139

A Communication System for Use in an Industrial Distributed Control System
M.G. Rodd, N.J. Peberdy, H.F. Weehuizen, D.P.A. Bean …141

Discussion …151

Session 8.   G.G. Wood, Chairman

Development and Analysis of Protocols for Distributed Computer Networks
E.A. Yakubaitis, Ya.A. Kikuts, S.V. Rotanov …153

Discussion …165

Ring Computer Networks for Real Time Process Control
A. Gościński, T. Walasek, K. Zieliński …167

Discussion …179

Session 9.   T.J. Harrison, Chairman

Panel Discussion

Bottlenecks in the Design and Implementation of DCCS and the Ways to Fight Them.
T.J. Harrison …181

Introductory Statements:
  S. Narita …183
  M. Maxwell …189
  G.G. Wood …191
  R.-R. Tavast …193

Discussion …195

Additional Papers

EMUNET-Emulator of Network Systems
H.G. Mendelbaum, G. de Sablet   201

The Communication Software on a Node of the RDC Network
Wu Zhimei, Zhang Wenkuan, Zhang Yingzhong, Cheng Yunyi   211

Author Index   217

WELCOME ADDRESS

T. Vamos

IFAC President

On behalf of IFAC Council I am happy to address this meeting not only because it is my duty and privilege but also because I can express my satisfaction with IFAC's directions of interest. To tell the truth, we would like to organize not only fashion directed meetings, but such ones which go somehow before the general fashion, which predict from the well-estimated trends, stimulate exchange experiences and after that summarise in order to provide something relevant to our control community.

If this symposium were the first of the distributed control series I would say did we sleep before? 1979 was just the right time to begin and, coming here, it is appropriate to speak about experiences about unsolved problems in more depth.

Let me tell you some personal views on the topic which I consider to be even more revolutionary than is generally advertized. Distributed control is much more significant in control philosophy than decentralized control was after the hierarchical centralized ones.

Decentralized control is mostly a system which is really de-centralized: that means the system as a whole, handled as a black box, is a resolved and reorganized centralized one, not losing the strong central control but only delegating some tasks and some information to the lower subjugated levels.

Distributed control is a much more liberal solution indicating a highly cooperative philosophy: a coalition of components arranged not hierarchically but in a very democratic coordinative way. Any centralization loses its rationality as we consider larger and larger systems, systems which have no rigorous physical limitations but can be augmented or dissolved adaptively. A very big system is even theoretically uncontrollable in the old sense, due to combinational explosion and due to randomness of parameters. It's nevertheless controllable as a coalition of autonomous system partners, which cooperate through a well-defined network of information and flow. Flow is understood as anything different from information, e.g. flow of energy, fluids, goods, people, etc. High information transmission technology, high level exchange protocols, reliable well-organized system components are basic requirements and that's the reason why we had to wait until now for this revolution. Several earlier systems like international telephone exchange and some power systems started to realize these ideas long before they were formulated. It's my firm belief that for the future the idea of distributed cooperative systems will be the dominating control principle, for every large-scale system which intends to survive and develop.

IFAC itself is a cooperative system and it is appropriate to express our gratitude to those who cooperate in IFAC. Distribution of information, call for papers and people are organized by our forty national organizations by voluntary, non-profit groups. Participation is free, without any kind of discrimination, independent of the transients of government policies. Our goals and aims are more global and more stable.

Thanks to the International Program Committee of this Workshop that shaped the program which is of interest not only for those who could come here at this time, but for all people who look at IFAC's events for information and stimulation.

Thanks to the local organizers headed by Vice-President of IFAC, Chairman of our Technical Board, Academician Boris Tamm.

Soovin koige paremat. Soovin edu.
Best wishes and big success.

WELCOME ADDRESS

B. G. Tamm

Workshop Chairman, Tallinn Technical University

On behalf of the National Organizing Committee and International Program Committee I am glad to extend a sincere welcome to everybody participating in the 4th Workshop on Distributed Computer Control Systems here in old Tallinn.

Just recently I received two telegrams, one from Dr. W. Gellie, Chairman of the sponsoring committee of this workshop, IFAC TC on Computers, the other from Professor P. Larsen, Chairman of the co-sponsoring committee, IFAC TC on Education both sending their best regards for the success of our workshop and asking me to transfer that to you.

Our workshop is already the fourth in the series of Distributed Computer Control after those held in Tampa (1979, USA) Ste. Adele (1980, Canada) and Beijing (1981, PRC), so we ought to have some kind of experience.

Nevertheless the topic of our interest is developing dramatically and I know the hard efforts of the members of the International Program Committee in selecting the best papers from among those submitted. I should like to thank everybody who sent contributions and congratulate the authors of the papers selected for the Final Technical Program.

As you know IFAC is a society of volunteers consisting of specialists in automatic control who are ready to undertake personal efforts besides their everyday jobs, in order to promote science and technology. In this respect I should like to thank every member of IFAC, especially the Chairman, Raul Tavast, Tom Harrison, who has had a hand in all four of the workshops, and L. Motus and E. Trachtengerts, members from the USSR NMO, for their outstanding contribution to this IFAC event.

Dear guests, I should also like to assure you that the volunteers from the Institute of Cybernetics, Academy of Sciences of the Estonian SSR and Tallinn Technical University have done their best to create a fruitful professional atmosphere as well as to ensure your joyful stay in Tallinn.

SOFTWARE DESIGN FOR MULTIPROCESSOR SYSTEMS COMPUTER CONTROL

E. A. Trakhtengerts and Yu. M. Shuraits

Institute of Control Sciences, Moscow, USSR

Abstract. The paper treats the specifics of the multiprocessor computer systems software design, namely the design of language means, translation and program parallelization means, operation systems and debug systems. Feasibility of parallel and serial program execution, the influence of the execution mode on the useful performance of a computer system and the reduction of execution time for a group of tasks or mean service time for a flow of queries are considered.

Keywords. Multiprocessor computer system; parallel computations; automatic parallelization; program branch; usefulness of parallelization; multiprocessor computer system performance.

INTRODUCTION

The advent of multiprocessor computer systems capable of performing parallel computations put forward new requirements to the software which realizes control algorithms. These requirements may be verbalized in a single phrase: parallel performance of the computer process.

The process of parallelization imposes specific requirements on
- language means of programming for multiprocessor systems which should make parallel computations possible and provide their timing;
- means of translation which should provide automatic parallelization of the object program;
- means of debug controlling asynchronous performance of certain tasks and localizing errors in running the program on certain processors;
- operation system distributing computational resources and providing its own parallel operation;
- design of algorithms taking full advantage of parallel computations;

The first part of the paper treats various ways to realize the first four requirements.

The second part discusses rational combination of the parallel and serial forms of program execution, the effect of program parallelization upon the useful capability of a multiprocessor system and some problems pertaining to the design of parallel algorithms.

PART I

1. Language means

The language means in multiprocessor systems are intended for the organization of serial - parallel computations. They differ from the "traditional" programming languages in that they incorporate some additional units providing parallel realization of program fragments and its timing.

These include
- the introduction of vector and matrix operations and means of masking the operations on elements of the vectors;
- the apparatus for the creation of sections of parallel program execution which are further referred to as branches, and
- the apparatus for branch synchronization.

The expressions on arrays (vectors, matrices) generally employ the same operations as scalar expressions. Usually subarrays of various kinds and operations on them are specified. A logical conditional statement permits only those operations with the elements of the arrays to be performed which correspond to the TRUE value of the logical expression for conditional statement. Thus masking of the operations with vectors is performed which is generally done by hardware.

Special statements for branch description and initialization are introduced for parallel branch operations. In any point of the program one or several branches may be initialized for parallel execution. Usually static and dynamic definitions of parallel branches are given. In the latter case the number of branches in a given point is dynamically obtained in the course of the program execution.

The body of a branch is specified statically i.e. in the process of translation and, generally, may not be formed dynamically. The begining and end of each branch are specified by special statements.

The synchronization of the computing process was attained through the introduction of variables or arrays of the "event" type and wait and event termination statements. The operands for these statements were the variables or arrays of the "event" type. To provide processing of the same data array by several branches statements of the "semaphore" type were introduced. To raise the level of synchronizing primitives the mechanisms of the conditional critical intervals [4], monitors [2,3,5], sentinels [6], control expressions [1] and rendezvous (in the Ada language) were created. One should note that the above rather complicated synchronization designs may be expressed through the semaphores as well, but the use of these designs increases software "reliability" and lowers the probability of errors in the program when complex interactions are described.

2. Translation means for parallelization

Program analysis and its parallelization in the course of the translation may be carried out as follows.

At the first step linear sections and simple cycles are isolated.

A linear section of the program is a part of it whose statements are executed in the natural order sequentially or in the order determined by unconditional branch instructions. A linear section is limited by the start and end statements.

A start statement of the linear section is the statement for which at least one of the following requirements is satisfied: either it has more than one direct predecessor or its direct predecessor has more than one direct follower.

Using these definitions one may easily construct the linear section search algorithm.

A simple cycle is a fragment of the program consisting of one or several cycle and cycle body statements not containing transfer-of-control beyond the cycle boundaries. These boundaries are found by formed indicators of cycle description used in the corresponding programming languages.

This terminates the process of the analysis of a program graph.

Parallelization inside linear sectors, design of ordered linear sequences (OLS) and parallelization of simple cycles may be executed in parallel.

2.1. Parallelization of linear sections

Parallelization inside a linear section is carried out by statements; inside the statements parallel execution of arithmetic expressions is possible.

The variables processed by the statements P_i of the linear section may be categorized in four groups:
1. Read-only denoted as W_i; 2. Write-only denoted as X_i; 3. Write-after-read denoted as Y_i; and 4. Read-after-write denoted as Z_i. When two statements P_1 and P_2 work in parallel memory cells or, which in this case is the same, identifies of the variables read by statement P_1 should not be affected by writing into them statement P_2. Thus $(W_1 \cup Y_1 \cup Z_1) \cap (X_2 \cup Y_2 \cup Z_2) = \phi$.
Changing P_1 and P_2 symmetrically we obtain
$$(X_1 \cup Y_1 \cup Z_1) \cap (W_2 \cup Y_2 \cup Z_2) = \phi$$

If I_i denotes input data (i.e. the variables and constants of the righthand part of the i-th asignment statement) and O_i, output data (i.e. the variables in the lefthand part of the i-th asignment statement) the above requirements of informational independence of statement i, j may be written as
$$I_j \cap O_i = \phi, \quad I_i \cap O_j = \phi, \quad O_i \cap O_j = \phi$$

Proceeding from these necessary conditions various algorithms for parallelization of linear sections were suggested. The transformation of scalar arithmetic expressions for parallel computation is in reducing the number of steps necessary to compute the arithmetic expression. For example the computation of the expression a + b * c + d requires two steps. At the first step (b * c) and (a + d) are computed. At the second step the results obtained at the first step are summed up.

This parallelization of arithmetic expressions and parallel execution of information-uncorrelated linear section statements is possible in the cases when the computer system permits realization of pipeline processing and/or is provided with special arithmetical-logical devices for tracking, multiplication, shifting etc. Thus the computing process may be significantly speeded up.

2.2. Parallelization of cycles in the process of translation

For computer systems employing vector registers or sets of processor elements vector computations prove highly effective. The operations with the elements of vectors in such computations are performed one order faster than the same operations with scalars. Consequently the transformation of cycle bodies of sequential programs into vector operations may make the program execution essentially faster.

In the course of transformation of the cycle body of a sequential program into a vector operation the latter should be executed (resulting in parallel computations) on those elements of the vector the coordinates of all points of which are parallel to some plane. For instance, such that the condition $\Sigma a_i I_j =$ const holds. The value of the constant should change after each execution of the cycle body until all points of the cycle are not looked through.

Lending themselves for parallelization are normally not all the cycles but only those which satisfy some restrictions. Usually the following restrictions are imposed upon the body of the cycle:

A. It should not contain any input/output statements.

B. It should not contain any transfer-of-control outside the cycle.

C. It should not contain any references to the subprograms and functions whose parameters are generated variables.

D. It should satisfy certain restrictions on the form and order of index expressions.

The structure of the computer system greatly effects the cycle parallelization technique. Thus for systems of the ILLIAK-IV type one may employ the reference technique [8], for systems with a set of asynchronously operating processors, the hyperplane technique [8] or the method of parallelipipeds [14] etc. These methods differ both in the technique of parallelization and in the strength of restrictions imposed upon the cycles to be transformed.

In a parallel cycle body execution one should determine the range of feasible values for each index variable in which the vector operation may be executed. In doing so one should provide equivalency of the vector operation to the initial cycle. The solution to this problem is generally that of system of integer equations and unequalities [9]. Therefore the parameters the cycles to be transformed should be specified in terms of constants rather than variables. Then the entire preparation to parallelization should be carried out in the process translation rather than in the course of the program's execution.

It should be noted that the analysis of cycles used in FORTRAN programs has shown that depending on the structure of the computer system and, consequently, on the parallelization technique used, 30 to 60 percent of cycles in these programs lent themselves to automatic parallelization.

3. Determination of branches in programs

A way to reduce program execution time is parallelization of it in branches i.e. revealing such sections of it which may be executed simultaneously and independently. Program branch initialization takes a great deal of time therefore rational branching implies that individual branches be executed long enough. When greater program units than linear sections or simple cycles are analyzed for possible parallelization, the initial program is presented in the form of linear structures consisting of single-input-single-output nodes. In the given generalized graph of the program the so-called hammocks are isolated, that is, subgraphs with a single input and single output vertex. Note that a hierarchy of hammocks is admitted. Orderness and hierarchysity of linear structures permit one to reveal time limits of the execution of parallel sections and make their debug easier. We shall not go into the details of how hammocks are isolated. Note only that a number of algorithms are available for the purpose. Proceeding from the program's graph consisting of linear sections, simple cycles and OLS's are constructed which incorporate the above structures. When program branches are shaped all these OLS's are looked through starting with the first one. This is done after the elements of each OLS are allocated throughout local levels. The initial number of branches is determined by the content of the first OLS' first level. The subsequent analysis of the graph makes use of information and logical links to unite the branches and construct new ones.

In the process of branch formation the optimal structure of the program should be obtained i.e. the program of minimal execution time with given finite resources. This is achieved by means of uniting some branches together which results in the reduction of the branch formation time. The same problem may be stated in terms of mathematical programming. An algorithm solving it to a certain extent is designed.

4. The specifics of designing multiprocessor computers operations systems

Operation systems for multiprocessor computers fulfill the same tasks as in "conventional" computers but besides they [10]

- organize the interaction of parallel computational processes and their timing;

- schedule and dispatch computational processes with regard for their parallelization;

- reconfigure the system whenever necessary;

- dynamically reallocate the available resources.

Multiprocessor operation systems may employ only one specified processor. This makes the design of the system somewhat simpler but reduces its reliability since a breakdown of this processor results in a complete failure of the whole system.

Alternatively, an operation system may function with any of the computer's processors. This results in a more complicated design of the operation system, but increases its reliability due to the fact that under a failure of one processor the system may operate on another.

An operation system may, finally, be decentralized with a part of its functions performed on one of the central processors and the other part, on peripheral processors. In this approach central processors are free of performing the functions of the operation system which results in a higher efficiency of the computer.

Assigning individual modules of the operation system to different processors i.e. the distribution of the operation system functions among the processors depends on purpose of the system, the characteristics of the central and peripheral processors and on the topology and speed of performance of the interfaces which provide data exchanges between processors. Optimal distribution of functions among the processors may significantly increase the overall performance of the computer and its tolerance to hardware breakdowns.

5. The specifics of parallel programs debug

The major specific features of parallel programs which make their debug more difficult as compared to serial programs are

- asynchronous execution of sections of a parallel program, and

- physically simultaneous execution of operations on several elements of the array.

The first feature hampers reproduction of situations in which an error took place. Since parallel processes are asynchronous they may access the same data in different order. The order of processing may effect the result and the programmer is deprived of any means to restore the order of data processing for the localization of the error. This feature of asynchronous programs adversely affects complex debug of complicated programs.

The second specific feature offers no special difficulties. To localize an error in the course of the debug either all elements of the resulting array which were obtained physically simultaneously or only those specified by the programmer may be displayed.

PART II. THE EFFECT OF PARALLELIZATION ON THE PERFORMANCE OF MULTIPROCESSOR COMPUTER SYSTEMS

Parallel execution of some parts of a program reduces its run time. However as a rule parallelization requires certain additional operations to be executed which do not take place in its serial execution thus effecting the system performance. The usefulness of parallelization in multiprocessor computer systems (MCS) may be treated in three aspects:

a) its effect on the mean MCS performance;

b) change of execution time for a group of programs;

c) reduction of run time for a single program.

1. The effect on the MCS performance

The first question to arise is that of the conditions of measuring the performance. If one considers the average number of instructions executed by processors in a unit of time he may see that with parallel programs it does not change under fault-free operations of the processors. However if one considers only the number of executed useful instructions i.e. the instructions in a serial program such useful performance may be much less. This results from the need to execute some extra instructions for instance, those of branch origination and integration. The useful performance of a MCS may be estimated in the following manner. Let us observe I programs with the degree of parallelization [7] $E_i = t_i^l / q t_i^p$, $i=1,I$ where q is the number of processors and t_i^l and t_i^p are the execution times for a serial and parallel programs respectively. I is assumed large enough. Note that even without any additional operations E_i may be less than one. Let us estimate the useful MCS performance in execution of parallel programs. Following [13] denote β_{ik} as the part of time t_i^p during which i-th task uses k processors $(k \leq q)$. The total number of actions for the fulfillment of the program is

$$D_i = c t_i^p \sum_{k=1}^{q} \beta_{ik} k$$

The number of operations performed for all tasks is

$$D_p = \sum_{i=1}^{I} D_i = c \sum_{i=1}^{I} \sum_{k=1}^{q} \beta_{ik} k t_i^p$$

Assume that the execution of I programs both in the serial and parallel modes goes without idling of the processors.

In I programs there are $D^l = \sum t_i^l c$ useful instructions where c is the nominal performance of one processor. In a serial run of programs the MCS performance is

$$P_1 = \frac{D^l}{T_1} = \frac{c \sum_i t_i^l}{\sum t_i^l / q} = cq$$

In a parallel run the time of execution of programs with due regard for the expression for D_p is

$$T_p = \frac{D_p}{cq} = \frac{\sum_{i=1}^{I} \sum_{k=1}^{q} \beta_{ik} k t_i^p}{q}$$

In a parallel mode the useful performance of a MCS is

$$P_p = qc \sum_i t_i^l / \sum_i \sum_k \beta_{ik} k t_i^p \quad (1)$$

with $\beta_{iq} = 1$, $\beta_{ik} = 0$ for $k = 1, \ldots, q-1$ and $\forall i \in \overline{1,I}$

$$P_p = \frac{c \sum t_i^l}{\sum t_i^p} \quad (1a)$$

The change of MCS performance is

$$N_p = \frac{P_p}{P_1} = \frac{qc \sum_i t_i^l}{cq \sum_i \sum_k \beta_{ik} k t_i^p} = \frac{\sum_i t_i^l}{\sum_i \sum_k \beta_{ik} k t_i^p} \quad (2)$$

With a complete parallelization of programs

$$N_p = \sum_i t_i^p E_i / \sum_i t_i^p \quad (2a)$$

Thus the use of parallel programs decreases the useful MCS performance N_p times due to the increase of the number of operations executed in parallel.

<u>Example.</u> Estimate the loss in the performance of an MCS with 8 processors for a flow of similar tasks with the level of parallelization $\beta_1 = 0.04$, $\beta_3 = 0.27$, $\beta_8 = 0.69$ with that of all the rest tasks $\beta_i = 0$. Let $t^l = 5$ and $t^p = 1$.

Substituting these data into (2) yields

$$N_p = \frac{5}{1 \cdot 0.04 + 3 \cdot 0.27 + 8 \cdot 0.69} = 0.78$$

The overall useful performance is thus decreased more than 20 percent.

The above considerations were made on the assumption that no limitations were imposed on the volume of main memory. If program execution requires a larger volume of main memory waiting of data may result in idling of the processors thus affecting the performance of the MCS. Since in the serial processing mode a part of the main memory should be provided for each of q programs run on q processors while in the parallel mode the entire memory may be given to a single program if it activates all q processors a greater number of exchanges with external memory takes place in the former case and, consequently, the processors in the serial mode of operation may idle much longer waiting for data.

It should be added that concurrence of data pumping and program execution is more probable under the parallel mode which also results in the increase of MCS performance.

2. Execution of a group of programs

Consider two situations possible in execution of a group of programs:

a) the group consists of I programs. It is required to find the optimal version of program initialization providing the minimal execution time;

b) the flow of queries for each of the tasks is known. It is required to find the version of program execution providing the minimal mean waiting time.

Consider the first situation. Let the time of execution of each of the programs in the serial mode be t_i^l and in the parallel mode, t_i^p ($i = 1, I$). It is assumed that in the parallel mode all processors are busy. Introduce

$$x_i = \begin{cases} 1, & \text{if the parallel mode is used for the i-th program} \\ 0, & \text{otherwise} \end{cases}$$

If the task is performed serially denote

$$y_{ik} = \begin{cases} 1, & \text{if the i-th program is executed on the } k\text{-th processor} \\ 0, & \text{otherwise.} \end{cases}$$

Then the total time for the execution of I programs is

$$T^o = \sum_i x_i t_i^p + \max_k (\sum_i y_{ik} t_i^l) \quad (3)$$

Using the variable $T = \max_k (\sum_i y_{ik} t_i^l)$ the minimizing function (3) takes up the form

$$\sum_{i=1}^{I} x_i t_i^p + T \to \min \quad (4)$$

the following conditions should be met in this case

$$\sum_k y_{ik} + x_i = 1 \quad i = \overline{1,I} \quad (5)$$

$$\sum_i y_{ik} t_i^p - T \leq 0 \quad k = \overline{1,q} \quad (6)$$

$$x_i = 0, 1, \quad y_{ik} = 0, 1 \quad (7)$$

The problem (4)-(7) is the Boolean problem of integer linear programming. Using the well-known algorithms [15] its solution is quite easy. Note that parallelization of some programs permits the reduction of the solution time for a package of programs due to greater load of processors during the time T^o.

This may be explained with the following example. Let two processors be employed to execute three identical programs with the time of execution $t_i^l = 5$, $t_i^p = 3$. One may easily see that in serial processing the total execution time amounts to 10 units, in parallel 9 units while in the case of serial execution of two programs and parallelization of the third program the total time is 8 units.

Similarly one may treat the problem of minimizing execution time for a package of programs in the case when not all of the processors operate in the parallel mode ($q_i \leq q$).

Let the processors be presented by consecutive numbers. Then following [12] we may denote

$$y_{ijk} = \begin{cases} 1, & \text{if the i-th program starts its execution at the j-th instant of time and keeps busy all processors from } k\text{-th to } (k+q_i-1)\text{-th.} \\ 0, & \text{otherwise.} \end{cases}$$

The following function is to be minimized

$$T \to \min$$

under the constraints

$$\sum_j \sum_k y_{ijk} = 1 \quad i = \overline{1,I}$$

$$\sum_i \sum_k y_{ijk}(x_i t_i^p + (1-x_i)t_i^1) \leq T \quad k=\overline{1,q}$$

$$y_{ijk}(\sum_{i_1 \neq i} \sum_{j_1=j}^{j+x_i t_i^p + (1-x_i)t_i^1 - 1} y_{i_1 j_1 k}) = 0$$

$$x_i y_{ijk}(\sum_{i_1} \sum_{j_1=j}^{=j+x_i t_i^p - 1} \sum_{k_1=k-q_{i_1}+1}^{k+q_{i_1}-1} y_{i_1 j_1 k_1}) = 0$$

$$y_{ijk} = 0,1, \quad x_i = 0,1$$

The 3rd and 4th of these constraints determine the condition of concurrence of assigning the entire processor resource to the program. This problem is nonlinear, integer and its accurate solution is very hard to find.

The problem however will be essentially simplified if one assumes that the program is regarded to be executed when it is given qt_i^p of the processor time in the parallel mode and t_i^1, in the serial mode. Let us introduce the variable z_{ik} characterizing the processor time assigned to the i-th program on the k-th processor. The load time for the k-th processor is $T_k = \Sigma z_{ik}$. It is necessary to minimize

$$T \to \min \quad (8)$$

under the constraints imposed on the assignment of the processor time

$$\sum_k z_{ik} - x_i q t_i^p - (1-x_i)t_i^1 = 0 \quad (9)$$

The time of executing programs on each of the processors should not exceed T, i.e.

$$\sum_i z_{ik} - T \leq 0 \quad k=\overline{1,q} \quad (10)$$

Besides if the program in the serial mode may be executed only on one specified processor

$$z_{ik} \cdot \sum_{k_1 \neq k} z_{ik_1} = 0 \quad k=\overline{1,q-1}, \quad i=\overline{1,I} \quad (11)$$

The condition that the variables are integers leads to the following:

$$x_i = 0,1, \quad z_{ik} \in N \quad (12)$$

To find an approximate solution for problem (8)-(12) one may use the algorithm which brings the solution close to optimal. Its essence is in the following. Assign all the programs executed serially to the processors in such a manner that the most durable program takes up a free processor in the first place. Then on processor k^* with the greatest total program execution time τ^* transfer a program with the maximal value E_{i_1} into the parallel processing mode if the following inequality holds:

$$\sum_{k \neq k^*}(\tau^* - \tau_k) \geq q t_{i_1}^p (1 - E_{i_1}) \quad (13)$$

The program employing the parallel mode may be transferred onto free processors starting with the least busy. In doing so time τ^* decreases. This transformation of programs into the parallel mode may be carried out until the condition (13) is true for at least one program.

Example. If 5 tasks are to be fulfilled on three processors, each task of duration (t_i^1, t_i^p)= =(6.3; 10.4; 8.3; 15.6; 11.5). Employing the above algorithm we obtain the following assignment of serial tasks to the processors: (4; 5 and 1; 2 and 3) with the execution time (15, 17, 18). The degrees of parallelization of the 2nd and 3rd tasks realized on the third processor are 5/6 and 8/9. Condition (13) is true for the 3rd task since (18-15)+(18-17) > 3.8 /9. The execution of the 3rd task in the parallel mode requires 9 units of the processor time of which 7 units is provided on the third processor and 2 units, on the first. The total execution time for the package of programs is thus reduced by 1 and amounts to 17 time units.

Similarly one may treat the problem of time minimization for the execution of program packages in other statements. However the above statement sufficiently proves that one should thoroughly consider whether parallelization is really necessary in his stacked operation.

Consider now the second of the two situations given at the beginning of this part. If flows of queries are coming for execution of various programs (which is often the case in computer-aided management and control systems). The major attention has to be paid to the time of serving the queries. Let us analyze this situation from the viewpoint of possible parallelization of programs. Assume that all the programs have equal priorities. Let Poisson query flows reach programs with the time constant λ_i and the times of serial and parallel execution $t_i^l = 1/\mu_i^l$ and $t_i^p = 1/\mu_i^p$, respectively ($i = \overline{1,I}$). The results of the analysis of such system depend on the discipline of service accepted for the case when query queues occur.

However when a single program is served in either the serial or the parallel mode an analytical study is possible.

The parallel mode operation presents a situation similar to serving a flow of queries by a single device. Let us use the Pollachek-Khinchin formula [11] for the mean time of presence of a query in the system provided the service law is arbitrary:

$$E = b_1 (1 + \rho(1 + c_1^2)/(2(1-\rho)))$$

For an exponential time of service we shall have

$$E\omega_p = 1/\mu (1 + \lambda/(\mu - \lambda)) \quad (14)$$

For a permanent service time, with due regard for we shall have

$$E\omega_p = t(1 + \lambda t/2(1 - \lambda t)) \quad (15)$$

The serial program execution presents a situation similar to serving a flow of queries by q devices. For an exponential service time the expression for the mean time of presence of a query in the system is as follows [11]:

$$W_q = \frac{1}{\mu} + \frac{\lambda(\lambda/\mu)^q}{q\mu \cdot q! (1 - \lambda/q\mu)^2} \quad (16)$$

$$\cdot 1/(\sum_{n=0}^{q-1} \frac{(\lambda/\mu)^n}{n!} + \frac{(\lambda/\mu)^q}{q!(1-\lambda/q\mu)})$$

For a permanent service time the mean waiting time [11] is

$$W_q = \frac{q}{q+1} \cdot \frac{1-(\lambda t/q)^{q-1}}{1-(\lambda t/q)^q} \cdot \frac{q(\lambda t)^q e^{-\lambda t}/q!}{(q/t)(1-\lambda t/q)} \quad (17)$$

$$\cdot 1/(\frac{q-\lambda t}{1-e^{-\lambda t}\sum_{k=q}^{\infty}(\lambda t)^k/k!} + \frac{(\lambda t)^q e^{-\lambda t} \cdot q}{q!/(q-\lambda t)})$$

Comparing (14) to (16) or (15) to (17) depending on the distribution of the program execution time one may find out which of the service modes is preferable.

Example. Let two processors realize a Poisson flow of queries on some problem characterized by the exponentially distributed execution time with $(\mu_1, \mu_p) = (1, 1.5)$.
Substituting these values into (14) and (16) yields

$$E = \frac{1}{1.5 - \lambda}, \quad W_q = 1 + \frac{\lambda^3}{4 - \lambda^2}$$

The parallel mode turns to be better when $E < W_q$ i.e. approximately with $\lambda < = 0.77$. The systems load in the parallel mode is $\rho_2 = \lambda/\mu_p = 0.51$.

The limit value of the load also depends on the efficiency of parallelization $E_i = t^l/qt^p$. In the above example $E = 0.75$. Note that with $E = 1$ parallel execution mode is always preferable.

Analytical analysis in the general case of I flows of queries for programs is very difficult and modelling may be suggested instead.

On the whole it may be concluded that with equal priorities of programs the parallel mode works more effectively in systems with small loads where waiting times are little and the major part of service times are taken up by program execution. It should be said also that the parallel mode turns more effective for lengthy programs. However for programs with intensive query flows the more effective mode is serial.

When programs with different priorities are handled those with high priorities should be executed in parallel while those with low priorities serially since generally the system is little loaded by high priority programs and the major service time is the time of execution which should be reduced.

3. Parallelization of a single program

As it was noted above program parallelization decreases the performance of a MCS. However the useful speed of computation estimated in terms of the number of instructions of a parallel program executed by a MCS in a unit of time increases $S_p = t^1/t^p$ times. Let us enumerate the cases when, in our opinion, program parallelization is preferable:

- when service time for a program operating in control systems under time limitations is to be reduced;

- when the system's main memory or its reliability characteristics are restricted, and

- to provide a flexible load of processors with sharp variations in the flows of tasks.

To some extent the first and the third cases where treated in the previous parts of the paper. Reliability of individual MCS devices effects the selection of the program operation mode so that program restarts or duplication of computations may be necessary which may significantly increase the program execution time in the serial mode and reduce the overall efficiency of the MCS. Indeed if the failure probability for some device is nonlinear time dependent which is practically always the case the increase of program execution time may dramatically raise the probability of failure. This makes program restart necessary which takes extra time δ. In this case the efficiency of parallelization is $(t^1 + \delta)/qt^p$ which already may be close or even more than unity. The effective implementation of MCS's implies that a thorough analysis of the goals of parallelization should be made and criteria estimating the usefulness of this parallelization pointed out. It is desirable that alternate variants be developed with both a parallel and a serial version of the program execution and either of them be activated depending on the concrete situation in the MCS.

REFERENCES

1. Campbell, R.H., and Haberman, A.H. The specification of process synchronization by path expressions. Lecture notes in computer science, Springer-Verlag, Berlin, 1974.
2. Hansen, P.B. Operating system principles. Prentice-Hall, 1973.
3. Hansen, P.B. The programming language concurrent Pascal IEEE Trans. on Software Engineering, Vol. 1 (1975), No. 2.
4. Hoare, C.A.R. Towards a theory of parallel programming. Operating systems techniques. Academic Press. New York, 1972, pp.220-230.
5. Hoare, C.A.R. Monitors. An operating system structuring concept. Comm. of the ACM, vol. 17 (1974), No. 10, pp. 549-557.
6. Keller, R.M. Sentinels: A language construct for multiprocess coordination. Unpublished Memo, Dept of Computer Science, Univ. of Utah, Salt City, 1978.
7. Kuck, D.J., Maruoka Y., and Chen S.-C. On the number of operations simultaneously executable in Fortran-like programs and their resulting speedup. IEEE Trans. on Computers., Vol. C-21(1972), No. 12, pp. 1293-1310.
8. Lamport, L. The parallel execution of Do loops. Comm. of the ACM, vol. 17(1974), No. 2, pp.83-93.
9. Lebedev, V.G. and Shuraits, Yu.M. Parallelization of cycles with arbitrary steps. Avtomatika i Telemekhanika, No. 9(1979), pp.158-167 (in Russian).
10. Rabinovich, V.M. Specifics of the design of operation systems for uniform MCS's with variable structure. In "Problems of cybernetics. Multiprocessor computer systems with variable structures. VINITI, Moscow, 1981, pp.58-64 (in Russian).
11. Saaty, T. Elements of queueing theory with applications. McGraw-Hill, New York-Toronto-London, 1961.
12. Shuraits, Yu.M. Optimal resourse allocation in multiple execution of a complex of jobs. In "Urgent problems of the control theory and applications". Nauka Publ., Moscow, 1977, pp.122-125 (in Russian).
13. Trakhtengerts, E.A. Some estimates of the essectiveness of computation parallelization. In "Computer-Aided Control System Design". Statistika Publ.,Moscow, 1981, pp. 88-100 (in Russian).
14. Val'kovsky, V.A., and Kotov, V.Ye. Automatic design of parallel programs. Preprint No. 146, SO AN SSSR, Novosibirsk, 1979 (in Russian).
15. Wagner, H. Principles of operations research with applications to managerial decisions. Prentice-Hall, Englewood Cliffs, New Jersey, 1969.

DISCUSSION

Shileiko: At the beginning of your talk you stated that at present there were no discussions on the possibility of the realization of parallel processing.

Trakhtengerts: Well. I said that there were no discussions about whether parallel processing is necessary or not. Everybody agrees that it is necessary to create parallel processing and parallel processors. The question is how to create them.

Shileiko: Then I understand that parallel processing is impossible but necessary. I'll ask you a very simple question. Let us have a problem to generate a sequence of natural numbers, 1,2,3. Can you parallelize this process?

Trakhtengerts: I am not ready yet to answer this question, but I think it is possible. I can answer another question, that is, I know an algorithm for parallelizing the computation of all simple numbers, and how to create all integers.

Harrison: You started by looking analytically at a rather general case of determining parallelism in a given problem or expression. I think you demonstrated correctly that this is a very difficult problem. If you arbitrarily take all problems one may wish to solve and you want to parallelize them, that's very difficult. Later on you proposed and discussed a computer system consisting of separate multi-processors, stating that designing the operating system for that kind of system is difficult. But this is a much more constrained example than that of a general analytical expression. Have you applied your method to functions specifically related to operating systems? Have you taken an OS and attempted through your algorithmic method to determine where the parallelism exists and how it is best implemented?

Trakhtengerts: I think that up to now it has been impossible to create such an algorithm which would show when parallelism is useful and when it is not. The usefulness and effectiveness of parallelization highly depends on hardware, and it is possible to show that some algorithms that are very effective on one structure of a parallel system, are absolutely ineffective on another structure. It is possible to make a relevant analysis to understand whether it is useful to parallelize on some particular hardware, but no algorithm exists for that, either.

Inamoto: I'm looking at the computer system architecture. I'd like to ask a question concerning multiprocessing on some sort of a computer system. Have you any idea of the optimum number of the processors in case you parallelize, for instance, the FORTRAN compiler?

Trakhtengerts: I have no idea of how many processors are needed for parallelizing a compiling process. I think the problem should be stated vice versa: if we have a particular hardware, how to organize an effective compiler?

Shileiko: Can you tell us what the measure of effectiveness of the processes, either sequential or parallel, is?

Trakhtengerts: There may be several measures of effectiveness. One is very simple: the time to perform a task, not considering the cost. Another approach is well known - the cost of operations per second: the cost of the task divided by the time to perform the task. Usually two joined processors work less effectively than two separate processors. If the multiprocessor is a simple unity of processors it may be very ineffective. A multiprocessor contains several control processors and many ALU's. The latter are much cheaper than control devices. In this case cost per task time may be the proper measure of effectiveness. Besides, parallel processing is less universal than sequential processing.

Joudu: Will there be any difficulties in debugging parallel programs, so that we'll have to test all the branches, not depending on whether the program is for parallel or sequential work?

Trakhtengerts: I can see only one difficulty, but that's a big one: it is not impossible that after having made a mistake, one may have the same situation again. Parallel processing is asynchronous processing and, as soon as you parallelize synchronous processing, you will have the same situation as in sequential processing.

Dragunova: Does complexity of algorithms depend on the number of processors, and if yes, then how?

Trakhtengerts: It depends on the structure of the computer. For vector processors, for example CRAY-1 or PSP (Parallel Scientific Processors), the complexity of algorithms does not depend on the number of processors. For other multiprocessors, say, Burroughs 7000, that dependence may exist.

IEEE PROJECT 802: LOCAL AREA NETWORK STANDARD — A March 1982 Status Report

T. J. Harrison

Advanced Software Engineering Technology, IBM Corporation, Boca Raton, FL 33432, USA

Abstract. This paper describes Project 802 of the Computer Society of the Institute for Electrical and Electronic Engineers (IEEE)[1]. Project 802 is defining a standard for local area networks in which microprocessor controlled terminals and other devices are coupled on a peer-to-peer basis. The paper provides a brief history and rationale for the effort and an overview of the current draft standard.

Keywords. Project 802; local area network; LAN; telecommunication; token passing; CSMA/CD; OSI; reference model; communications computer applications.

Disclaimer and Caution

It is prudent to remind the reader of several cautions and conventions that apply to the current status of Project 802 and, in fact, to the tentative results of any standards-developing organization:

This paper is not an official publication of IEEE Project 802 or of any other organizations discussed, and it has not been approved by these bodies. It is based primarily on public information available in the literature and the minutes of meetings. The most recent meeting of the committee was held 1982 March 8-12 and minutes are not available as this paper is written. As a result, descriptions resulting from decisions at that meeting are based on the author's notes.

Opinions and speculation as to the content of the final standard, the current group consensus, and future actions are those of the author and do not necessarily represent the position of other committee members. In addition, since this activity is being done at the "working group" level where participants act as independent "experts," their positions, and those of the author, may not represent those of their sponsoring organizations.

The proposed standard still is undergoing change and this report is provided for current information only. It should not be used as the basis for decisions affecting future product design, specification, or sales activity. The final approval process has not yet begun and changes, perhaps major in scope, may be made before approval as an IEEE, American National or international standard is accomplished.

The author is not a member of the IEEE 802 committee but has been indirectly associated with the project and related projects through his involvement with a number of national and international standards organizations.

---

[1] This part of the paper borrows heavily from previous papers authored or co-authored by the author(1,2).

INTRODUCTION

Local area networks (LANs) are a product of recent advances in technology and the continuing evolution of computer systems. They are, in a sense, "an idea whose time has come." In the early days of computing, economics and the requirement for specialized support favored centralized systems in which the user came to the hardware to use it. As technology advances, competition, and mass production combined to reduce computing costs, the alternative of distributing computing power to the user became attractive and "distributed processing" (which had been around for a long time!) emerged as a popular concept. The introduction of the so-called "microprocessor" and "microcomputer" now promise the possibility of each user having significant computing power at the individual workstation (3). In addition, the microprocessor is spawning a whole new class of so-called "intelligent devices" that provide significant capability without utilizing the time-shared power of a central computer.

Information sharing, inherently provided by centralized systems, can be extended to microprocessor-based devices through the development of convenient communication protocols between such devices. Nonetheless, and even anticipating further spectacular advances in microprocessor performance and function, it is hard to visualize the day when the immense computing power of the large centralized "mainframes" will not be required as a part of the total computing environment. Distributed computing is but another computing capability; it does not replace, but rather complements, the use of centralized computers to solve information processing problems.

The need for LANs comes from the growing desire to communicate between small microprocessor-based devices at a cost consistent with the device cost. This need was anticipated by researchers as long as ten years ago and experimental local networks were established. With the tremendous increase in the number of available microcomputers, these networks are now of commercial interest.

Effective communication requires agreement on a number of technical points. First, for example, there must be agreement on the communication medium, such as the public telephone system, in which the characteristics of connectors, interconnecting lines etc must be such that transmitted and received signals are compatible.

There also must be agreement on certain protocols and conventions in order that information be exchanged. If one user speaks Chinese and the other speaks English, there probably must be agreement on which language will be used so that information, as opposed to mere signals, are exchanged. If the users do not share a common language, alternatives such as interposing a translator must be utilized, but this imposes a cost and, possibly, an undesirable loss of efficiency. Conventions, such as using "hello" to initiate information transfer and "good-by" to terminate it, must be agreed upon and observed by all users.

The number of necessary agreements is significant and the economic effects, such as the cost of a Chinese-English translator, must be understood and evaluated. In addition, future needs must be anticipated to provide for further advances in technology and to avoid, whenever possible, future costs of replacement and conversion of equipment and data.

The recognition of these needs and the technological environment led to the formation of the IEEE Local Network Standards Committee, formally known as Project 802, in February 1980. Their task was recognized as being very complex, very large, and commercially significant. An urgency was perceived to quickly establish agreement on some minimum set of requirements and protocols before a large number of incompatible devices emerged. Upon examination, it became apparent that no single standard could serve the needs of all applications. However, a limited "family of standards" which addressed a broad range of anticipated applications could result in economic savings through a reduction of unnecessary proliferation of incompatible designs and through the promotion of competition and mass production of large-scale-integrated electronic devices. In retrospect, the perception was accurate and the past two years have seen the commercial introduction of many LANs, some influenced by the IEEE work.

The urgency continues as demonstrated by the intensity of the committee work. Since the first meeting a little more than two years ago, the committee has met eleven times, typically for four or five days every two or three months. In addition, subcommittees often meet between committee meetings. Meetings of the committee and its subcommittees involve over one hundred people. This is a level of activity which ranks among the highest of any similar standardization effort. The most recent meeting was held 1982 March 8-12.

SCOPE AND ORGANIZATION

There was no generally accepted definition of a LAN and, indeed, there still is not[2]. Nevertheless, the group has evolved a definition as the basis for their effort. That definition, as incorporated into Draft B of the proposed (5) standard is:

> Local Network: A Local Network is distinguished from other types of data networks in that communication is usually confined to a moderate sized geographic area such as a single office building, a warehouse, or a campus. The network can depend on a physical communication channel of moderate to high data rate which has a consistently low error rate. The network is owned and used by a single organization. This is in contrast to long distance networks which interconnect facilities in different parts of the country or are used as a public utility. The Local Network is also different from networks which interconnect devices on a desktop or components within a single piece of equipment.

This definition implicitly includes the committee's concept of what the network will be used for and its concern for cost.

Specifically, its reference to an office building, a warehouse, and a campus, signals its primary use in the emerging "office automation" area in industry, although the evolving standard clearly has application in other areas, such as some industrial control applications and the research or university environment. Its statement about geographic area and its restriction to moderate-to-high data rates recognizes the costs associated with very high speed data transmission over long distances. The low error rate implies a highly noise-immune media, and extraordinary means to detect and correct errors at the data link and physical levels are avoided. Finally, the definition recognizes the existence of other types of data networks which have different goals and characteristics. In particular, but without specific reference, it differentiates its work from that of the IEEE committees working on a serial extension to the IEEE-488 bus and on several board-level microcomputer buses.

With regard to values for area, length, number of stations supported, and interconnection topology, the approved functional requirements specify a communication medium of at least two kilometers in length, supporting at least two hundred stations. The proposed standard supports several physical topologies and a variety of transmission rates.

It is also the committee's intent to focus on a minimum set of requirements that will allow devices to exchange messages at a very basic level. The committee recognizes that the exchange of _information_, as opposed to merely data bytes, requires agreements beyond those being discussed. These are not addressed since other committees are considering them at the national and international level.

The work of ISO/TC97/SC16 on "Open System Interconnect (OSI)" provides the framework for a common set of definitions and agreements organized as a set of "layers" that provide specific functions which, in the general case, are required for two users to exchange information. The Draft International Standard for the OSI Reference Model can be obtained from ANSI (5). The model consists of seven layers as shown in Fig.1. The boundaries between the layers are defined by agreed-upon parameters and procedures that are necessary for one layer to exchange information with its adjacent layers. The parameters and procedures that allow the two corresponding layers in two communicating systems to effectively exchange information are called "protocols."

---

[2] The international committee ISO/TC97 on Computers and Information Processing avoided a precise definition at their December 1981 meeting and merely noted: "The primary differences between Public Data Networks and Local Area Networks are that the Public Data Networks are not restricted to the user premises and that the protocols and facilities are not under user control" (4).

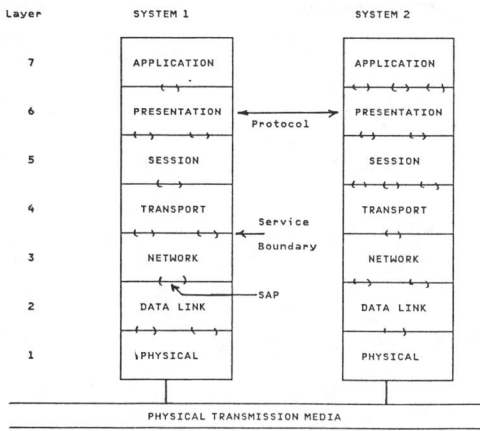

Fig.1: OSI Reference Model

The IEEE effort is concerned with the functions represented by the two lowest layers (Physical and Data Link) and the services provided to the third (Network) layer. It is the feeling of the committee that standards implementing layers three through seven will require little, if any, modification solely for the purposes of satisfying local network requirements. Liaison has been established with the major groups dealing with the higher layers to ensure, to the degree possible, that the 802 standard will be consistent with other emerging standards.

In light of the structure of the OSI model, the natural organization, and that adopted by the committee, mirrors the model itself. The overall committee, chaired by Maris Graube of Tektronix, has been organized into three subcommittees:

> The Media Subcommittee, chaired jointly by Chris Wargo of RCA and Nathan Tobol of Codex, has a scope roughly corresponding to Layer 1, or Physical Layer, of the OSI Reference Model.
>
> The Data Link and Media Access Control (DLMAC Subcommittee), chaired by Gerald Clancy of Honeywell, corresponds to Layer 2 (Link) of the OSI model.
>
> The High Level Interface Subcommittee, chaired by William Ladinsky of International Harvester is responsible for understanding the effect of local networks on the higher five layers of the OSI model.

REQUIREMENTS AND GOALS

The committee and subcommittees are guided in their work by a set of functional requirements and goals which evolved and were approved during the first few meetings. These are included as an appendix to Chapter I of the Draft B proposal (6).

The key requirement is to provide a layered architecture which:

- Corresponds to the ISO Open Systems Interconnection Reference Model;
- Permits efficient interconnection of moderately priced devices; and
- Can itself be implemented at moderate price.

The standard is also to provide broad functional applicability and should support applications ranging from word processing and electronic document distribution to digital voice transmission and industrial/process control. In addition, the standards should provide for the attachment of a variety of devices ranging from computers and terminals to gateways to other networks.

The subcommittees also established some specific goals early in its work which have guided their effort. Specific goals of interest in this discussion are the use of HDLC(7) as a model (if, and to the extent, it is appropriate to do so), and independence from topology, transmission rate, and specific transmission media. In varying degrees, these goals are being met by the evolving standard. The diversity of the goals, however, has led to the requirement for a number of options in the proposed standard which may seem large at first glance. However, not all options are independent. For example, selecting the CSMA/CD media access control method narrows the choice of media and transmission method to two out of the more than six included in the standard. The proposed standard requires that the manufacturer specifically state which options are incorporated in a given device. It also defines the criteria for claiming compliance to the standard.

A result of these options is that two conforming pieces of equipment may not be able to communicate with each other. In some cases, however, they may be able to coexist in the same physical network. That is, they can both be connected to the physical media, communicate with other compatible devices, and not interfere with devices having incompatible options.

OVERVIEW OF THE PROPOSED STANDARD

The OSI Reference Model (OSI/RM) provides a framework for understanding the 802 activity. Project 802 is concerned with the lowest two layers of the OSI/RM, the physical and data link layers. The OSI/RM recognizes and allows the creation of sublayers by grouping related functions. This has proved convenient for Project 802, primarily due to the different options for media access control (MAC). The characteristics of the MAC options, however, are such that one of the sublayers includes functions associated with both the physical and data link layers of the OSI/RM. Thus, the sublayer straddles the boundary between Layers 1 and 2 in the OSI/RM.

The mapping of the OSI/RM physical and data link layers into the three layered

LAN/RM is shown in Fig.2. The three LAN/RM layers are called the (Local Network) Logical Link Control (LNLLC), Media Access Control (MAC), and Physical Layers. The multiple LLC Service Access Points (L-SAPs) provide logical ports at the 3/2 layer boundary with the OSI/RM Network Layer. Single SAPs, called the MAC-SAP and the P-SAP, provide logical ports between the LLC and MAC layers, and the MAC and Physical Layers, respectively.

The LNLLC layer provides two types of link services to the Network Layer. The <u>Type 1</u> or <u>Connectionless</u> service allows the exchange of protocol data units between two L-SAPs without the requirement for establishing a logical link. These frames are not acknowledged in the LLC layer (although they might be at a higher layer) and no error recovery or flow control are provided.

<u>Type 2</u> or <u>connection-oriented</u> service requires the establishment of a logical link between two L-SAPs prior to the exchange of user data. When in the data transfer mode, frames are transmitted (or delivered) in sequence. Both error recovery and flow control are provided. A conforming device may provide either Type 1 or Type 2 service; however, in the latter case, it must also provide Type 1 as well.

A third type of link service, currently being referred to as <u>Type 1.5</u>, is being considered but is not yet included in the proposal. It would provide a <u>connectionless-acknowledged</u> service that provides a simple acknowledgement without error recovery (e.g., automatic retransmission) or flow control[3].

The Medium Access Control (MAC) layer fundamentally provides an application-entity with the ability to "capture" the physical medium for its use. Three control methods are provided in the proposed standard: Carrier-Sense Multiple-Access with Collision Detect (CSMA/CD), Ring Token-Passing, and Bus Token-Passing.

The Physical Layer provides the functions necessary to transmit and receive bits between peer-entity P-SAPs. Several transmission schemes are provided in this initial standard, including both broadband and baseband coax, and shielded twisted pair systems. Consideration is being given to a fiber optic medium as well.

---

[3] In the ring token-passing media access method discussed later, this type of service function is provided by the "copy" and "address recognized" indicators.

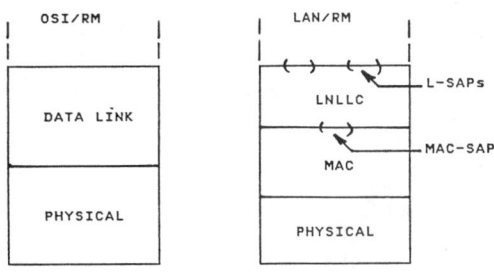

Fig.2: Mapping of OSI and LAN Reference Models

LAN FRAME STRUCTURE

Many of the characteristics of the proposed LAN can be discerned from the fields in the transmission frame. As stated earlier, a goal of the DLMAC Subcommittee was HDLC conformance to the degree possible. A comparison of the HDLC and the proposed LNLLC frames is shown in Fig.3.

In comparing the HDLC and LAN frames, a number of differences can be identified. The first is that HDLC frame boundaries are delimited by an F or "flag" field consisting of a unique sequence of bits. In order to avoid a data or address bit sequence identical to the F field, HDLC-compatible devices provide "bit stuffing," a technique that inserts bits in a prescribed manner whenever necessary to prevent the flag pattern from occurring within the frame. If this were not done, an erroneous frame boundary would be detected.

After considerable discussion, the subcommittee reached the conclusion that the HDLC flags are undesirable in the high-speed environment envisioned for the LAN. This stems from the need for non-linear coding during the transmission and receipt of data and its

HDLC FRAME

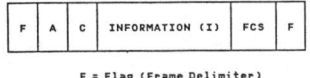

F = Flag (Frame Delimiter)
A = Address (Destination)
C = Control
FCS = Frame Check Sequence

CSMA/CD FRAME

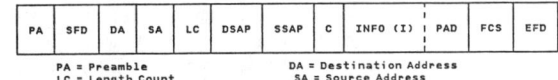

PA = Preamble   DA = Destination Address
LC = Length Count   SA = Source Address
PAD = Padding Octets   DAP = Destination SAP Address
SFD = Start Frame Delimiter   SSAP = Source SAP Address
EFD = End Frame Delimiter

TOKEN-PASSING METHOD FRAME

| SFD | AC1 | AC2 | DA | SA | DSAP | SSAP | C | INFO (I) | FCS | EFD | IDL |

AC1 = Access Control 1 (Ring Only)
AC2 = Access Control 2
IDL = Idle Time Fill (Ring Only)

Fig.3: HDLC and 802-LAN Frame Structures

implications in terms of hardware performance, implementation, and susceptibility to error. The patterns adopted for starting frame, ending frame, preamble, and idle delimiters take advantage of the encoding techniques utilized for transmission. They provide, together with a 32-bit FCS, a Hamming distance of at least 4.

A second difference noted in comparing the HDLC and LNLLC frames is the inclusion of additional address fields for the LAN. HDLC provides for point-to-point links, where there are only two stations on the link, and multipoint unbalanced links, where the transmitting station is always either a primary station or a secondary station responding to a primary. In the peer multipoint environment provided by the LAN, permission to transmit is not required and multiple sources can be concurrently sending messages to the same destination station. As a result, the receiving stations must be told the identity of the sending station in order to sort out the transmissions. In addition, the concept of concurrent access to multiple higher-layer protocols through the use of SAPs requires a means to identify the protocol (and, hence, the SAP address) being utilized by the source and destination stations. The result, therefore, is the addition of the SA, DSAP, and SSAP fields as, if you like, extensions of the HDLC addressing concept.

Addressing

Source and Destination Addresses

Destination adddesses may be either _individual_ or _group_ (multicast) addresses. Individual addresses uniquely identify stations on the link and, conceptually, remain unchanged from HDLC addressing. Group addresses identify more than one station as the destination of the message. A one bit in the first position of the first octet of the DA indicates a group address and zero identifies an individual address. A special group address (all ones) is designated as an "all station" address, recognized by all connected stations. Since a group address has no meaning as a SA, the first bit of the first octet is always zero so as to maintain consistency in address assignment. In addition, group addresses cannot be used in the connected class of service (Type 2) and group transmissions are not acknowledged at the LNLLC level.

The SA and DA are of equal length and may be either 2 or 6 octets long. All stations on any given link must use the same length addresses. Conforming equipment is not required to implement both 2 and 6 octet addresses.

With a 6 octet address, there are 2^{48} possible addresses. The rationale for this huge addressing range comes from the desire to accommodate the "Universal Product Code (UPC)" addressing concept introduced by Ethernet[4]. The idea is that every piece of equipment could have a unique address throughout the world! The exact manner in which this will be administered has not been determined, but is being pursued by the committee. The second bit of the SA and DA is set to zero to indicate a "universal" address and to one to indicate locally administered addresses. Effectively, therefore, there are two distinct address spaces, each containing 2^{46} addresses. The large locally administered addressing range also offers the user the possibility of partitioning the address for some purpose such as associating a block of addresses with a particular building in a multibuilding complex.

Service Access Point (SAP) Addresses

In accordance with the OSI/RM, SAPs are identified by addresses. In the DSAP, the first bit provides the Individual/Group indicator and the second is set to binary 1 and reserved for future, but unspecified, committee use. This leaves six bits for SAP addresses that could correspond to individual protocols at the boundary between the DLC and Network Layers. For the SSAP, the individual/group concept does not make sense so the first bit is used as a Command/Response (C/R) bit, as in HDLC.

Control Fields

The single-octet C control field shown in Fig.3 is used to designate command and response functions invoked at the LLC layer. In addition, it contains frame sequence numbers when required. The general formats of the field are identical to those used in HDLC although there are differences in the defined command functions. Three formats provide for numbered information transfer (Type 2 service), numbered supervisory transfer, unnumbered control, and unnumbered information transfer (Type 1 service). The available commands are a subset of those associated with HDLC with addi-

---

[4] _Ethernet_ is a trademark of XEROX CORPORATION.

tional commands unique to the LAN. They include such things as receive ready (not ready), reject, and exchange identification.

The AC and LC fields can be considered as specialized control fields for the token-passing and CSMA/CD media access methods. CSMA/CD requires a minimum frame size for proper operation so that some padding of the information field is necessary when the number of data bytes is small (or zero). The LC field indicates the total number of octets in the DSAP, SSAP, C, and I (Information) fields so that the receiver can strip off the padding bits. Although shown as a separate field, the PAD field is really a variable (including zero) part of the I-field.

The AC fields provide control information for the MAC sublayer in the token-passing media access methods. The ring token-passing method utilizes both AC1 and AC2, each of which is a single octet. AC1 contains the token state indicator and the optional priority and reservation bits. A single bit is reserved for use by the optional centralized monitor function.

The AC2 field in the ring token-passing method contains an extension bit (reserved for future use), a bit indicating if the frame is to be used by the MAC or LNLLC sublayer, and 6 bits to encode MAC commands. In the bus token-passing method, AC2 is used for control functions that are similar, but not all identical, to those used on the ring.

The ring method requires a station to transmit for a minimum length of time approximately equal to the propagation and processing delay (latency) of the physical ring. For short frames or a large ring, this requirement is satisfied by transmitting an idling signal as indicated by the optional trailing IDL field.

The FCS (Frame Check Sequence) is four octets long in all the access methods. The check sequence being utilized is a cyclic redundancy check (CRC) and utilizes the generator polynomial $x^{32}+x^{26}+x^{23}+x^{22}+x^{16}+x^{12}+x^{11}+x^{10}+x^8+x^7+x^5+x^4+x^2+x+1$.

MEDIA ACCESS CONTROL METHODS

The committee decisions regarding access control mechanisms represent a classical example of how standards are, and must be, developed. Early in the activity, two potential candidate methods were proposed. Each had its advantages and disadvantages, its proponents and its deprecators. The methods were orthogonal in the sense that they could not be merged into a single proposal due to fundamental technical and philosophical differences. The committee became polarized into two groups, neither of which could "out vote" the other.

Despite a strong desire to have a single method, it became clear that the advantages of each method were important for particular applications. Fearing that further delay would only result in a proliferation of incompatible equipment offerings, the committee adopted both methods as a pragmatic compromise. It was felt better to agree on two methods than to delay or defeat the entire standard. Fortunately, the decision to sublayer the Link Layer of the OSI/RM allows both methods to utilize the same LNLLC, so that a high degree of commonality still exists.

Carrier-Sense Multiple Access with Collision Detect (CSMA/CD)

The CSMA/CD method is based on the design for Ethernet [4] as announced by DEC, Intel, and Xerox (8). The basic concept for sharing the physical transmission media is that a station listens before transmitting to determine if the line is idle. If it is not, the station defers until the line is idle for a certain period of time to ensure that the media and stations have recovered. Once the station begins transmitting its frame, it continues to listen for a similar period of time to make sure that a signal "collision" has not occurred. This period is essentially equal to the maximum round trip propagation delay plus delays in the collision detection hardware. A collision will occur during this "window" if two stations determine the line is idle and begin transmission, only to have the transmissions collide between stations. The need for the station to remain active during the entire collision window also establishes a minimum frame size, giving rise to the requirement for the LC (Length Count) field. The minimum frame size is dependent on both the transmission rate and the physical extent of the network.

If a collision is detected, the station reinforces the collision signal for a brief period of time (the "jamtime") to ensure that all stations detect it. Each transmitting station then quiesces and delays subsequent transmission attempts by a pseudo-random time period. The set of numbers on which the randomization is based doubles the time to each retry attempt (known as binary exponential backoff).

CMSA/CD Properties.

The important characteristics of CSMA/CD are that it is completely fair in the sense that every station equally contends for access; it is a simple and easily implemented access algorithm, and it is quite efficient when traffic loading is low[5].

CSMA/CD also has some characteristics which may be undesirable in some applications. The method cannot be unfair so that, if it is necessary to shed load, it cannot be discriminatory, a desirable feature when some traffic is of higher priority. Access time and, therefore, response time is statistical rather than deterministic. This is often considered unacceptable in a real-time environment. The response time saturation characteristic, furthermore, is nonlinear, becoming asymptotic to infinity at a relatively low loading value[6]. Concern is also expressed by some that the system performance dependence on propagation delay may limit the system evolution as future enhancements to increase distance or transmission speed are considered. This is because the minimum acceptable frame size increases with either an increase in the medium length or the transmission rate. CSMA/CD is directly applicable only to bus topologies and, therefore, does not satisfy applications where other topologies may be more appropriate.

Despite the simplicity of the access method when compared to the procedural aspects of the token approach, physical implementation of CSMA/CD may place additional demands on equipment and installation practices. The listen-while-transmitting requirement means that the dynamic signal range must be limited. In the case of the DEC/Intel/Xerox definition, this limit is about 10-12 dB. This, and the need for impedance matching to avoid reflections, leads to: 1) the use of 50 ohm coaxial cable for baseband transmission rather than the more commonly used 75 ohm cable; 2) the requirement to place transceivers only at prespecified locations along the cable; 3) a relatively stringent set of recommendations regarding cable coupling and cable segment lengths; and 4) the need for signal amplification down the drop from the cable tap point. This latter requirement may complicate installation requirements if, for example, the tap point is located in a ceiling or wall cavity since it implies the need for active electronic components at the tap location.

Token Access Methods

Conceptually, the token access methods are as simple as CSMA/CD, although the media management functions are more complex and imply greater logical capability in the attached device. The basic concept is that the stations on the LAN are formed into a <u>logical ring</u> whether or not the physical topology is a ring. Control of the transmission medium is passed between stations in an ordered sequence by means of a "token". The token represents a "permission to transmit" that grants exclusive use of the transmission medium for a maximum time determined by a "token-holding timer." Depending on the token-holding time and the frame length, the station may transmit multiple frames during this transmit "window," as opposed to the single frame inherent in the CSMA/CD method. At the completion of the window or when the station has transmitted all waiting frames (possibly none), it passes the token to the next station in the logical ring.

The control of a token access LAN is more complex than in the case of CSMA/CD. Specifically, means are required to initialize the token passing sequence in the case of non-ring topologies, as well as for administrative (monitoring) functions such as error control to detect and correct faults; e.g., the loss of the token or the existence of multiple tokens. The standard provides that the monitoring function be replicated in all stations capable of passing the token, although centralized control is not precluded.

Bus Token-Passing Operation

The details of the token-passing access methods differ depending on the physical topology of the network. In the so-called "broadcast" or non-physical-ring topologies such as the bus and star, the MAC layer maintains a record of the predecessor and successor nodes in the logical ring. It also has means for creating the logical ring, admitting or deleting nodes with the necessary adjustment of

---

[5] Due to the statistical nature of the method, efficiency calculations depend on the statistical assumptions regarding loading, message lengths, etc. Although there is continuing debate over these calculations, at least one source (9) suggest good efficiency up to a traffic loading of about 40% of the total bus capacity.

[6] Under a set of stated assumptions, the asymptote occurs at approximately 55% of offered load at 2 km and 10Mbps (9)

predecessor and successor station addresses, and handling faults such as lost or multiple tokens. All stations are connected in parallel across the line and receive all transmissions. Stations not having token-handling capabilities can coexist on the physical network and can respond to queries from the token-holding station as long as the response is within the primary station's token holding time. In addition, its response must not confuse other stations. This allows a primary-secondary type of relationship.

A station may be in one of several modes: Watching the medium, using the token, passing the token, initializing the ring, claiming the token, or sending an immediate response to the token holder. When in the watching mode, the station listens to activity on the medium. When the station receives a valid frame addressed to it, the MAC functions process the frame either directly or by passing the information to the LNLLC layer. If the received frame requests an immediate response, the station enters the sending-response mode. If the received frame is token, the station enters the using-token mode.

If the station goes for a prolonged time (greater than a defined timeout period) without hearing any activity, the station may infer that recovery or restart of the logical ring is needed. Either the station claims the token and restarts it around the old logical ring or it attempts to reinitialize the logical ring, depending on its state at the time the silence occurs. The node which just passed the token has primary responsibility to reclaim the token if it is lost. If the token is not recovered, other stations will attempt reinitialization.

When a station receives the token, it performs some logical ring maintenance functions (such as accepting new stations into the ring) before it transmits any information. These maintenance functions are not charged against its "token holding time." When all of the information has been sent or the maximum token holding time is exceeded, the station passes the token to the next station in the logical ring. The token is passed by sending an abbreviated frame (SFD AC2 DA SA FCS EFD) that explicitly transfers control to the successor station. The station then watches for activity on the line. As soon as anything is heard, it resets its responsibility for recovering the token and reenters the watching mode. If the token pass is unsuccessful, the node attempts to recover the remaining linkages. If, while in possession of the token, the station hears another station claiming to have the token, the station must "drop the token" and enter the watching mode. In this way, duplicate tokens are eventually eliminated.

Procedures are provided for initializing the logical ring, assigning monitoring (administrative control) responsibility, and adding and deleting stations. The initialization process involves a "contest" in which the winning station has the right to be the initializer. Adding stations involves the use of a "demand window" through which stations can request entry into the logical ring and a subsequent algorithm which sorts out simultaneous requests by multiple stations.

<u>Bus Timed-Token Bandwidth Allocation</u>. At the March 1982 meeting, the committee incorporated an optional method whereby a set of stations is guaranteed a certain percentage of the maximum number of octets that could be transmitted on a rotation of the token around the logical ring (10). Due to the presence of the token-holding timer, the maximum token rotation time (or maximum octets allowed per rotation) is known and fixed. The corresponding maximum octets allowed can be considered as the bandwidth of the ring. In the adopted method, three classes of stations are allowed and these are designated as Synchronous (S), Asynchronous 1 (AS1), and Asynchronous 2 (AS2).

An S station may transmit up to a predefined number (NS) of octets whenever it receives the token. By knowing the maximum number of S stations allowed and the sum of their NSs as compared to the bandwidth (in octets), it can be seen that the S stations are guaranteed a minimum percentage of the bandwidth. Furthermore, the bandwidth allocated to the S stations will be shared equally among them.

The remaining bandwidth, plus that guaranteed to the S stations but not actually used due to a lack of octets to send, is shared among the AS1 and AS2 stations. The sharing algorithm is essentially a counting scheme whereby AS1 stations are assigned a number NAS1 which is the maximum number of octets each station can send when allowed to do so. The AS2 stations are assigned a similar number NAS2. AS1 stations are allowed to send FS = NAS1 - TFS octets, where TFS is the total number of octets sent by all stations (including itself) on the previous token rotation. If FS is zero or negative, the station may not send any frames. Similarly, AS2 stations may send FS = NAS2 - TFS octets if FS is greater than zero.

The effect of this algorithm is to divide the bandwidth available to AS1 and AS2 stations according to the ratio NAS1/NAS2. The fractional bandwidth assigned to AS1 stations is shared equally among them. Under heavily loaded conditions, however, some AS2 stations may not receive a fair share of the bandwidth left for the AS2 stations depending on their position in the logical ring. This is not considered a serious problem since heavy loading presumably will occur only infrequently and AS2 service requirements are less demanding than those of AS1 and S stations.

Ring Token-Passing Operation

The operation on sequential (i.e., physical ring) media is somewhat different. The stations are connected in series with the medium and frames not intended for a given physical station are relayed to the next physical station after a short processing delay. A bypass relay function is provided for inactive (i.e., offline) stations to preserve circuit continuity. Tokens are passed from one station to the next without explicit knowledge of predecessor and successor station addresses since these are inherent in the physical connection. Due to the repeating action, an unclaimed message would circulate on the physical ring forever so frame removal procedures are necessary. The frame removal responsibility belongs to the transmitting station but fault detection algorithms remove frames if the token-holding station malfunctions.

In the ring token-passing method, AC1 contains the token which may be either free or busy. When optional preemptive priority is not used, a station may transmit available frames whenever it receives a free token. The station merely sets the token indicator to busy, appends the remainder of the frame (DA ... EFD), and sends it to the next station on the physical ring. The transmitting station may send multiple frames, all marked with a busy token, up to a limit imposed by the token-holding timer (counter).

Intermediate stations between those identified by SA and DA receive each message, check it for errors, and retransmit it to the next station after a few bit-time delays for processing. If the intermediate station detects an error, it sets the last bit of the first octet of the EFD to one before it is transmitted.

When the frame arrives at the DA station, the station _copies_ the frame (assuming it has the necessary buffer space available), sets two redundant "address recognized" (A) bits and two redundant "copied" (C) bits in the second octet of the EFD to one, and retransmits the frame to the next station on the ring. If the DA station is congested and cannot copy the frame, it sets only the A bits. The A, C, and E bits provide the SA station with information needed for error recovery procedures. The A _and_ C bit combination essentially provides the Type 1.5 connectionless-acknowledged service mentioned earlier. The use of redundant bits for the A and C indicators provides error protection since the EFD is not covered by the FCS.

In the meantime, the token-holding (SA) station continues sending frames or an "idle" pattern of bits until at least the first frame it sent traverses the ring. When it recognizes is own SA in this frame, it may send a "free token" abbreviated frame (SFD AC1 AC2 EFD) to the next station on the ring. Optionally, it may wait until it receives the _last_ frame it sent. This option reduces the ring capacity but provides earlier assurance that all subsequent frames were sent and received correctly.

Preemptive Priority Option. At the March 1982 meeting, the committee adopted an option for priority preemption in the ring token-passing method. Two three-bit fields are provided in AC1 to designate the current priority of the token and to reserve the next free token at a higher priority level. A station having a higher priority than the token-holding station sets the reservation bits to its higher priority level as the frame passes through the station. When the AC1 frame returns to the token-holding station, the next free token is issued with its priority bits set to the reserved priority level. Stations of lower priority are prohibited from claiming the token so it passes to the requesting station or an intermediate station of equal priority that now has frames to send.

The original station which upgraded the priority level has the responsibility to downgrade the token to its original priority when all higher priority stations have sent their available frames. When the original station sees a free token at the higher priority, it assumes that the higher priority stations have been satisfied and it downgrades the token priority before passing it to the next station. This guarantees that the token pass is restarted at the point it was interrupted. Since there are eight levels of priority, the token may be upgraded a number of times before it returns to its original value and is restarted around the ring.

Token Method Properties

The important properties of the token-passing access methods are that they can, with the provision for priority or bandwidth allocation, be fair or unfair, as the situation requires. They are also deterministic so that they can provide predictable response, thus making them more attractive for some real-time applications. The methods promise to be very efficient, with the effective capacity of the system perhaps exceeding 90% of the raw capacity of the medium. Furthermore, the efficiency is relatively insensitive to both the number of stations and the length of the medium.

On the negative side, the token-passing methods are unquestionably much more complex than CSMA/CD. Committee members experienced in the design of complex VLSI chips, however, believe that the design of suitable chips is achievable in current technology. The design evolving as the standard does not represent an existing product, although many similar systems have been in use for several years. As a result, a significant effort to verify the design through simulation and prototyping probably will be required prior to widespread industry acceptance.

PHYSICAL LAYER IMPLEMENTATION

The physical layer is concerned with the details of transmitting bits from one system to another. It involves signal levels, modulation techniques, testing circuits, etc and protocols for their control. Some of these details are independent of the actual transmission medium but many others depend on the exact type of medium and transmission technique. In addition, some transmission systems are better suited for particular media access methods. For example, the broadband transmission schemes are not applicable to the ring token-passing method.

It is the intent of the committee to provide a variety of media and transmission rates where it makes sense to do so. Due to the effort required to prepare the specifications, however, the initial standard will not include all the selected alternatives. It is anticipated that the first version of the standard will include the following:

- For CSMA/CD

    A baseband system using 50 ohm cable at a 10 megabit per second (Mbs) transmission rate. Future extensions will provide additional rates of 1, 5, and 20 Mbs.

    A broadband system similar to that used in CATV systems using 75 ohm coaxial cable at a 10 Mbs rate. The specification for this system will be issued for technical review in May 1982 and may or may not be in the initial standard.

- For the Ring Token-Passing Method

    A baseband system using 150 ohm shielded twisted-pair cable at rates of 1 and 4 Mbs.

    A baseband system using 75 ohm coaxial cable at 4, 20, and 40 Mbs. The latter rate is not in the original 802 requirements and its inclusion has not been approved by the 802 Committee, although the Media Subcommittee has approved it.

- For the Bus Token-Passing Method

    A single channel, phase-continuous FSK system utilizing 75 ohm coaxial cable for rates up to 1 Mbs.

    Five and 10 Mbs single channel, phase-coherent FSK systems using 75 ohm coaxial cable.

    A broadband system using multilevel duobinary AM/PSK at 1, 5, 10, and 20 Mbs on 75 ohm coaxial cable.

In addition, the committee is pursuing, in cooperation with Electronic Industries Association (EIA) committees, fiber optic media for star and ring topologies. The transmission rates have not been determined.

STATUS AND PLAN

The purpose of the March 1982 meeting was to consider comments received as the result of the committee ballot on Draft B and to continue refining the document. Major changes were made at the meeting, such as the incorporation of specific options for bandwidth allocation and preemptive priority in the token-passing methods and the decision to use the LC and PAD fields in CSMA/CD. It is the intent of the committee to issue the next revision (Draft C) in May 1982 for a 30 day committee ballot. As implied earlier, not all sections are considered ready for a formal ballot and these will be included for technical review only.

Comments resulting from the Draft C ballot will be discussed at a meeting in August 1982. Resulting revisions will be balloted with the expectation that the required 75% plurality will be achieved at the committee level. The proposed standard will then be sent to the Technical Committee on Computer Communications (TCCC) for their approval prior to its submission to the IEEE Standards Board. Unless some significant aspect has been overlooked, one would expect that the version sent to the TCCC, except for changes of an editorial nature, will be the initial version of the resulting IEEE standard. The committee intends to extend the standard through revisions which, for example, will detail requirements for transmission rates and media not included in the initial standard.

It is expected that the standard will be considered by the ISO (International Organization for Standardization) through its subcommittee on data communications, ISO/TC97/SC6. Its introduction into ISO probably will be through one of two routes. The European Computers Manufacturers Association (ECMA) currently is considering a set of documents derived from the CSMA/CD sections of the IEEE 802 proposal. As a liaison member of ISO, ECMA could request consideration of its documents. It is also expected that ECMA will consider the token-passing methods at its future meetings. As an alternate route, Draft C will be distributed to members of American National Standards Subcommittee X3S3. This subcommittee, as the body responsible for U.S. participation in ISO/TC97/SC6, could submit all or part of Draft C (or its successors, if any) as a U.S. contribution. Whichever approach is used, it can be expected that the proposal eventually will be approved as an ISO Standard.

ACKNOWLEDGEMENT

The author would like to acknowledge the assistance of Gerry Clancy, Bob Donnan, Rich Fabbri, Tom Phinney, and Gary Robinson in reviewing an earlier version of this paper. The errors that remain, and there may be some, are the result of the author's incomplete notes taken during a typical, but very hectic, meeting of the 802 Committee.

REFERENCES

(1). Gerald J. Clancy, Jr. and Thomas J. Harrison, "Local Area Network Standards -- A Status Report," 3rd IFAC Workshop on Distributed Computer Control Systems, Beijing, China, August 15-17, 1981 (Proceedings to be published by Pergamon Press).

(2). Thomas J. Harrison, "IEEE Project 802 Local Area Network Standard," Share 58, Paper C212, Share Inc., (111 E. Wacker Dr., Chicago, IL 60601), Los Angeles, March 17, 1982.

(3). T. Manuel, "Microcomputer-sized units handle one million instructions per second," Electronics, February 24, 1982, pp.39-40.

(4). "Resolutions Taken at TC97 Meeting," Paris, 2-4 December 1981, ISO/TC97/n1046, Secretariat for TC97, American National Standards Institute, New York.

(5). "Data Processing - Open Systems Interconnection - Basic Reference Model," ISO/DP 7498, Secretariat ISO/TC97/SC16, ANSI, New York, August 6, 1981.

(6). IEEE 802 Local Network Standard - Draft B, Institute of Electrical and Electronic Engineers, New York, October 19, 1981.

(7). "Data Communications -- High-level Data Link Control Procedures - Frame Structure," ISO Standard 3309-1979 (2nd Ed.), ISO Central Secretariat, Geneva, Switzerland.

(8). "The Ethernet, A Local Area Network: Data Link Layer and Physical Layer Specification (Version 1.0)," jointly published by Digital Equipment Corporation, Intel Corporation, and Xerox Corporation, September 30, 1980.

(9). W. Bux, "Local-Area Subnetworks: A Performance Comparison," IFIP Conference on Local Networks, Zurich, Switzerland, August, 1980.

(10). R. M. Grow, "A Timed-Token Protocol for Local Area Networks," To be presented at IEEE Electro '82, Boston, May 1982.

DISCUSSION

Narita: Dr. Harrison, will you make a similar speech at the next meeting or do you expect any major changes in the content you have just mentioned? Secondly, when will there be the first silicon chip for the token-passing method?

Harrison: I'll answer very briefly to both: I don't know the answer to either! I have the feeling that the committee itself is anxious to complete its work. In fact, I spent many hours writing a paper before the March 1982 meeting, figuring that I wouldn't have to change it, and then after that meeting I spent three rather hectic days totally rewritting the paper, because the changes were major. I don't expect that there will be further changes of that magnitude, but I would not recommend to anyone that they go out and start, as we say, pouring silicon, based on what is in the May draft. If I am back next year, perhaps I can give a report on the final standard and I hope there will not be too many changes.

As far as chips are concerned, I have no information on that. To my knowledge nobody has publicly announced their plans with respect to a chip which will implement a media access method, with the exception of the ETHERNET-like chip from DEC, INTEL and XEROX. Within the committee, however, the discussion is such that one would believe that there are several other semiconductor manufacturers (in the United States and Japan, in particular) who are very interested in the activity. Whether or not they are designing something, they have not stated publicly.

Vohandu: I have two questions. Do you make your ring token priority downgrading automatically or by free will? Doesn't that game with AS1 and AS2 seem to be like poker where three different classes of people are playing and only the people in money can play it really?

Harrison: Number one - on the token downgrading. The ring token algorithm is such that the station originally interrupted is responsible for downgrading the token. It does it automatically when it receives a frame with a free token at the level it had assigned. For example, if priority 4 is higher than 3 and I am the token holding station, and at level 3 I'm asked to release the token at 4, that's what I do. Now that token may go around and get jacked up to 5 and 6 but then is returned to 5 and then back to 4. When I see it again at 4 and it is free, I know that all the four and higher priority requests have been satisfied. So I grab that token, return it to 3 and send it on to the next station. It may, because of transmission time, be immediately jacked up again to 4. But the algorithm says that when I see a free token at the level I assigned to it, then I grab it and bring it back down to where it was. So it is automatic, but it may go through several steps before you get there.

Your second question was about the synchronous AS1 and AS2. Well, In Phoenix I sat down and attempted to write an analytical expression to predict the probability of any given station getting service under any given conditions. It is a non-linear function because every time the token comes to you, you reset your counter. If you take a six-station case you have six non-linear functions or resettings taking place. The only way I could predict ahead of time how my network would act is to write a simulation program which by the way, the author of the scheme R.M. Grow has done. He, however, has used the system so much that he has a heuristic sense of how it works. I found it took me several days of playing with charts of numbers, not having my computer with me, before I really began to understand how it worked. I might point out that R.M. Grow is presenting a paper this week at IEEE Electro '82. It is the last reference in my paper, and he has a complete explanation of his scheme, using somewhat different terms from mine. I had to simplify it for this brief presentation.

Rodd: You mentioned just in passing the monitor function. I think from the process control point of view the big worry is what happens when a token gets lost. How and where does your monitor function get exercised?

Harrison: As I indicated, the monitor function can be centralized or non-centralized. The exact way it works depends on whether it is the bus or the ring scheme. Each station has a monitoring facility, so in the bus scheme, for example, when sending the token on to the next station, I'm responsible for that token until I hear the next station come on line, take it, and do something with it; either send a message or pass it on to the next station. At that time I drop my monitor responsibility. However, if I send it and after a time period I hear no response, I invoke the monitor function and assume that the token has been lost. I then attempt to reinitialize the logical ring.

As another example, if a station gets on the line and hears no activity for some period of time, the monitor function is activated and it assumes that the token is lost and will attempt to restart the token. If two stations attempt to start the token simultaneously and they hear each other, they are both obligated to back off. So, eventually somebody is going to start the token again using a system whereby you essentially go out and ask who wants to be in the logical ring. This distributed monitor function is defined by state diagrams in Draft C of the proposed standard. It is worked out in quite some detail.

STANDARDIZATION WORK FOR COMMUNICATION AMONG DISTRIBUTED COMPUTER CONTROL SYSTEMS

G. G. Wood

European Market Research & Product Planning, Foxboro Yoxall, Redhill, Surrey RH1 2HL, UK

ABSTRACT: The current position of some standardization groups working in local area communications are discussed and compared with the needs of Distributed Computer Control Systems in industry. The standards groups are International Electrotechnical Commission (IEC), International Standards Organization (ISO) and Institute of Electrical and Electronic Engineers (IEEE).

The paper then introduces the following issues which may affect further evolution of the work on these standards:-

(1) The need and direction for services and protocols in the higher levels.

(2) The impact of fibre optics which favour a change from the passive "T bus" to a network of point-to-point links for communication.

(3) The economic arguments for the application areas with numerically small markets to adopt subsets of the standards developed for large markets.

KEYWORDS: Local Area Networks, Communications, PROWAY, IEEE 802, IEC 625. Industrial Data Highways, Communication sub-systems.

Disclaimer. This paper is not an official statement by any standardization body. The information is drawn from individual discussions and working documents which are subject to change, consequently it should not be relied upon for decisions about product design or commercial activity. The opinions are those of the author and do not necessarily represent his employer, any other committee member or their sponsoring organizations.

The author has been a member of the IEC Working Group on PROWAY since its inception and Chairman of the Coupler Sub-Group. He is a liaison member of the IEC Working Group on Interface Systems for Programmable Measuring Instruments.

INTRODUCTION

The age of the microprocessor has arrived and microprocessors are being interconnected for many different application purposes and environments. The most critical factor in these multi-micro systems is the communication structure. User concern for this subject is reflected in committees seeking to produce standards for communication among distributed microprocessor based devices. These standards are unusual in that they seek to define a standard before industry usage has created a "de facto" standard which can then be tidied up and documented. Several of these evolving standards could support industrial Distributed Computer Control Systems (DCCS). These applications are characterised by up to 100 microprocessor based stations separated by distances up to 2 KM. Such systems are the industrial equivalent of a Local Area Network (LAN).

This paper covers the author's view of the present status among standards which could support a DCCS or Industrial LAN in the future:-

- Process Plant Communications (PROWAY[1] and Process Application formats)

- Laboratory Communications:
  (Serial extension of IEC 625-1)[2]
- Office Communications:
  (IEEE 802[3] and Ethernet)[4]
- Post and Telegraph Communications
  (CCITT X 25)[5]

Following a summary of present status among these standards, the paper discusses some areas which are not covered by the committees and some of the trade-off decisions which must be made.

PRESENT STATUS OF STANDARDS

The IEC PROWAY Committee has defined functional requirements for an industrial data highway. Table 1 lists some of these requirements and reasons for their choice.

1. Multiple master stations sharing the highway, with passive connections from each station to the data highway. This ensures that loss of one station or loss of power in one part of the plant will not interrupt communication among other plant areas.

2. Single wire, serial, multi-drop connection among stations. This gives a simple message path and economic benefits in cabling and connections. Options include redundant paths.

3. Variable length message frame with no restriction on the bits and bytes contained in a user's application message.

4. Operation of the communication system must be deterministic to support calculation of worst case access time and guaranteed response figures.

5. Automatic acknowledge, within the communication sub-system, for each message sent. This is a safety feature to support fast fault recovery, ensure messages are not lost and keep them in sequence.

TABLE 1. GUIDELINES FOR GOOD INDUSTRIAL DCCS

The PROWAY standards group is working to meet the Table 1 criteria by defining a standard for Layers 1 and 2 of the OSI model[6].

The 625-Serial and IEEE 802 groups are working on several standards with various optional choices in Layers 1 and 2. Some versions will meet all the criteria of Table 1 and others which meet only criteria 1, 2 and 3.

The X25 standard includes Network Layers of the OSI model and can support criteria 2 to 5 of Table 1. Multi master is not easily provided by X25, however a number of manufacturers have developed products using X25 components with non-standard enchancements to provide multi master capability.

AVAILABILITY

PROWAY and 625-Serial are still in the definition stage and products meeting these standards are not expected in the market place until 1985 or later.

IEEE 802 is in the final stages of definition and products for some of its subsets are now available in the market place.

DEVELOPMENT OF UPPER LAYER FEATURES

The application of small DCCS projects will involve a single, local bus. In such a case the functions of a network layer are not needed. Also for small systems the maximum frame length and task or session connections are usually fixed at installation time and absorbed into the application programs. The small DCCS system effectively has its application layer directly interfaced to the Link Layer.

For larger projects with multiple independent busses, Network and transport layers are needed to free the application from concern about message length, absolute addresses and connections.

For ease of system expansion from a small DCCS to a large Network DCCS the interface logic or state machines should be the same for each interface above the Network layer. If this goal is achieved, then low cost single bus applications can be built without the intermediate layers (345). However, they can easily be expanded into larger networks by inserting the extra layers and changing the application address references from absolute to logical.

PASSIVE "T" CONNECTION OR STORE AND FORWARD

The industrial community have requested passive "T" junctions to increase the integrity of their DCCS installations.

PROWAY, Serial-625 and IEEE 802 each embody versions which meet this aim. However, this requirement will cause problems when optical data links are applied in industry.

The technology of optical point-to-point data links is now well established and offers significant opportunity to eliminate such traditional problems as:
- electrical interference
- lightning effects
- ground loops
- spark hazard in explosive environments

Passive "T" optical connections are not expected to be economic in the near future so other methods must be used to ensure link integrity when power fails at a station. Three solutions can be considered:-

i) Battery backup to sustain the active retransmission elements in each station.

ii) Optical switching relays which bypass the station active elements when power fails.

iii) Multiple point-to-point links giving a network solution for rerouting and bypassing a failed station.

The choice among these will depend on a balance of first cost and installed maintenance cost. Future development of LSI components for Networking Protocols will probably favour the third solution.

ECONOMICS

Table 2 gives an approximate estimate of the market for station couplers and logic in DCCS applications.

Industrial Uses (e.g. PROWAY) 10^5 Units per year
Laboratory and Light Industry (e.g. Serial-625) 10^6 units per year.
Office Automation (e.g. IEEE 802) very large
PTT and CCITT (e.g. X25) very large.

TABLE 2. MARKET ESTIMATE FOR DCCS STATIONS

For economic reasons the development of PROWAY and Serial 625 standards must consider adaptions of X25 or IEEE 802. This allows the economics of large scale productions.

In some situations a DCCS requires links through the PTT for a remote sub-system or management computer. The use of X25 adaptations should simplify such interaction with PTT systems.

REFERENCES AND NOTES

1. PROWAY is being developed by the International Electrotechnical Commission Technical Committee 65C, Working Group 6. This group was previously named TC65A, WG6 and drafts parts of PROWAY were circulated for national comment by TC65A.

   Another working group TC65C WG1 is working on standards for message formats in industrial data highways.

2. International Electrotechnical Commission Technical Committee 65C, Working Group 3. This group was previously named TC66 WG3 and defined a short distance, bit serial, byte parallel data bus which has been published as an IEC standard number 625. This group is now investigating a long distance, fully serial bus which will have maximum compatability with 625. This paper refers to the proposed serial extension as "625-Serial".

3. Institute of Electrical and Electronic Engineers, Project 802. Draft documents were published for comment at the end of 1981. The drafts include many sub options some of which are very close to the Ethernet standard.

4. The Ethernet, A Local Area Network: Data Link Layer and Physical Layer Specification. Jointly published by DEC, Intel and Xerox Corporations. Also referred to as the DIX Data Highway.

5. International Telephone and Telephony Consultative Committee. Recommendation X25 covers physical, Data Link and Network functions.

6. International Organization for Standardization ISO TC97 SC16 "Reference Model for Open Systems Interconnection". ISO Draft Proposal 7598.

DISCUSSION

Harrison: The only comment that I would make is that, first of all, we are operating in real-time and in parallel in the sense that those two committees (IEEE-802 and PROWAY) met basically at the same time. Graham (Wood) did have access to my paper but, I guess, really not about the same time he wrote his.

I'm not trying to defend IEEE-802, I'm trying to report on IEEE-802. I'd make the observation, however, that in his necessity to provide comparison charts in a very brief way Graham does tend to mix advantages and disadvantages within the same chart in a way which is confusing. You cannot say that it is a disadvantage for CSMA/CD, but then in the same column put down the advantage of a token, because they are fundamentally non-comparable. You really need to take, go through and sort out the advantages and disadvantages of the ring token scheme which, in fact, does have immediate acknowledgement, as compared to the bus token scheme which does not yet have it as stated on your chart, compared to CSMA/CD which has no hope of ever getting it, for the present, and then also have a separate column for CSMA/CD. IEEE-802 has three standards. Now what?

Wood: I had tried to make the assumption that I was comparing PROWAY with the bus version of IEEE-802, not the ring, because PROWAY at the moment is aiming for a bus only, not a ring. I had assumed that we'd be comparing with the token version of IEEE-802.

Harrison: Well, that's true, but you also have shown the disadvantages of active taps, and truly, the only active tap is CSMA/CD. The other is the one we required to bend the cable and bring it in. But in a bus you really don't. That's a passive tap.

Wood: That is the first time I've heard that statement because I think you've also said the physical layer was identical across all versions.

Joudu: How can you prove the very high level of reliability of one undetected error for a thousand years?

Wood: It is very difficult. There have been a number of papers but what we are assuming is that the line noise is known. The noise is going to give us one residue faulty bit per 16^6 received at the link level. Once you have made that assumption, then for a given average message length and a given data rate it is simply a matter of statistics. We have been able to show that, with the CRC method that is available and with the bit stuffing situation, you can detect most of the bit faults that occur with the probability of eventually 3×10^{-15}, that is one undetected fault in a thousand years. I am not sure if there has been a published paper, but there were a lot of papers presented and argued inside the PROWAY committee on that subject.

Tavast: Would you comment on the choice of the HDLC frame for PROWAY.

Wood: The economic situation that I was arguing was, the choice of IEEE-802 or HDLC. We had been analysing the HDLC frame for two or three years, evaluating all the error probabilities, and how we could extend it. So we felt we understood that frame reasonably well and we had satisfied ourselves that we could use it. If we are to start using the IEEE-802 frame, we'll have to do all the analyses again, if you like. There are still some questions about IEEE-802 applicability for some of the functions that PROWAY believes are important. We have found ways of using the HDLC to satisfy all of those functions. So, with some question marks about parts of IEEE-802, our present position is to stay with HDLC. And of course there are plenty of LSI chips coming out that were aimed for HDLC. But we understand that the important bits that PROWAY wants to add on can be added as micro code on the chip. So we don't require inventing a new chip - which is a big economic problem.

Natiello: I'd also like to make a comment in that aspect. As a member of the industrial community I emphasize that we badly need data communication standards for use in process control. We don't necessarily want to stop or control developments of standards, but things which stay in the committee for ever and ever are not very useful to us. We are in many cases under time pressure.

Wood: Another point I could add is that I think at least 50% of the people who are supplying industrial communications systems have to develop and build their own systems. They don't intercommunicate, but many of them have used at least the HDLC frame chips which are available as the minimum building blocks of the present systems. We have found it easier to move up to larger, more integrated versions of an HDLC-related structure.

Narita: What is your personal opinion of the very slow, sluggish, out-of-date speed of one megabit per second of PROWAY? Do you think it is enough for medium or large scale applications or is there any plan to raise the speed, say, to 5 or 10 Mbit/sec.

Wood: Yes, based on the data rates that we've evaluated, we feel that many industrial plants will be satisfied with the 1 Mbit rate. We are fairly happy as a committee that we can let the IEEE-802 and the IEEE-625 committees pioneer the physical technology to run it with 10 Mbit and 20 Mbit. Because we are using a layered structure, we see no problem to shift to that at a later date. But following Lou's (Natiello) comment, we are quite happy to stay with 1 Mbit rate to get something quickly using existing technology. We expect eventually to change the physical layer to optics which has other implications as well. If we want to get something published and something out, we are deliberately making the higher layers independent, so that they won't be affected if we do change the bottom layer.

Vohandu: You are speaking on the lower level protocols. What about use of standard data formats, e.g. ISO data transmission coding tables in the higher (application) level protocols of your standard?

Wood: I think PROWAY is treating the data field as completely transparent. This is what I was getting at when I mentioned the IEC TC65A WG7 (now IEC TEC65C WG1), trying to agree standards for the application format which is the data content. They will be considering the different data tables. There are some recommended formats used by the laboratory people. I have tried to say that for a floating point number you put your floating point this way and your mantissa goes this way. It is fairly arbitrary to make a decision, so we don't all invent our own different internal company standards for higher level protocols.

DEVELOPMENT AND QUANTITATIVE EVALUATION OF DISTRIBUTED SENSOR BASE MANAGEMENT SYSTEM

T. Muto, C. Imamichi, A. Inamoto and S. Kato

Computer System Works of Mitsubishi Electric Corp., Kamakura, Japan

Abstract. The sensor-base-management-system (abbr. SBMS) is a standard software package, designed to perform the control data aquisition and the control output data initiation, in place of application software. Furthermore, the distributed SBMS was newly developed to support distributed system architectures, upon success of the conventional SBMS. This distributed SBMS is assumed to be constituted on a distributed computer control system, where remote process input/output subsystems together with front-end-processors will be installed as local stations on the plant floor.

In order to give the practical aspects of this distributed SBMS, the model system applied to a hot-strip-mill control system is outlined. The quantitative evaluation studies on this distributed SBMS, focussing the interests on data communication cost, data processing cost and system performance loss are done with reference to the above-mentioned model. The objectives of these studies are to give the system designers adequate suggestions on where is the balancing point between a distributed and a non-distributed architecture, what is the optimal number of the local stations, and hopefully more.

Keywords. Data aquisition; data processing; large-scale system; computer applications; computer evaluation; steel industry; hot strip mill.

INTRODUCTION

In the field of industrial computer systems, some of the major trends in these years are that computer systems are getting larger in scale and more complex in configuration and that many of so-called large-scale systems have become common. This means that industrial computer systems are now being much more heavily relied and that the functions of computers are being widely expanded. In addition to that, the control range of computer systems is now geographicly expanding and computer systems now control not limited to each of the local parts of the plants but totally control the plants.

To fit to those needs or trends, the power of industrial computers has been drastically and continuously up-graded with respect to CPU processing capability, capacity of mass storage, speed of I/O channels and so on. However, it has been proved that power-up of individual computers alone may not cater for all the requirements on system scale, but that the final solution may be placed on system structure. (1) To share CPU tasks between two or more main computers and (2) To distribute the load of host computers to front-end-processors (abbr. FEP), are two of the major approaches. The former is the concept of multi-processing and the latter is the one of distribution. In this article, the discussion will be focussed on this distributed system architecture.

Meanwhile, the most important function of industrial computer systems is no other than to control the plants, and for this purpose, tremendous amount of data processing on process input/output is required. Recently, the data processing on process input/output has become much larger in scale and more complex in procedure, as the control algorithms become more advanced and more complicated. To develop these data processing software for each system exclusively may be too much burden to application software. The sensor-base-management-system (abbr. SBMS), a standard software package, has been developed in order to remove such burden from application software. So, the major functions of the SBMS are data aquisition of process input data, initiation of process control output data, filing of historical or time-serial process data and the other functions concerning process data handling. The details concerning this SBMS will appear in section 2.

As mentioned before, distributed system architectures have become extremely common in industrial computer system fields, so that this SBMS together should have the distributed architecture. The distributed SBMS, described in this paper, has been newly developed to respond to these requirements, of which design concepts are:
(1) Despite that physically the system may be distributed, the users of SBMS need not be conscious of the distributed architecture;
(2) The data processing load on host computers should be distributed to FEPs;
(3) The system should have the support facilities for multi-computer system architectures.

The detailed descriptions on this distributed SBMS will be given in section 3.

As for the practical applications of this SBMS, the model will be presented, where the distributed SBMS is applied to a hot strip mill line control system of the steel mill plants. In section 4, the system configuration, system scale and utilization aspects of this SBMS at the hot strip mill application model will be given.

The quantitative evaluation studies on this SBMS, the major framework of this paper, will be tried and presented in section 5 on the model of the hot strip mill application. The advantages of distributed system architectures have been qualitatively reported in the variety of articles and papers, but not quantitatively. However, the quantitative evaluation data concerning the trade-off between distributed architecture and conventional architecture would be so much helpful to the system designers, when the decision whether distributed or not should be made at the system design stage. At quantitative evaluation studies in this paper the attentions will be concentrated on the three items, say (1) data communication cost; (2) data processing cost; and (3) system performance loss. The special formula and premises on each of three items, will be introduced, to offer the numerical results.

THE SENSOR BASE MANAGEMENT SYSTEM

System Configuration

The sensor base system (abbr. SBS) is a kind of control data base system, to be constituted on a process input/output hardware system. The main objectives of the SBS are:
1) To offer user programs (application programs) logical access-methods on process input/output hardware system;
2) To keep off user programs from complex procedures on historical data processing (high-speed data scanning or long-term data storing);

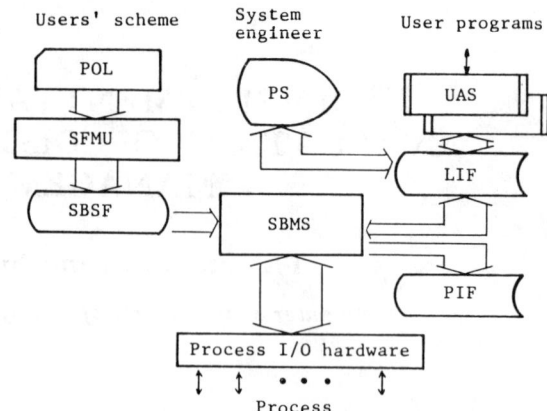

Fig. 1. System configuration of the sensor base system

3) To control and to process all of process input/output data systematically;
4) To standardize data-processing, such as limit checking, digital filtering and the other error checkings.

In other words, if the SBS is introduced, application programmers need not know about complex hardware addressing of process input/output mechanism, nor about scanning and filing procedures, concerning time-serial process data. Also, in case of hardware interface change, the modification will be made on the SBS only and no maintenance will be necessary on application programs.

The SBS, of which basic configuration is illustrated in Fig.1, is composed of process input/output hardware, physical image files, logical image files, an SBMS, data-base scheme files and users' access-subroutines. The general descriptions on each of these main components of the SBS are given below.

Sensor-base-management-system (SBMS). The SBMS, the principal part of the SBS, is composed of data processing and system management programs. The logical image files and physical image files of the SBS are all under control of this SBMS.

Physical-image-file (PIF). The hardware image of the process input-output system is maintained and updated periodically in this PIF.

Logical-image-file (LIF). At the data aquisition procedure, the SBMS processes and arrange the physical data in PIFs and saves the processed logical data in the LIFs. At the output procedure, the process is vice-versa.

Sensor-base-scheme-file (SBSF). The processing procedures or the schemes concerning the SBMS are stored on these SBSFs, which include logical-to-physical-conversion (LTPC) procedures, physical-to-logical-conversion (PTLC) procedures, and process data specifications. The SBMS performs its data processing jobs, with reference to these SBSFs.

Scheme-file-maintenance-utility (SFMU). This utility program allows the SBS users to do maintenance works on the scheme files, such as registration, elimination and modification. The problem-oriented-language (POL) is provided for this utility program.

Users'-access-subroutines (UAS). These subroutines are provided for interface between the user programs and the SBMS.

Process-simulator (PS). This is a CRT (Cathode-ray-tube) based support facility associated with the SBMS, which supports system test, simulation and monitoring. The further descriptions are coming in the succeeding subsection.

Functional Description

From the view-points of functional configuration, the SBMS could be divided into following three subsystems.

Logical process I/O subsystem (LPIO). This subsystem will furnish not physical but logical access methods on the process I/O hardware system. This subsystem performs basic functions of the SBMS and supports instantaneous process data handling. Interface with hardware system, engineering unit conversion, malfunction checking and logical conversions are included in this subsystem.

Short-term-quick-scanning (STQS). The SBMS will initiate high-frequency but short term process data scanning, when it gets an event from the process. The time-serial process data, corresponding to each of the process events, are offered to the user program. The user program can retrieve these short-term time-serial data, by specifying an event number as a key, from the logical files of the SBS.

Long-term-slow-scanning (LTSS). The SBMS periodically aquires the specified process data and does some data-processing such as digital filtering, averaging and getting maximum & minimum. The acquired and processed data are to be stored on the large-scale disk memory files, for long term use. The user program can demand the retrieval of these time-serial process data to the SBMS by specifying a set of time and logical point number as a key.

From the view-points of file size, scanning period and number of logical points in normal application, the above-mentioned three subsystems are compared to be arranged in tabular form, as in TABLE 1. The functional configuration of this SBMS has been designed and developed, so that this SBMS would be applied in the wide range of industrial computer system application field.

Here, some quantative comparison work will be done on these three subsystems of the SBS. In Fig.2, the required response time or scanning period of the process data is taken as the X-coordinate axis, while the system scale evaluated by the number of logical points is taken as the Y-coordinate axis. The normal application range of each subsystem is evaluated to be shown as an enclosed area in XY-plane of Fig.2.

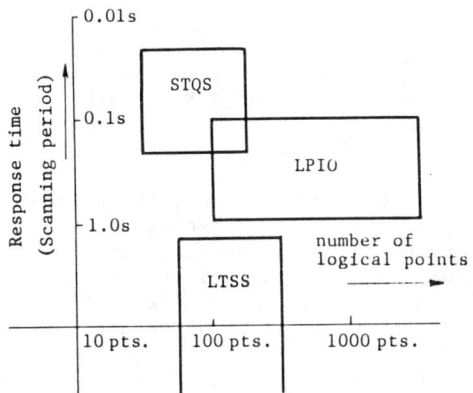

Fig.2. Application range of three subsystems.

Process Simulation & Monitoring

In the industrial computer system application field, the support function, associated with the SBMS is one of the most important points for qualification, when application of the SBMS is actually studied or planned. As associated support functions, the SBMS furnishes the functions on system test and maintenance, of which man-machine interfaces are totally CRT-based. The detailed functional descriptions on these support functions are given immediately after.

TABLE 1 Comparison of SBMS Three Subsystems

Subsystem	LPIO	STQS	LTSS
Scanning period or response time	100 ms ∿ 1.0 s	20 ms ∿ 200 ms	1.0 s ∿ 60.0 s
number of logical points	100 pts. ∿ 2500 pts.	30 pts. ∿ 200 pts.	50 pts. ∿ 300 pts.
data processing	instantaneous	time - serial (short term)	time - serial (long term)
file size	∿ 2.5 K words	∿ 10 K words	∿ 1 M words
file holding period	instantaneous	50 s ∿ 100 s	1 week ∿ 1 month

Process simulation. In simulation mode, the SBS is isolated from the process I/O hardware system, so that the input process data can be fed by a CRT key-board manually and the output process data from the user programs can be monitored by a CRT viewer, but not put to the process. To facilitate this process simulation function, the SBMS has two alternative modes, on-line mode and off-line mode (= simulation mode).

Process I/O signal test. Process input/output signal testing could be done by means of a CRT keyboard, while the system is connected with the plant.

Process monitoring. While the system is operating in on-line mode, the status of the process and the control output signal of the users' control programs could be monitored on a CRT viewer.

THE DISTRIBUTED SENSOR BASE MANAGEMENT SYSTEM

System Configuration

In the previous section, the general descriptions of the conventional or centralized SBS, have been presented. In this section, the newly developed distributed sensor base system (DSBS) with distributed SBMS will be described. The general system configuration of the DSBS is illustrated in Fig.3, which corresponds to Fig.1 of the conventional SBS. In case of the DSBS, the total process I/O hardware system is localized and distributed among the local stations as local process I/O. The each of the local stations, which will be installed on the plant floor, closer to the controlled equipment, is composed of a front-end-processor (FEP) as intelligence and a data communication device, in addition to a local process I/O subsystem. The local sensor base system, constituted on each of FEPs, as illustrated in Fig.3, is composed of a local management system, physical image files, logical image files, scheme files and users' access subroutines, and has almost identical structure as the host station sensor-base system has. In normal situation, very limited scale of process I/O hardware (limited to interface within the host computer room) is provided in the host station.

The fundamental philosophy of this distributed sensor-base system is "Distribution of process I/O hardware and data processing CPU load". However, this DSBS ensures to its users "Transparency" as a distributed system. This "Transparency" is described more in detail.

Transparency about process I/O hardware distribution. The users need not be conscious of "Distributed system architecture". In other words, although the process I/O hardware is distributed and localized, from the view points of DSBS users who have interface with process I/O only by the access subroutines, the total system seems to be one unique centralized system.

Transparency about location of user programs. Regardless of the location of the user program, the identical access methods (subroutine) are offered from the DSBS.

Transparency about multi-computer system architecture. Even if one resource of the local process I/O hardware is shared between the user programs on more than two host stations, the resource management is completely implemented by the SBMS. No burden will be imposed to the user programs, concerning resource sharing of process data.

In order to realize the above-mentioned transparency, the local SBMS in local stations and the SBMS in host stations are designed to have organic coupling in this DSBS. To be more concrete, this organic coupling requires the data communication measures, with high speed and high performance. Normally, in this case, dataway system seems to be the most suitable for the data communication measure, and this DSBS is to be based on the dataway system.

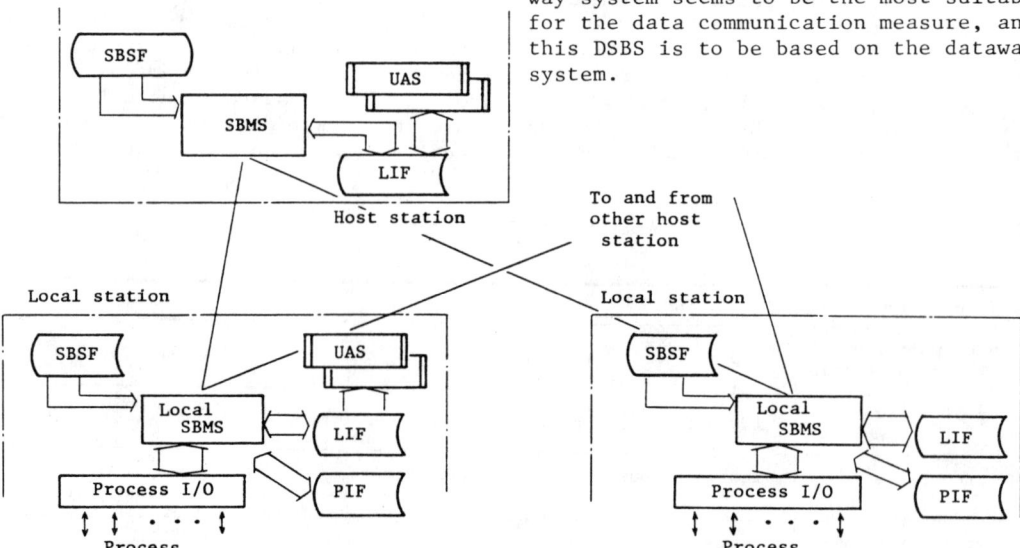

Fig. 3. System configuration of the distributed sensor base system

Functional Description

Described in this subsection is how the main three subsystems of the SBMS are implemented in the newly developed distributed SBMS.

Logical process input/output (LPIO). First of all, as for logical input function, the physical image of each local process input is made-up at each local station. The logical process input image after processing by the local SBMS is transfered to each host station, to build-up an identical copy of the logical input image made-up in the local station. At the host station, the logical image from each local station is put together, to be built-up as a total system image. This logical input mechanism is shown in Fig.4(a).

Next, as for logical process output, the logical image is first built-up at each host station, the copy of which is transfered to the local stations. At the local station, the copies of logical output image from the host stations are merged and resource-control concerning process outpout is managed at that time by the local SBMS. And after that, the local logical image is converted into the physical image, which is given to the process output hardware. The above mechanism is illustrated in Fig. 4(b). The functions concerning on-line and off-line (simulation) mode are the same as in the original SBMS and the process simulation and monitoring support are the same as well.

Short-term-quick-scanning (STQS). In this case, the most parts of data processing job are done in the local station. The time-serial data acquired by the local SBMS are saved in LIFs on the local memory for a certain period. Every time when the short-time scanning is terminated and the data are ready in the local file, an event signal is transmitted from the local station to host stations, to inform that the corresponding process data are ready. The user program in a host station can access those data in the local LIF by the access subroutine. The data flow concerning this STQS in the distributed system is illustrated in Fig.5.

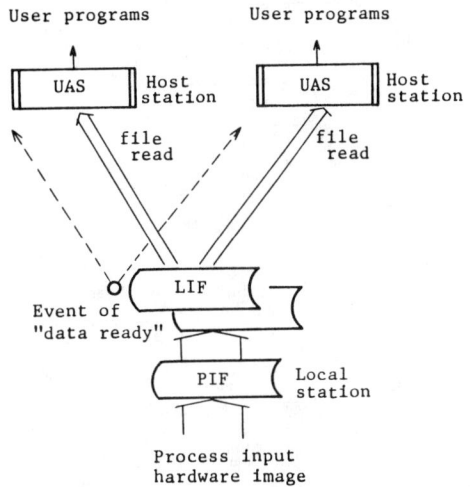

Fig. 5. Data flow of short term quick scan

Long-term-slow-scanning (LTSS). The large-scale logical files associated with this LTSS are reserved on the host computer disk memory, since these file could not be reserved on FEP memory. As shown by the data flow of Fig.6, the host computer SBMS processes the logical input data transmitted over from local station SBMS and builds up the long term data files, with reference to the scheme file SBSF in the host computer. So, this LTSS is designed to be constituted over the distributed LPIO system.

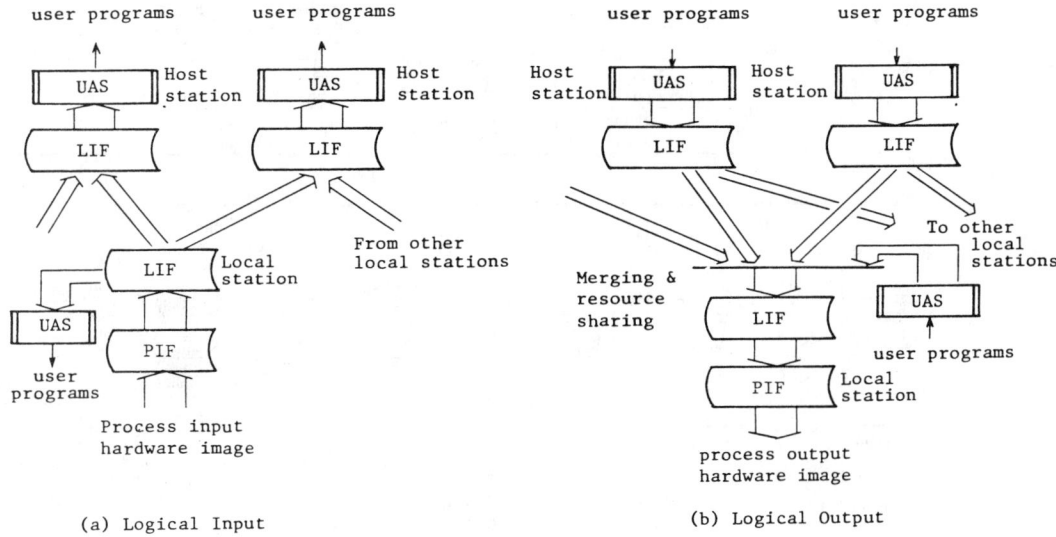

(a) Logical Input   (b) Logical Output

Fig. 4. Data flow of logical process I/O subsystem

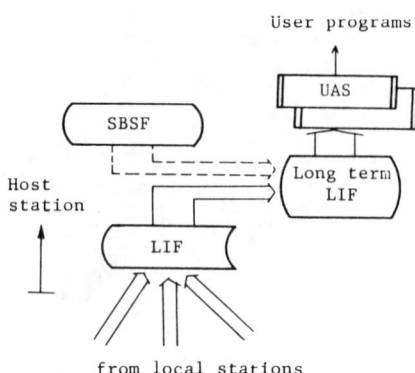

Fig.6. Data flow of long term scan

APPLICATION

In order to make the discussions and descriptions on the distributed sensor base system more concrete, here in this section, the practical aspects of the distributed SBMS will be offered by introducing a model system applied to a hot strip mill control system. A hot strip mill has been proved to be one of the most respresentative application fields of the distributed sensor base system, so that the model system may give the relevant image of the distributed SBMS.

Hot Strip Mill Computer System

A hot strip mill very naturally requires a distributed computer control system, according to the following reasons :

(1) Hot strip mill plant is a complex and large-scale plant and needs large-scale computer control system;
(2) It is relatively easy to decompose the overall control system into several subsystems;
(3) Physical range of a plant is so wide (normally 500m to 1,000m).

The distributed computer control systems have been introduced to the hot-strip mill from the early development stage of the distributed computer system, in advance to the other plants application.

System Configuration

The distributed computer control system configuration shown in Fig.7 is a respresentative system configuration on hot strip mill application, where the distributed sensor-base system could be applied. In this case, the total mill line is thought to be divided into five zones, for each of which an intelligent local station with FEP is to be installed. On the other hand, in the host station, two sets of main computers are to be applied, one of which is for mill line control and the other is for furnace area control. These two main computers are loosely coupled, to form a loosely-coupled multi-computer system.

In this system configuration, not only process input/output hardware, but also peripheral devices are distributed. However, in this paper most attention will be concentrated on the sensor base system.

The general hardware specification of the processors and data-ways involved in this system configuration is outlined in TABLE 2, where the hardware equipment of Mitsubishi Electric Corp. happen to be applied.

CRT: Cathode ray tube
OTW: Output typewriter
FEP: Front-end-processor
VAS: Voice annaunciator
LP : Line printer
STW: System
FDD: Froppy-disk-drive
DAD: Dual access disk

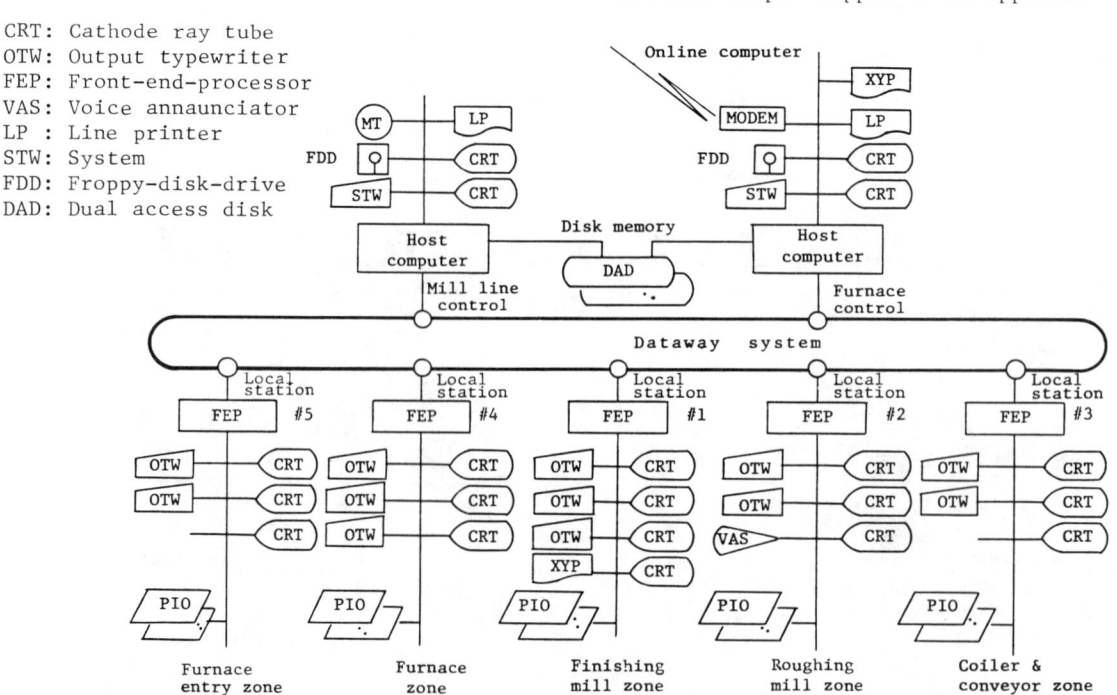

Fig.7. The representative hot strip mill computer control system

TABLE 2 Hardware Specification

Component	General Specification
Host computer	MELCOM 350 - 50 (A2500) add time = 0.2 μm memory cycle time = 0.65 μs/64 bits maximum memory size = 1 M words number of instructions = 250
Local computer	MELCOM 350 - 50 (A2100) add time = 1.2 μs memory cycle time = 0.60 μs/16 bits maximum memory size = 192 K words number of instructions = 164
Data way system	MDWS - 30S transmission rate = 15 M bps maximum no. of stations = 256

TABLE 3 Physical Scale of SBMS

No.	Local station identifier	Number of PI/O points	Distance from host station (m)	
			CASE-1	CASE-2
1	Finishing mill	2000	20	100
2	Roughing mill	600	250	350
3	Coiler & conveyor	700	200	300
4	Furnace	750	350	450
5	Furnace entry side	900	500	600

TABLE 4 Logical Scale of SBMS

No.	Local station identifier	Number of logical points		
		LPIO	STQS	LTSS
1	Finishing mill	700	100	40
2	Roughing mill	100	50	20
3	Coiler & conveyor	250	30	20
4	Furnace	400	10	200
5	Furnace entry side	250	10	0

Distributed Sensor Base System

The system scale of the distributed sensor base system applied to the host strip mill computer control system will be shown. First of all, regarding physical scale, the number of process input/output hardware points and distance from the host station to local station are shown in TABLE 3, where distance is not direct distance but cable length. As for the distance data, the following two cases are taken into account. In CASE-1, the host station is assumed to be installed inside the plant floor, while in CASE-2, the host station is to be installed in the plant-office detached to the plant floor.

Next, from logical point of view, the system scale will be shown as the numbers of logical points, at every local station. These numbers as to LPIO, HSQS & LTSS subsystems are listed up and tabulated in TABLE 4, indexing a station. The data in these tables will be utilized as basic data at the quantitative evaluations of the distributed SBMS in the next section.

In addition to those quantitative figures concerning the distributed SBMS, to show how each of three subsystem, i,e. LPIO, STQS and LTSS actually fulfills its functions, the practical aspects of these subsystems at the hot strip mill control system application are arranged in tabular form as in TABLE 5.

EVALUATION STUDIES

Quantitative Evaluation

Concerning the system model for evaluation analysis, the DSBS applied for a hot strip mill computer control system, which has been outlined beforehand in the previous section, will be adopted. Further, in order to make evaluation results visual and understandable, the following expression techniques will be applied. As one of the parameters of evaluation on each case, system scale will be chosen. This means that the cross-point or the conflict point between the distributed architecture and the conventional architecture may be obtained, by increasing system scale. Here, on increasing the system scale as an index, it is temporarily assumed that the number of the local station should be increased sequentially starting from local station No.1 to local station No.5. The increasing sequence assumed here is just the temporary sequence and so that sequence itself is trivial.

Besides, the system scale as an index will be appropriately arranged and normalized to convenience of every evaluation case, so that the curves of the conventional architectures may be linear or quadratic.

TABLE 5 Practical Aspect of SBMS

Subsystem	Use
LPIO	o Operator's desk interface. o Interface with electric drive system. o Interface with sensors specially tracking sensors. o Miscellaneous process I/O.
STQS	o Data aquisition for mill control model calculation. o Data aquisition for learning control or feed forward control. o Data aquisition for precise quality control.
LTSS	o Data aquisition for furnace combustion control and energy saving control. o Data aquisition for long term quality control and long term learning of models.

Process Data Communication Cost

As far as the hot strip mill computer control system is concerned, the data communication cost on the process input/output interface, say mostly the hardwiring cost, may occupy approximately 10 to 30% of the total system cost, so that the reduction of hardwiring cost may be the serious problem for the system designer.

If the distributed system architecture will be adopted, the amount of the hardwiring and consequently total wiring cost may be reduced, while the cost of both a dataway hardware system and local stations containing FEPs may be increased. In order to evaluate these data communication cost, some of the formulation techniques are required. Here, in this paper, two cost functions are introduced, one for the cabling cost estimation in the conventional architecture and the other for data communication hardware in the distributed architecture. Those cost functions are detailed below.

Hardwiring cost. Cost function of hardwiring of process I/O system could be approximated and be formulated as :

$$C_W = \sum_{i=1}^{n} (\alpha * P_i + \beta * L_i * P_i), \quad (1)$$

where C_W : total hardwiring cost;
P_i : number of process I/O points;
L_i : length from host station;
α, β : parameters;
i : local station index.

In other words, it is supposed that the hardwiring cost may be proportional both to the number of process I/O points and to the number of process I/O points multiplied by cable length.

Data communication hardware. Cost function of data communication hardware could be formulated as:

$$C_d = C_{DW} + n * C_{FEP}, \quad (2)$$

where C_{DW} : dataway system initial cost;
n : number of local stations;
C_{FEP}: cost of FEP hardware cost;
C_d : total data communication cost.

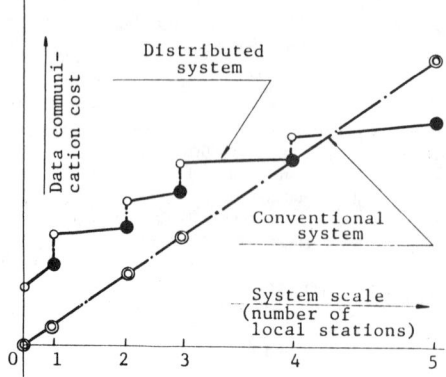

Fig.8(a). Evaluation of data communication cost in CASE-1

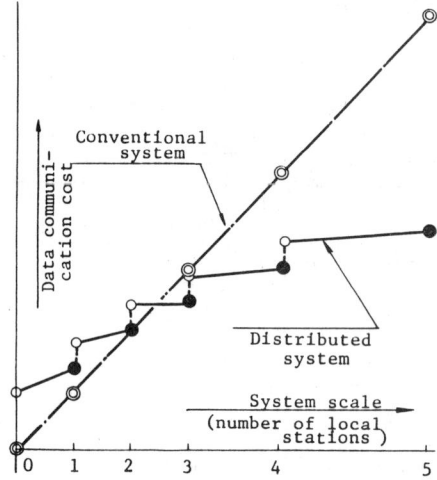

Fig.8(b). Evaluation of data communication cost in CASE-2

In the TABLE 3, the basic data concerning data communication cost were shown in two cases, CASE-1 that the host station is installed within the plant-floor, maybe next to the finishing mill pulpit and CASE-2 that the host station is inside the plant-office detached to the plant floor. The results of the data communication cost evaluation, are shown in Fig.8 (a) and Fig.8 (b) for CASE-1 and CASE-2 respectively.

In CASE-1, the conflict point between a distributed and a conventional system seems to be at the point of three or four local stations, saying in system scale. A distributed system may get more advantage in CASE-2, where the conflict point may be a couple of local stations.

Data Processing Cost

The evaluation measure on this CPU load distribution is how the total system cost of data processing is reduced by distributing system duty of host computers among FEPs. As for the data processing cost, the unit processing cost at FEPs should be set much low, compared with the unit cost at host computers, of which hardware cost could be naturally high. On the other hand, the data processing cost for data communication between stations should be taken into account at the distributed architecture.

The following premises should be made for the data processing cost evaluation.

Premise-1. The data-processing cost for a particular job at each station is proportional to the average CPU duty ratio (%) occupied by that job.

Premise-2. The data-processing cost is also related to the required response time (or turn-around time). On those jobs which require quick service of CPU, high processing cost may be imposed. This relation between CPU cost and response time may naturally differ, depending on whether a host station or a local station. The processing cost vs. response time curves at each station shown in Fig.9 are thought to be practical.

Premise-3. The data processing cost at each station is proportional to the processing hardware system cost of that particular station, where the processing hardware system includes a CPU, a main memory unit and a set of disk-memory-drives.

Premise-4. The data processing cost related to the SBMS contains the processing cost of (1) Interface with the process I/O hardware system; (2) Engineering unit conversion; (3) Logical to physical or physical to logical conversion; (4) Error checking; and (5) Image data communication between stations, if necessary.

Under the above-listed premises, the actual numerical calculations are done on the hot strip mill system application, of which results are presented both in Fig.10 for the LPIO subsystem and the STQS subsystem. Reviewing these figures, it may be concluded that the distributed system architecture may favor to the conventional architecture, supposed that the system may include three or four local stations or more.

System Performance Loss

On evaluating the system reliability of the sensor-base system, it seems that to offer the expected amount of system performance loss may be more practical, rather than just to give the reliability data of the total hardware system and/or the individual hardware components. According to this philosophy, the concept of the expected system performance loss (abbr. ESPL) will be introduced,

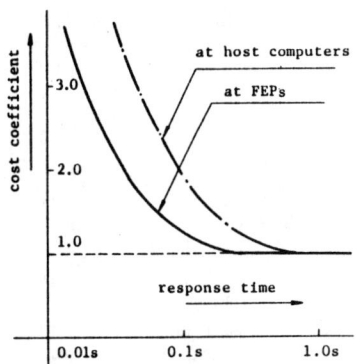

Fig.9. Cost coefficient vs. response time

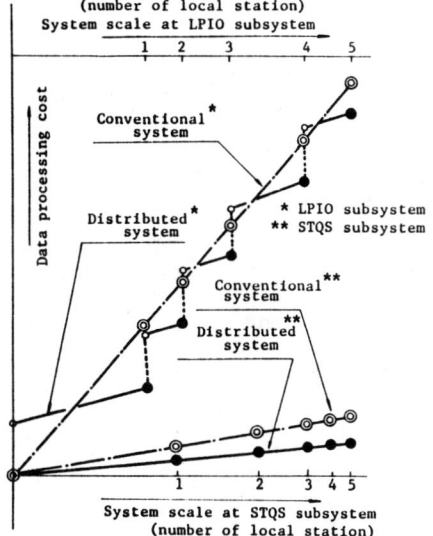

Fig.10. Evaluation of data processing cost on LPIO & STQS

of which the definitions are:

Definition. The system performance loss, associated with the fault of a particular hardware subsystem should be assessed as the sum of the hardware cost of all the hardware subsystems of which performance or functions are lost due to that fault. The ESPL should be defined as this system performance loss multiplied by the down-time ratio (down-time ratio = 1.0 - availability) of that hardware subsystem.

The following premises are required.

Premise-1. The hardware cost of a host station is well balanced as in the following formula:

$$C_{HOST}(j) = C^*_{HOST}(j) + \sum_{i=1}^{n} C_{FEP}(i,j) + \frac{C_{DW}}{2}, \quad j = 1, 2 \quad (3)$$

where $C_{HOST}(j)$: cost of a host station(j) at the conventional architecture;
$C^*_{HOST}(j)$: cost of a host station(j) at

the distributed architecture;
$C_{FEP(i,j)}$: cost of FEP(i), imposed on a host station(j)
C_{DW} : cost of a data way system;
n : number of local stations.

Namely, the original hardware cost of a host station at the conventional architecture, is thought to be divided into the hardware cost of a reduced host station, FEPs and a dataway system, at the distributed architecture. Besides, since an FEP is shared by two host stations, the hardware cost of an FEP must be shared as well, depending on how much the particular FEP performs for each of the host stations. And the cost of a dataway system is to be equally shared by two host stations.

<u>Premise-2</u>. The down time ratio of a certain hardware subsystem is proportional to the hardware cost of that subsystem.

In other words, as the hardware cost of a particular subsystem increases the number of components or elements of that subsystem will be increased correspondingly, and the down time ratio may be increased, consequently.

By introducing the basic data concerning system hardware cost into the above-mentioned premises, the total ESPL could be easily calculated for both of the system architectures. The results of these numerical calculations are shown in **Fig.11**.

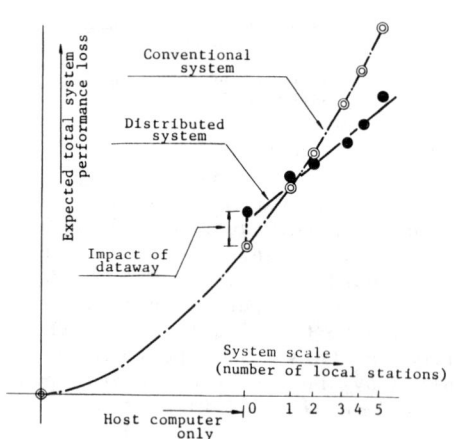

Fig.11. Evaluation of expected system performance loss

CONCLUSION

In this paper, to begin with, the review of the conventional SBMS was given on system structure, functions and other features. Next, the newly developed distributed SBMS was introduced and its design concepts were discussed, where the concept of "Transparency" was considered to be one of the most important points. How this "Transparency" is actually realized, was also described, by means of the data flow charts, and as for how this distributed SBMS is practically applied, the system model constituted on a hot strip mill computer control system was introduced.

In the latter half of this paper, the emphasises were concentrated on the quantitative evaluations. The evaluation criteria were developed on each of these evaluation items: (a) data communication cost; (b) data processing cost; and (c) system performance loss. The numerical calculations were done, on the basic data obtained from the model.

As for the quantitative evaluation, it seems that the approaches and evaluation criteria have not yet been completely established, and also the lack of basic data may be serious. However, the results obtained here may hopefully offer some of the benificial suggestions on the feasibility studies whether a system should be distributed or not and what is the optimal number of local stations to maximize the system performance.

Future works may be on the further establishment of the quantitative evaluation approaches and refinement of the basic data, so that these approaches proposed here may be applied for the assessing studies of a distributed system or the system optimization design, widely in the variety of application fields.

REFERENCES

Kato, S. and Nasu, S. (1980). A distributed industrial computer system. <u>Mitsubishi Electric ADVANCE</u>, <u>11</u>, 17-19.

Scherr, A.L. (1978). Distributed data processing. <u>IBM Syst. J.</u>, <u>17</u>, 325-335

Jenkins, D.W. (1978). Distributed or not ? The choice between distributed and central computing control. <u>Control Eng.</u>, <u>25</u>, 61-64.

Satyanarayanan, M. (1980). Commercial multiprocessing systems. <u>Computer</u>, <u>13</u>, 75-95.

DISCUSSION

<u>Vamos:</u> Can you give figures on the percentage of the total control instrumentation investment, compared to the total investment of the plant? Also, what is the ratio of the control equipment investment to that of the other part of the instrumentation? Besides, what can you say about the instrumentation installed on your plant, especially the signal transmitters. Is it conventional or something newly designed?

<u>Inamoto:</u> My figures concerning the computer system investment of the hop strip mill are the following: 10 or 20 per cent is the cost of the electrical equipment: 10 or 20 percent of the cost of the electrical equipment is the cost of the computer installation. Thus, the computer installation is 2 to 3 per cent of the total investment of the plant. At the same time, the cost of the computer system tends to be 2 to 3 times higher than that of the other instrumentation, not including electrical equipment. As to your third question then I must say that I'm not familiar with signal transmission equipment, but most of the installation in the steel mill is of a new type. People in Japan mostly do not want to have conventional systems, they want the most advanced technology.

<u>Rosza:</u> In your presentation you mentioned that a part of your program deals with process simulation. Would you elaborate on that part too?

<u>Inamoto:</u> We skipped over this part of the description because of the time limitation. The system has CRT-based simulation support functions. We call this facility process simulator. This sensor-based system has a special mode for simulation in which the signals can be fed by the CRT keyboard in order to simulate the plant. You can test the application program by manipulating the keyboard. You can monitor the output signal from the user programs on this CRT. The system is detached from the plant and you can test the program in off-line mode.

<u>Wood:</u> On your last slide you were working with the assumption that adding extra hardware makes for less reliability. If you put the extra hardware forward as a redundant system then that tends to reduce the probability of a failure. How does that improve the curve?

<u>Inamoto:</u> I think that is a most interesting point. We are now trying to evaluate how the redundancy of the sytem affects this curve. Our study is not yet completed, so I cannot say how much the system redundancy will be reduced.

<u>Narita:</u> I guess the design and implementation of the DCCS is what we would call a multiobjective mathematical programming problem. In your last slide the optimum number of the subsystems was something like 2 or 3. If you take, for example, the cabling cost or some other criteria, then the optimum number of the subsystems is 4 or 5, so maybe the designer will be confronted with the problem of which figure to choose.

<u>Inamoto:</u> I have no exact solution for that kind of a multipurpose problem. At this stage it depends on the system engineer which criteria he will pay attention to. We shall have to solve this problem in the future.

<u>Narita:</u> I didn't understand what the definition of the system performance was. The system performance loss was defined as the sum of the hardware cost. Is this the hardware capital cost or what? Can you give me a simple example?

<u>Inamoto:</u> Suppose there is a failure in the CPU of the local station and maybe some of the control functions are lost. In this case the system performance loss is the total cost of the local station. Or, if the host station CPU is down, then the system performance loss will be the total cost of the host station, not including the front-end processors and the dataway systems.

THE NOVA CONTROL SYSTEM — GOALS, ARCHITECTURE, AND SYSTEM DESIGN*

G. J. Suski, J. M. Duffy, D. G. Gritton, F. W. Holloway, J. E. Krammen, R. G. Ozarski, J. R. Severyn and P. J. Van Arsdall

Lawrence Livermore National Laboratory, Livermore, California 94550, USA

Abstract. The Nova laser fusion facility is presently under construction at LLNL. It is a large glass laser, generating 150 trillion watt pulses of infrared light at a wavelength of 1 micrometer (μ). Nova will achieve these energy levels using 10 amplifier chains, each focused onto fusion targets a few millimeters in diameter. Conversion of the 1 μ light to shorter wavelengths of 0.5 μ and 0.35 μ is expected to allow Nova to perform experiments in the physics regime of thermonuclear ignition.

The control system for the Nova laser must operate reliably in a harsh pulse power environment and satisfy requirements of technical functionality, flexibility, maintainability and operability. It is composed of four fundamental subsystems: Power Conditioning, Alignment, Laser Diagnostics, and Target Diagnostics, together with a fifth, unifying subsystem called Central Controls. The system architecture utilizes a collection of distributed microcomputers, minicomputers, and components interconnected through high speed fiber optic communications systems.

The design objectives, development strategy and architecture of the overall control system and each of its four fundamental subsystems are discussed. Specific hardware and software developments in several areas are also covered.

Keywords. Computer applications; computer control; data acquisition; digital systems; hierarchical systems; process control; programming languages.

INTRODUCTION

Nova is a 150 terawatt, ten arm laser fusion research facility currently under construction at the Lawrence Livermore National Laboratory (LLNL) (Simmons, 1982). As the world's most powerful glass laser system, Nova will provide researchers with an important new tool in the study of inertial confinement fusion (ICF). A principal objective of Nova is to demonstrate the feasibility of generating power from controlled thermonuclear reactions.

The predecessor to Nova was the highly successful, 30 terawatt, twenty arm infrared Shiva laser system (Figure 1). It was the first large laser system at LLNL to employ a comprehensive computer based control and data acquisition system. The control system for Nova uses an evolutionary design based upon the proven Shiva control system (Suski, 1979).

*Work performed under the auspices of the U. S. Department of Energy by the Lawrence Livermore National Laboratory under Contract No. W-7405-ENG-48.

Fig. 1. The twenty arm Shiva laser is Nova's predecessor.

An intermediate laser system called Novette is also under construction. It is scheduled for completion by early 1983 (approximately two years before Nova). Novette will utilize two beams and a control system of the Nova design to provide 30 TW of light on fusion targets. It will provide important data on laser operation and target performance in preparation for full scale Nova experiments.

The Nova system's ten laser beams will be capable of concentrating 100 to 150 kilojoules (kJ) in 3 nanosecond pulses of infrared light on a fusion target a few millimeters (mm) wide. The system also will generate light at shorter pulse lengths in power bursts up to 150 TW. Recent results have shown that shorter wavelength light is much more efficient at driving target implosions. Therefore, Nova will incorporate the ability to frequency double or triple its fundamental output, allowing target irradiation at 1.05 micrometers (μ) wavelength (infrared), 0.53 μ (green), or 0.35 μ (ultraviolet).

All arms of Nova must deliver their individual 15 kJ pulses to the target simultaneously (within $\pm$ 5 picoseconds). To achieve this objective, a single 100 microjoule pulse is selected from a train of such pulses. It is amplified to approximately 50 joules in a single pass through a nine stage preamplifier, and then it is split by partially reflecting mirrors into ten parallel chains of power amplifiers, each consisting of 15 cascaded laser amplifiers. Each pulse emerges from the output of its 180 meter long chain with a beam diameter of 74 centimeters. The pulses are subsequently reflected by large alignment mirrors, converted to shorter wavelengths, and finally focused onto a fusion target inside a 5 meter diameter aluminum vacuum vessel.

OVERVIEW

This paper will discuss the control system for Nova (Figure 2). First, the requirements and criteria established for the control system will be summarized to assist the reader in understanding the motivation for its design. A discussion of the architecture and commonly used components developed for the Nova Control System will then be presented. Applications of this system to achieve major control requirements conclude this paper.

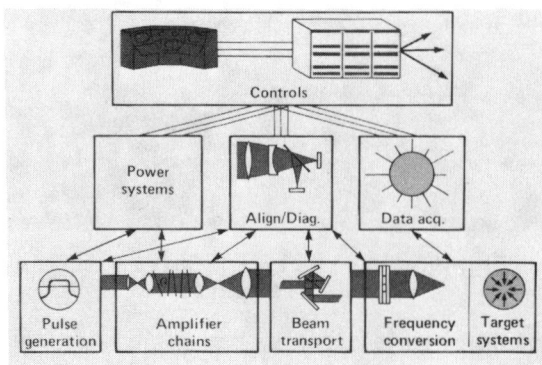

Fig. 2. Nova is operated through a hierarchical control system.

CONTROL REQUIREMENTS FOR NOVA

The Nova control system must satisfy control and data acquisition requirements in four fundamental areas:

Power Conditioning Subsystem

Functions. During a single laser "shot", over 60 megajoules of energy, stored in 25 kilovolt (kV) capacitor banks, is released through computer controlled ignitron switches to fire over 8600 flashlamps and 60 magnetic field devices. The discharge from the lamps imparts energy to neodymium doped glass rods and disks. This energy is released when infrared (1.05 μ) light passes through the glass and is amplified. The magnetic field devices are called Faraday rotators. These devices use the transmission properties of paramagnetic glasses in intense magnetic fields to prevent reflected light from propagating in a reverse direction through the amplifiers which can cause component damage.

Requirements:

- Control of seven 1.5 megavolt-ampere (MVA) power supplies and twelve 100 kVA supplies.

- Firing of ignitrons to discharge capacitors with microsecond precision - switching 20 to 30 megamps in 1 ms.

- Generation of triggers to synchronize diagnostics and components.

- Monitoring of safety and abort systems.

- Acquisition and analysis of current waveforms and the actual firing sequence.

Alignment Subsystem

Functions. The light pulses travel approximately 250 meters from the point of origination until target impact. Along this path, there are four principal locations where the pulsed beam is turned, pointed, centered, divided, expanded, focused, and diagnosed. These are the Master Oscillator where the initial pulse is formed, the Splitter Array where the pulse is divided into 10 paths, the Amplifier Chains where the pulsed beams are amplified and expanded, and the Output/Target Area where the beams are aimed and focused on the target. Alignment sensors generate video images of beam positions and profiles by viewing light split off from the main paths with partially transmitting mirrors or partially reflectory optics. Charge coupled device arrays (CCD's) are used to convert the beam information to video information which can be viewed on black-and-white television monitors, or digitized for computer analysis.

Requirements:

- Local and remote control of more than 1000 stepping motors to manipulate mirrors, lenses, shutters, and other alignment components at rates up to 300 steps/second (Nova standard) and, in a few cases, 5000 steps/second.

- Perform automatic alignment of the main oscillator, splitter arrays and amplifier chain components.
- Record alignment configurations for subsequent re-use and analysis.

Laser Diagnostics Subsystem

Functions. Accurate characterizations of beam quality, energy, and temporal properties are required at key locations throughout the laser system to interpret target results. Such measurements also aid in preventing damage to laser components due to excessive energy densities.

Laser energies ranging from 70 joules (J) to 10 kJ and time durations ranging from 100 picoseconds (ps) to 10 nanoseconds (ns) must be diagnosed. Beam characterization utilizes absorbing glass calorimetry and high speed photodiode detectors to measure energy. Transient digitizers and streak cameras record time discrimination. Digital solid state memories attached to CCD based video sensors capture and store pulsed images.

Requirements:

- Setup, control, and acquisition of data from approximately 100 sensors producing over 300,000 bytes of data for each laser shot.
- Acquisition of 160,000 bytes of data from several digital memories attached to streak cameras in distributed locations.
- Post shot storage, display, and analysis of results.

Target Subsystems

Function. Over 120 instruments acquire data on a single shot to characterize the results of the target implosion. A large portion of this data is collected as temporally resolved waveforms using transient digitizers and high speed streak cameras. Diagnostics include x-ray and particle detectors, calorimeters, photodiodes, photomultipliers and other instruments with resolutions ranging from 10 picoseconds to several seconds. The target system requires extensive reconfiguration of instruments, their sensitivities and triggering specifications, between successive target shots. The target chamber and several diagnostic instruments are serviced by individual vacuum systems. These require interlocked operation to prevent incorrect configurations of their more than 500 valves, leak detectors, and pressure detectors.

Requirements:

- Setup, control and acquisition for approximately 120 instruments generating a total of 10^7 bytes of data.
- Maintenance of vacuum integrity through interlocked control of valves according to known permissible configurations and pressure sensor readings.
- Support of major reconfigurations of diagnostics instruments between target shots.

DESIGN OBJECTIVES

In addition to the fundamental requirements summarized above, two additional objectives motivated the design of the Nova Control System:

- Reduction of development and long-term maintenance costs by employing a consistent overall architecture and a set of common components.
- Ensuring the flexibility required to optimize designs according to the unique requirements of the individual subsystems.

DESIGN CRITERIA

The design criteria established for the Nova Control System are representative of those found in large distributed control systems of this type. In addition to the primary criterion of cost, principal criteria include:

Reliability. The precise specification of reliability for control systems such as Nova's can be elusive. When matters of personnel safety are not involved, reliability criteria are often subordinate to other performance goals and cost constraints. However, two criteria were clearly established for Nova.

The control system must function normally in the high voltage, high current pulsed power environment which generates harsh electromagnetic interference (EMI) conditions. In addition, no single point failure of the control system should preclude overall operation of the laser system. (Manual or reduced levels of automated control are acceptable.)

Adaptability. Requirements evolve as Nova nears its final stage of construction, and will continue to evolve through the years of its operation. The control system must be adaptable to changes and extensions to the laser system configuration and its operational needs.

Performance. Simply stated, control system performance (speed and functionality) should not constrain the shot rate of the laser.

NOVA CONTROL SYSTEM DESIGN

Functional Organization. Nova's control system employs a distributed, computer based architecture which evolved from the successful Shiva laser control system. It is organized functionally according to the four fundamental subsystems; Power Conditioning, Alignment,

Laser Diagnostics, and Target Diagnostics. A fifth, unifying subsystem called Central Controls centralizes, augments, and coordinates the other subsystems' functions (Figure 3). The criteria of reliability and adaptability are met by the computer based, extendible nature of the system. The flexibility required to optimize individual subsystem architectures is provided by the inclusion of the Central Control subsystem. It establishes a single point at which compatible interfaces for command, control, and data interactions are established.

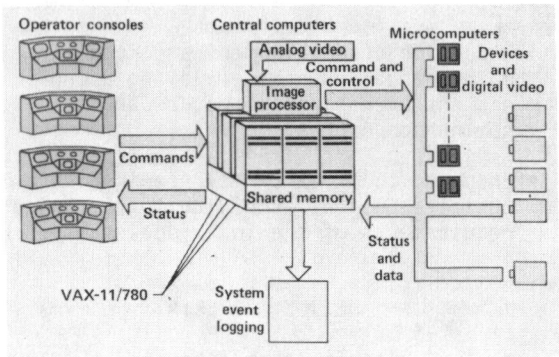

Fig. 3. A single subsystem integrates and centralizes distributed functions.

Development Strategy. The development strategy for the Nova control system capitalizes on the distinction between subsystems to accelerate the implementation cycle. The development teams are organized according to the five subsystems. Initial effort concentrated on the development of common products and techniques which would later be employed across the four functional subsystems. Developers of these products subsequently migrated into lead roles in the subsystems which employ those "tools". This reduced the need for a separate team of developers, and simplified the task of technology transfer into the subsystem efforts. Actual subsystem development was organized according to the following guidelines:

- Parallel Groups - The development efforts for the four fundamental subsystems proceeded in parallel using four distinct teams.

- Common Products - The products acquired and developed for common functions became resources in the development and integration of the subsystems.

- Central Support - A central facility, supported by a VAX-11/780, provided a central, well maintained, resource for software development and global system information.

- Distributed Testing - Hardware and software particular to each subsystem are tested in distributed locations configured with appropriate facilities. This has reduced unnecessary interference between the subsystems, resulting in an acceleration of the total development effort.

A Distributed Control System. In this hierarchically structured system approximately 50 Digital Equipment Corporation (DEC) LSI-11/23 microcomputers provide localized control and data acquisition capabilities in geographically distributed locations throughout the laser fusion facility. Data from these front end processors (FEP's) is collected, analyzed and integrated at the Central Control level with three redundant Digital Equipment Corporation VAX-11/780's minicomputers. Remote command and control capabilities, higher level control functions (e.g., automatic laser alignment), and high volume data storage and manipulation, are implemented at this level.

The physical distance between devices to be controlled suggests a distributed architecture. However, there are additional factors motivating the use of distributed microcomputers for device control:

- The real time nature of device control establishes requirements for computer interrupt handling capacity, low level device polling, and fast responses to simple control and status inputs. It is inappropriate, from a cost and performance perspective, to meet these needs in large central computer.

- A requirement for simple device control capabilities located physically nearby is efficiently achieved by providing the locally situated microcomputers with attached control panels.

- Device control can be optimized through individual microcomputer configurations which are not constrained by the requirements of unrelated devices.

- Independent local control functionality supports individual device maintenance while the remainder of the control system remains operational.

The large central minicomputers in Nova support control functions which cannot be effectively implemented in the distributed microcomputers due to cost considerations or technical constraints. For example, the design of the central controls architecture was strongly influenced by the need to perform alignment adjustments on over 230 laser components utilizing video images from 55 distributed sensor locations. Replicating expensive video processing capabilities at several locations is not cost effective. Therefore, automatic alignment functions are concentrated in the central computers which can efficiently support this speed and memory intensive task.

Development and Off-Line Analysis Support. The development and operation of a computer based control system are incompatible tasks for a single shared resource computer. Experience with the Shiva system indicated that a computer system configured and optimized to perform continuous control activities could not give adequate service to developers. For example, the need to provide high availability of operational controls precludes all but intermittent testing activities. In addition, software development activities such as editing and compilation significantly deteriorate control system performance.

In order to maintain high development and maintenance productivity, a separate system was established for Nova on which software and hardware development activities could occur without interfering with normal control system operations. This computer system is configured with several access terminals, a large volume of mass storage, a variety of analysis software, and specialized output devices (e.g., graphic hardcopy units).

This computer is connected into the control system network. This feature allows experimental data to be transferred to it for long-term analysis and storage, an important requirement for successful Nova operation. Large data bases used intermittently by the control system, such as the relatively static system configuration information, are also maintained on this system.

Development Strategy. The general architecture of the control system was established early in the Nova project, necessarily in advance of final specifications for the four fundamental subsystems. Early development, therefore, concentrated on developing the set of hardware and software packages which would be used to construct, support, and integrate the subsystems.

The next portion of this paper discusses the central architecture of the control system and packages developed for common uses. The distinct subsystem architectures which employ these tools to suit their individual requirements are then described.

CONTROL SYSTEM ARCHITECTURE

The hardware architecture of the Nova Control System is illustrated in Figures 4. The LSI-11/23 microcomputers communicate to each other, to remote devices such as CCD cameras, and to the three VAX-11/780's primarily via fiber optic communications links. Command and control information flows from the central computers into the distributed controllers and devices, with status and data typically returning. The VAX-11/780's are configured with four megabytes of local memory each, plus approximately two megabytes of shared memory. Attached to the central computers is a high speed array processor and a video digitizer. This package can analyze images from any of the video-based sensors by utilizing a computer controlled video switching network. Typically, the three central computers share tasks to provide a satisfactory performance level, but the system can be operated at reduced performance levels from a single VAX-11/780 if necessary. Four fully programmable operator consoles provide central control capabilities to any of the four subsystems. An Event Logger implemented on a separate computer maintains a time annotated record of important control system actions.

Central Operator Consoles. The principal operator interface to Nova is provided through four, fully programmable central operator consoles. The design for each console (see Figures 5 and 6) is optimized for use by a single person who can view status and initiate control functions for any of the four fundamental control subsystems. Three mid-resolution (512 x 640) 19" color displays are driven by a Ramtek RM9400 graphics controller. Two peripheral displays provide status information. A centrally mounted third screen displays operator command templates. It is overlayed with a transparent touch panel which translates operator selections into screen coordinates. The touch panel is augmented by a force operated joystick for stepping motor control and a numeric keyboard for data entry.

A software package treats each console as an integral control unit. This package supports the graphical display of control system elements as "entities" which may be selected via the touch panel to indicate control commands. Alternatively, shapes or colors can be changed via an executing program to indicate status. Designed to be control system oriented, this package has two important characteristics:

- High performance - Operator selections are immediately acknowledged. Screen updates lag data by less than 0.2 seconds.

- Control Oriented Programming - Engineering users prefer to program control system functions as opposed to writing computer graphics software. Therefore, the software interface to the operator console allows reference to names and functions of control system elements in preference to details of screen coordinates and graphics controller characteristics.

This high performance package for on-line control applications is augmented by general purpose graphics packages (Brown, 1981) to satisfy requirements of lower performance but higher functionality in post shot data display and analysis.

Image Processing. The control system must be capable of analyzing video beam images generated by alignment system cameras once every one to three seconds. These analyses determine 1) the center coordinates of soft-focused, noisy, laser beam profiles and 2) the offsets between diffraction patterns of overlapped alignment cross hairs. The VAX-11/780, with its virtual memory capability, provides the address space required for complex image analysis. It is augmented by a high speed auxiliary image processing system. This consists of a Quantex video frame digitizer with 256 x 256 addressability interfaced directly to a Floating Point System FPS-120B array processor. The array processor

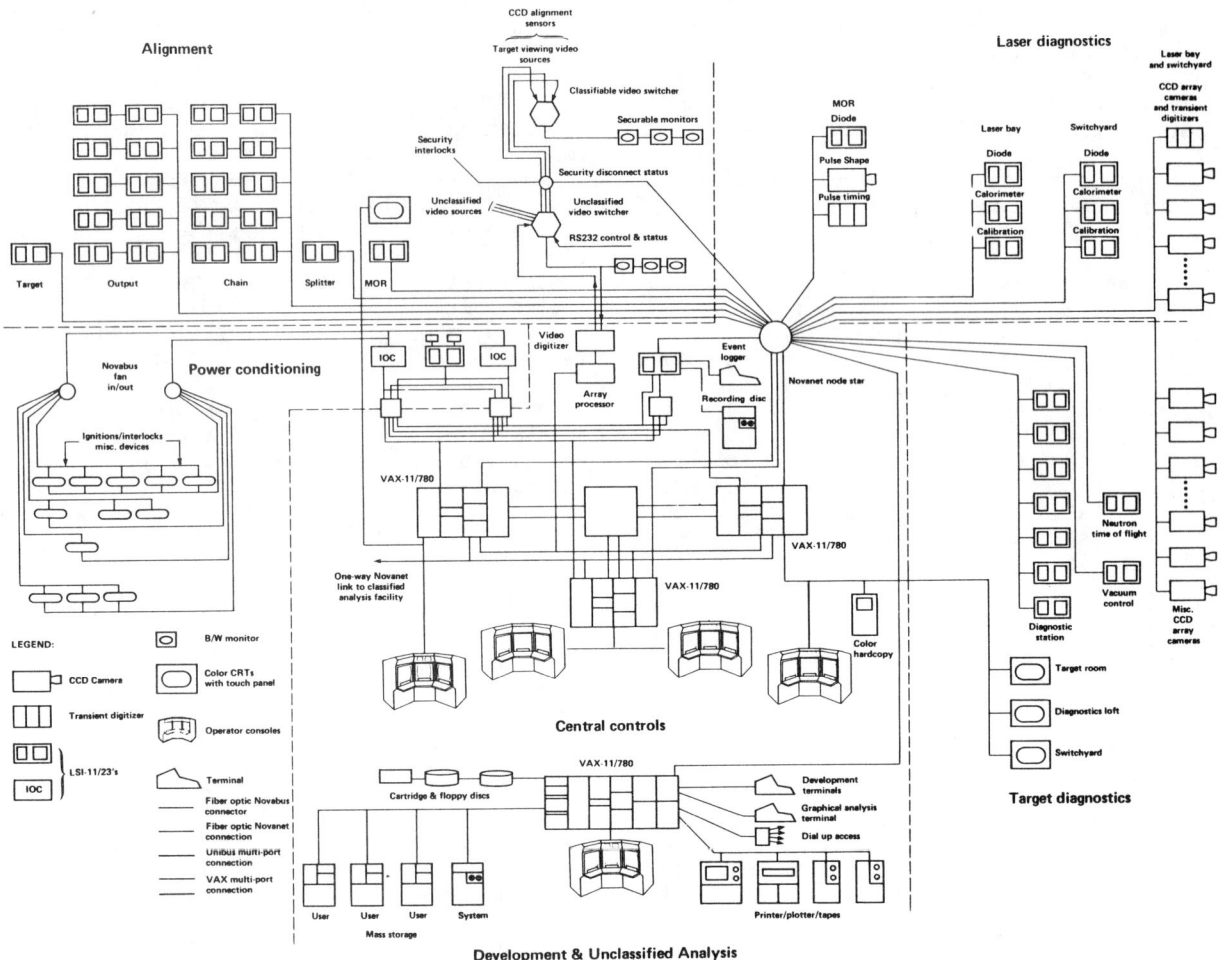

Fig. 4. The Nova control system hardware architecture.

Fig. 5. The Nova central operator console.

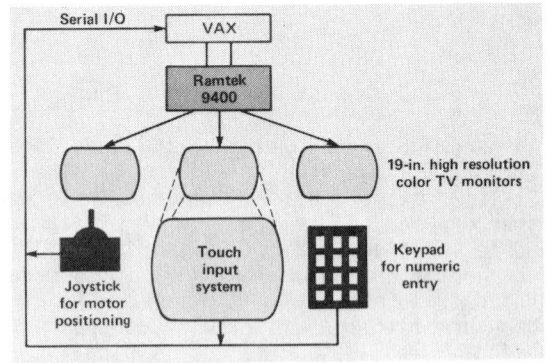

Fig. 6. Operator console I/O system.

is interfaced to a central VAX-11/780 computer through its I/O bus (UNIBUS). This auxiliary system increases the speed of image processing by a factor of 100, allowing closed loop image analyses for the entire facility to be performed by a single set of hardware. Video sensor images to be analyzed are selected through a computer controlled video switching system.

Support of Distributed Communications

In a distributed computer control system (DCCS), well-defined methods for communicating commands, status, data and synchronization between computers, devices, and software processes are required. The performance of the communications media, as measured by computer loading and speed of transfer, should not constrain the control system's ability to service the facility. In addition, restrictions on data format and content must be minimized. These latter criteria are motivated by the concept that the general purpose "tools" of the control system developers should enhance, not hinder, their ability to meet their objectives. Nova uses shared memory and fiber optic transmission links to satisfy such communications requirements.

Shared Memory. Communication between processes residing in the same or physically adjacent computers in Nova employs shared data structures in commonly accessible memory. This technique is particularly well suited to providing identical status information to more than one process at a given instant. The use of shared memory on an extensive basis across control subsystems yields data flow and access paths similar to those encountered in distributed point-to-point computer networks. Management of shared data structures, access control, and a "network analyzer" utility to assist in diagnosing communications activity in shared memory are provided by a software package called NSM (Network Shared Memory). NSM supports management of common data structures across subsystems, labels collections of these structures (called "tables"), sets up access control (Read, Modify, Write) on a per program basis, provides the semaphores necessary to interlock table updates, and signals suspended processes when table changes occur. NSM minimizes process overhead while waiting for data changes. It also improves the maintainability of the control system by managing the topology of the shared memory network.

Novanet. The primary medium for distributing command, status, and control information to remotely located computers is the Novanet fiber optic communications system. A collection of intelligent network hardware devices (Novalink) provides network functionality up through the Physical, Data Link, and Network Layers (including routing) as described in the ISO OSI reference model for Open System's Interconnections (Tanenbaum, 1981). Software supports the required functions of the Transport and Session Layers (blocking/deblocking and process to process connections).

The objective of the Novanet development effort was to provide a high performance method of interconnecting computers with computers as well as computers with devices. This combination is an intriguing feature in control systems since it implies that any computer in the DCCS network may access any device (such as a remote digital image memory) connected into the network. This results in simplified programming, more flexibility in architecture, cost reductions, and higher performance by the elimination of intermediate processors.

Principal characteristics of Novanet are:

- 10^7 bits/second using fiber optics.

- Devices connect directly into network.

- Node to node routing under hardware control.

- Assignment of network mastership to nodes under hardware control.

- Asynchronous word and block transfers.

  Word by word error checking and transfer acknowledgements performed by hardware.

- Network dynamically segmented into subnetworks to increase total available bandwidth.

The Novanet protocol (Figure 7) employs a 50 bit serial packet to transfer data, assign network mastership, and send control information to the network interfaces. Each packet is acknowledged by the receiving node's hardware. Control of network mastership is via token passing. Several modules comprise the Novanet support hardware. A Master/Slave Controller (MSC) plugs directly into a computer bus. The MSC utilizes microprogrammed state machine control to implement data serialization, data verification, arbitration of network mastership, and direct memory access (DMA) transfers to and from computer memory. A Multiple Device Interface (MDI) is used to connect remote digital image memories directly into the Novanet network. MDI's operating in slave mode to MSC's can transfer 160,000 bytes of image memory in approximately 2 seconds. The Q-bus Device Interface (QBDI) is functional similar to the MDI, but provides an LSI-11 bus (Q-bus) compatible connection to the network. Programmed operation, DMA, and interrupt functions of standard Q-bus interfaces are supported.

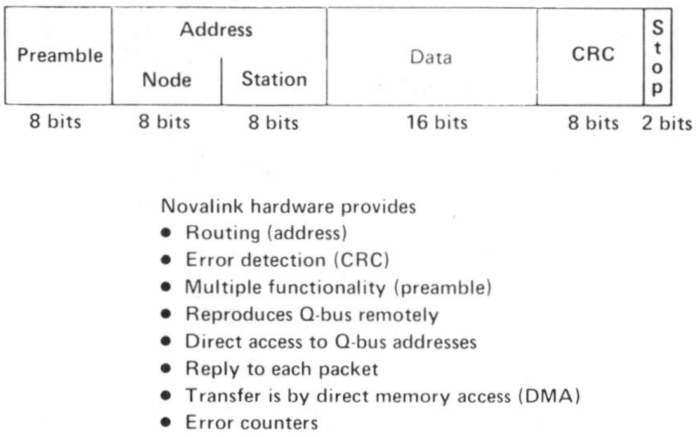

Preamble	Address		Data	CRC	Stop
	Node	Station			
8 bits	8 bits	8 bits	16 bits	8 bits	2 bits

Novalink hardware provides
- Routing (address)
- Error detection (CRC)
- Multiple functionality (preamble)
- Reproduces Q-bus remotely
- Direct access to Q-bus addresses
- Reply to each packet
- Transfer is by direct memory access (DMA)
- Error counters

Fig. 7. Novalink serial packet.

The central communications element for the network is the Node Star (Figure 8). This unit broadcasts incoming data to all processor and device nodes, exclusive of the originator. The Node Star converts the network, which consists of physical point-to-point fiber optic connections, to a logical "bus". The use of the Node Star was motivated by limitations in the number of fiber optic "taps" allowed in multi-drop systems. It has the ability to segment the network into distinct subnetworks during high volume data flow conditions. Configured as a collection of smaller Node Stars which are selectively

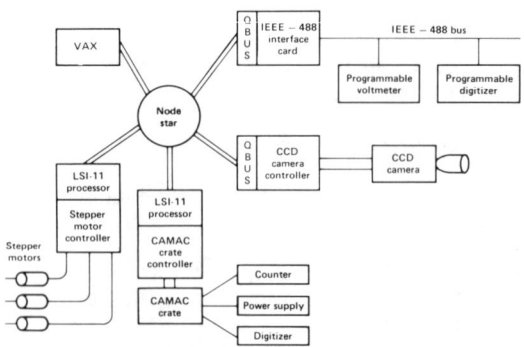

Fig. 8. Novanet connects devices and computers.

coupled together under program control, a single point failure in the Node Star will affect only a small subset of the network nodes. Up to 64 subnetworks can be created in a 256 node Novanet system. The Node Star also provides self-diagnosis capability for Novanet by supporting detection and isolation of faulty nodes. The final element of Novanet hardware, a network analyzer, captures and displays selected network traffic as a diagnostic aid for hardware and software.

Novabus. Several tasks in Nova require relatively simple device control functions, but must be closely synchronized at shot time. Many of these functions, such as ignitron firing and interlock sensing, are associated with the Power Conditioning subsystem. Another such function is providing of preshot triggers to the diagnostics subsystems. The Novabus system was developed to provide a fiber optic based communications link, driven from a central FEP, for these functions. Novabus uses the same fiber optic technology as Novanet, but utilizes a simplified protocol (Figure 9) optimized for device control from a single bus master.

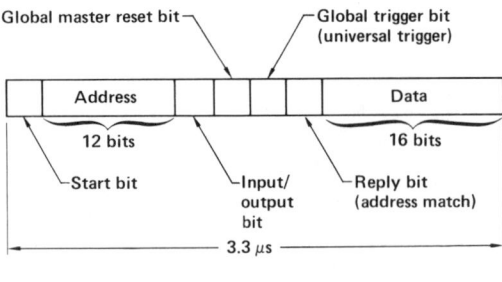

Fig. 9. Novabus serial packet.

The principal hardware modules in the Novabus system include the Fiber Interface Controller, which connects to the LSI-11 Q-bus, a Node Star capable of accommodating sixteen lines, and a Node Drop for each group of physically local devices to be connected to the fiber bus. Each Node Drop supports connection of devices to the network using a standarized bus which is logically similar to the Q-bus. In addition, it repeats the optical signal to allow chaining of Node Drops in series. All 110 drops in Nova are contained within twelve such chains. Novabus is implemented redundantly, with each device serviced by two distinct Novabus systems. This architecture increases reliability and availability. In addition, when both Novabus systems are fully operational, control functions are shared in order to improve control system performance.

Event Logging. A DCCS typically will have many functions operating concurrently. An important method of tracking operations activity, and analyzing the cause and effect of control system malfunctions, is examination of an event log. An event, in this sense, is an action or status change which has significance at the operational level. Examples in Nova include operator activation of an alignment control sequence, the completion of a remote mirror gimbal motion, and the opening of a safety interlock. For Nova a prototype event logging facility, employing a DEC PDP-11/34 mini-computer, has been connected via Novalink and multi-port memories to the control system. It provides an independent station at which significant, time stamped control system events can be displayed and stored.

Praxis: A Language for Control Systems Implementation. The maintainability of a DCCS is dependent on the standarization and documentation of its hardware and software components. Techniques for achieving these characteristics in hardware are well known and established. In software, maintainability is affected by its readability - the clarity with which the source statements represent their function. In a large DCCS effort, techniques to assist the software engineer in ensuring compatibility between interdependent modules and data structures are also required. The ADA programming language development effort, sponsored by the United States Department of Defense, is attempting to achieve this and other goals in control system applications. The Nova project, through the Department of Energy, has fostered the development of the Praxis language (Evans, 1981) to achieve some essential goals of ADA on a schedule useful to the Nova project.

Praxis is a modern, high-level computer language designed for the efficient programming of control systems applications. It is a strongly typed, block structured language in the tradition of Pascal, with much of the power of the forthcoming ADA. It supports the development of software composed of separately compiled modules, user defined data types, exception handling, detailed control mechanisms, and encapsulated data and routines. Direct access to computer facilities, efficient bit manipulation, and interlocked access to critical regions are provided.

Praxis contains high-level constructs which enhance readability and portability in control system applications. It also allows controlled access to machine dependencies, enabling the use of high performance, real time control capabilities specific to a particular computer architecure. Praxis supports separate compilation of modules while enforcing strong data type checking between such modules at compile time. This reduces the time nescessary to debug software at execution time.

The Praxis compiler is written in the Praxis language. It executes under the VAX/VMS and RSX-11/M operating systems, generating optimized code for VAX's, PDP-11's, and LSI-11's. Over 150,000 lines of Praxis source code now exist in the Nova control system. An example of a Praxis language interrupt driven clock routine for the LSI-11/23 is found in Figure 10.

```
Declare
    Vector is structure
      Routine : interrupt procedure () initially service
      Status  : logical initially 8#340
    Endstructure

    Clock  : volatile location (8!100) vector
    Ticks  : static integer initially 0

Enddeclare

Interrupt procedure service ()
    Ticks *= + 1

Endprocedure [service] ()
```

Fig. 10. A Praxis interrupt driver clock routine.

Configuration Management. In the development of a DCCS, two types of configuration management must be addressed, management of software configuration and management of control system configuration. The primary objective in software configuration management is to maintain multiple versions of software modules. This allows establishment of working "baselines" and revision levels which ensure access to proven software while development continues. The need for this capability increases with the number of developers in a DCCS. Nova uses the standard VAX-11/780 file directory system to maintain software revisions from the initial development to the on-line stages. A software librarian has responsibility for maintaining quality assurance and baseline control. In general, a more flexible software library system is desirable, but until recently, has not been commercially available for the VAX-11/780.

The second category, management of the control system hardware configuration, is required to assist in its operation and maintenance. In particular, device data which changes frequently must be easily accessible and modified. Such data includes the network configuration, predefined configurations of laser system components, and predefined sequences of operations. The principal support for such functions is provided by a vendor supplied relational data base system (RDB) for the VAX-11/780 named Oracle (Oracle, 1981). RDB's prove well suited to DCCS applications. Flexibility in the control system software is frequently gained through functions driven by tables, which can correspond directly to equivalent tables in an RDB.

THE SUBSYSTEM ARCHITECTURES

The four fundamental subsystems which employ these and other common components, have each developed architectures uniquely suited to their requirements (previously summarized). These architectures are now examined.

Power Conditioning. The Power Conditioning subsystem must perform short programmable sequences of low-level command and control for several hundred distributed devices at precisely spaced intervals to microsecond accuracy. One method of achieving this capability is to connect all devices to the I/O bus of single microcomputer, such as an LSI-11/23 FEP. This simplifies the task of synchronizing control sequences. However, the devices to be controlled are in geographically distributed locations in harsh EMI environments. The solution chosen in this subsystem was to extend the LSI-11/23's I/O bus using the high speed fiber optic Novabus, developed specifically for this application. All devices in the Power Conditioning system are controllable from either of two redundant LSI-11/Novabus systems to provide fault tolerance.

Timed control sequences are maintained in multi-port memories which connect the two Novabus FEP's to the central VAX-11/780's. A third interconnected LSI-11 supports a separate maintenance control console which can, if necessary, complete a full system shot sequence independent of the central subsystem.

Several standardized interfaces for device control at each drop on the Novabus are used. For example, the Capacitor Bank Interface fires high voltage ignitrons at precise intervals using an on-board timer triggered over Novabus. The Interlocks Interface provides the ability to monitor the hardwired, triply redundant, safety interlock system. The Power Supply Interface controls and monitors the megavolt-ampere power supplies which charge the capacitor bank.

Software in the Power Conditioning subsystem is functionally balanced between the central computer and its FEP's (Figure 11). An operator interacts with the central computers through a series of menus on the central operator console. The central computers respond by placing commands for control devices in the memory shared with the FEP's. Alternatively, commands can originate from two centrally resident programs; the shot scheduler program controls events during shot preparation and the sequencer program completes the time critical firing of the system. Actions initiated at the operator console can result in the loading of a shot "configuration", including voltage levels, timing information, and device selections, from a RDB. The FEP's route those commands to the remote hardware through Novabus, and regularly poll this hardware to update a status table in shared memory. The control computer thus has visibility of the system status at all times. FEP's also log changing status data

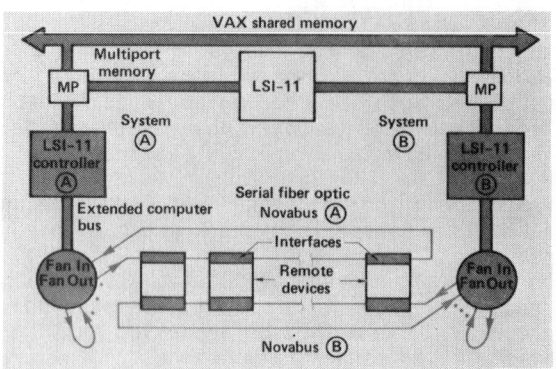

Fig. 11. Dual Novabus application in Power Conditioning subsystems.

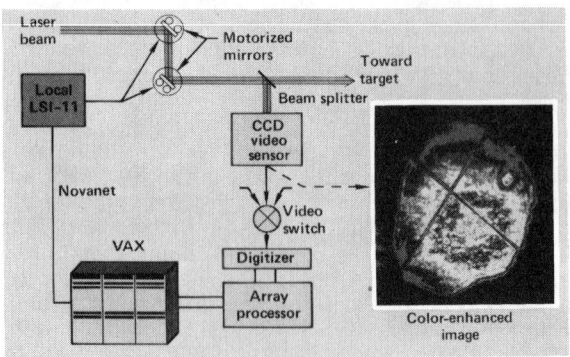

Fig. 12. Automatic alignment through image analysis.

at very high rates to provide a history of voltage levels and actions in the seconds prior to and immediately after a system shot.

Alignment Control. Approximately twenty-three LSI-11/23 based stepping motors controllers (SMC's) are used to control movable alignment subsystem components driven by nearly 1000 stepping motors. The Nova SMC's provide substantial functionality at each location:

- Variable rate control of 60 stepping motors.

- A programmable alphanumeric control panel.

- Programmable backlash correction.

- Crosscoupled control of beam pointing and centering devices.

- Support of optical position encoders.

- Connection via Novanet to the central control subsystem.

SMC's provides device oriented (as opposed to motor oriented) command and control functionality. For example, a lens mounted on a three axis, three motor gimbal is treated as a single controllable device. Predefined combinations of device positions, referred to as configurations, are loaded into the SMC's through Novanet. These configurations are maintained in a central RDB which can be updated by operations personnel.

Automatic laser alignment, crucial to maintaining a high shot rate, is performed by software resident in the central computers. Position information (derived using beam image analysis) is translated into correcting stepping motor commands, which are sent to the SMC's through Novanet (Figure 12).

Laser Diagnostics. This subsystem (Ozarski, 1981) acquires, centralizes, and analyzes over 200 simultaneous measurements at shot time. Three distinct classes of devices are serviced. Very high speed transients recorded by digital streak cameras are stored in distributed locations in 160,000 byte digital memories. Arrays of data from transient digitizers (~2000 bytes each) and small quantities of data from single value, time varying detectors (e.g., photodiodes and calorimeters) are also processed. The large volume of data produced by the streak cameras and transient digitizers precludes economical processing in local FEP's. Therefore, the streak cameras connect directly in Novanet using MDI's. Transient digitizers connect to Novanet through IEEE-488 interfaces connected to QBDI's. Data from these devices is transferred directly into the high capacity central control computers for storage and analysis. Data can be retained in the digital memories of each device until the central system acquires it.

Six FEP's are used to process the low volume data. As in the Alignment subsystem, substantial local control capability is provided. A typical FEP (Figure 13) contains sufficient logic to support it as a 128 kiloword microcomputer, communication hardware for extended triggers and Novalink, interfaces for optional local terminals and floppy discs and command and control logic for its diagnostics sensors.

The flexibility afforded the subsystem designer to reduce cost and optimize control via the FEP's is shown here in the Data Interconnect and Power Distribution (DIAD) system. A controller connects the FEP to the DIAD which in turn distributes commands and collects data from up to 20 separate sensors. Up to three DIADs may be connected to an FEP. Communication between sensors, DIADs, and the FEP is over 125 kilobit/sec., optically isolated lines.

Each sensor may have associated variable gain amplifiers configured with shot dependent parameters. These configurations, stored on a RDB in the central computers, are transmitted to the amplifiers through Novanet and the FEP.

Target Controls. The Nova target diagnostics system (Figure 14) must uniquely configure over 100 diagnostic instruments which characterize target performance on each shot. Widely varying signal levels and bandwidths dictate a flexible subsystem architecture which carefully treats issues such as instrument isolation, grounding, crosstalk pre-

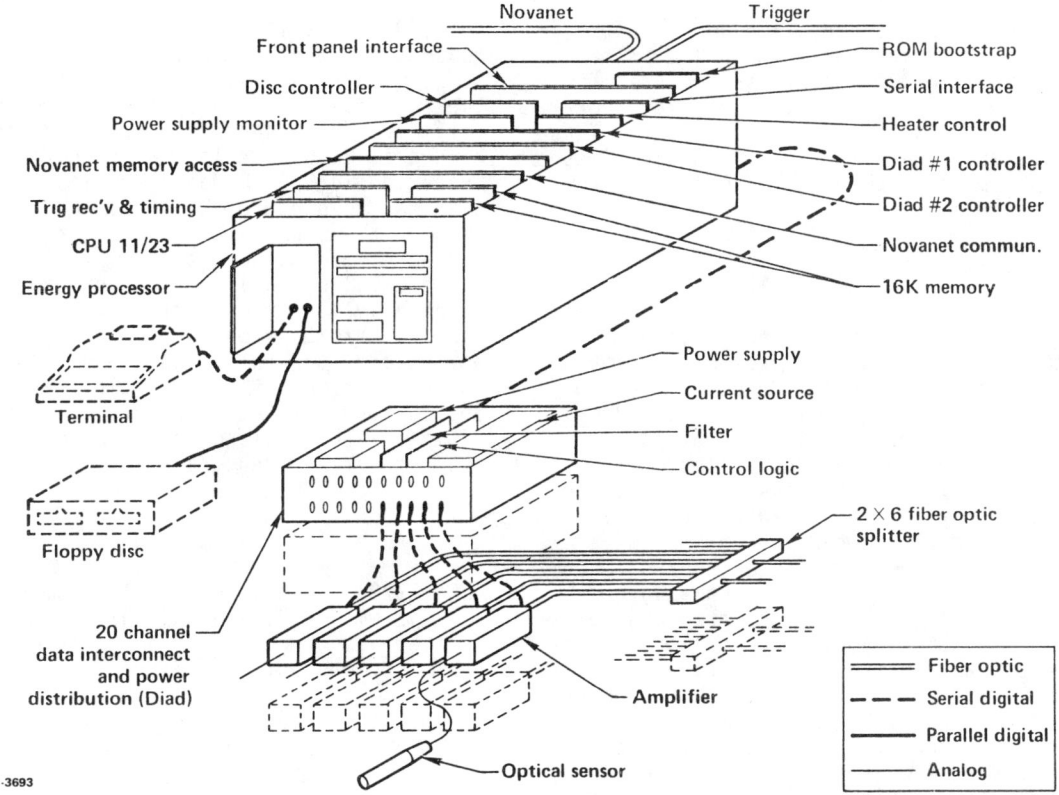

Fig. 13. LSI-11 based laser data acquisition unit.

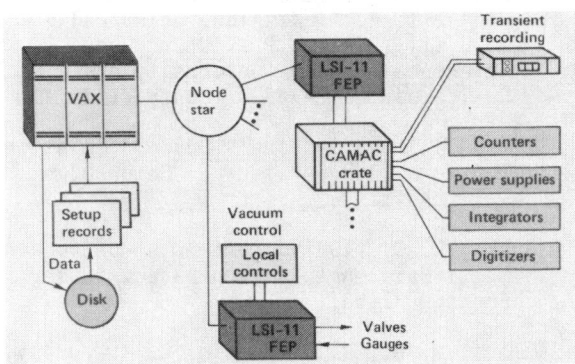

Fig. 14. Nova's target subsystem allows frequent instrument reconfiguration.

vention, and signal degradation. To assist in meeting these requirements, all target diagnostics sensors interface through CAMAC (AEC, 1972) equipment. Geographically distributed FEP's minimize analog signal cable lengths to maximize the signal quality.

Optically isolated, single screen extensions of the Nova operator consoles provide local control capability identical to that available in the central control room. Since the technology used for control is identical, only a single set of hardware and software for both local and remote control is required.

A diagnostic station comprises an electrically isolated group of racks, at least one FEP, at least one CAMAC crate (CAMAC bus and module rack), and the digitizers and equipment required to support an associated diagnostic detector. Supported through the CAMAC system are transient digitizers, charge integrators, calorimeters, and high voltage diagnostic power supplies.

Instrument locations, calibrations, and gains used by the FEP are stored in a central computer RDB. Analysis of the one megabyte of data acquired on a single shot requires this configuration information. Therefore, after each shot, a single data base consisting of both raw data and instrument configurations is created and stored for subsequent long-term analysis.

Subsystem Synchronization. The use of a DCCS for a facility such as Nova requires methods for synchronizing distributed control activities to varying tolerances. Requirements on flexibility, accuracy, and speed of synchronization directly affect its cost. Nova synchronization utilizes two hardware systems augmented by software to support the changing balance of requirements between flexibility and accuracy as shot time is neared.

Until one minute before a target shot, most synchronization is accomplished through operator and software communication. During the time the banks begin to charge until 10 ms. prior to the shot, devices are triggered over the Novabus to an accuracy of $\pm$ 200 sec. During this time the Power Conditioning FEP's are dedicated to executing the sequencer script and checking for abort status conditions. At 10 ns. prior to the shot, a universal trigger is broadcast over Novabus to preset timers. These timers deliver synchronization signals accurate within

± .5 sec. to arm diagnostics and fire the ignitrons. At 1 sec. prior to pulse propagation, the final timer starts the Fast Timing Control system to generate up to 128 programmable triggers with ± 1 ns. precision. These triggers are propagated over metallic wires or fiber optics according to their destination.

CONCLUSIONS

Multi-level control systems can utilize microcomputers as front end processors to provide low level local control and acquisition functions while larger mini-computers provide centralization and high level control functions. The development philosophy of using parallel teams to develop controls for parallel subsystems has been an effective approach in Nova and the preceeding Shiva system. Development and inter-subsystem compatibility is enhanced by having a large central facility to support software development across subsystems. The goal of an integrated command, control, and data acquisition system is realizable by establishing a central subsystem with which the fundamental control subsystems interact. Subsystem synchronization needs, however, may require more direct paths for integration. These requirements can be satisfied in a manner consistent with the overall system architecture.

SUMMARY

The control system for Nova meets several criteria, in particular flexibility and reliability, through the use of a distributed computer based control system. The system is organized into a hierarchy of computers grouped according to the four fundamental control and data acquisition tasks for Nova. The flexibility to optimize control subsystem architectures for individual requirements has resulted in four relatively distinct and innovative architectures. A fifth, unifying subsystem is utilized to integrate and centralize control functions. A high degree of programmability, extending from low level microcomputer up through the central operator consoles, ensures the ability to meet new control system requirements as they are identified. The control system survives the intense EMI environment of NOVA through the use of fiber optic isolation, careful grounding and shielding, and fault tolerant designs.

ACKNOWLEDGEMENTS

In addition to the authors, a number of people have contributed substantially to the Nova Control system effort. L. Berkbigler developed the Novabus system; R. Calliger and D. Kroepfle helped to develop the target systems; D. Christie developed the safety interlock system; T. DeGroot and J. Smart developed the power conditioning software; J. R. Greenwood led the Praxis effort with BBN, Inc.; J. R. Hill was responsible for the Novanet software; C. Humphreys developed the image processing system; D. McGuigan has led the relational data base efforts; D. Myers developed the laser diagnostics acquisition hardware; J. Oicles led the synchronization system design; S. Ruddock developed the Node Star for Novanet; J. Wilkerson has developed much of the console and laser diagnostics software; and G. Aune and T. Lynch of Digital Equipment Corporation have contributed substantially to our understanding and effective use of VAX/VMS. The authors also wish to acknowledge the consistent support of J. Holzrichter, beneficial criticism of R. Godwin, and the technical guidance of W. Simmons. The patience of C. Gardner in preparing this manuscript is gratefully noted.

REFERENCES

AEC (1972). CAMAC Modular Instrumentations System for Data Handling. *Atomic Energy Commission Committee on Nuclear Instrument Modules.* TID-25875.

Brown, M. D. (1981). *Tellagraf User's Manual.* Integrated Software Systems Corporation, San Diego, Ca.

Evans, A., Jr., C. R. Morgan, J. R. Greenwood, M. C. Zarnstorff, G. J. Williams, E. A. Killian, J. H. Walker, J. M. Duffy (1981). *Praxis Language Reference Manual.* University of California UCRL-15331, Contract 3634909.

RSI (1981). *Oracle User's Guide.* Relational Software Incorporated, El Segundo, California.

Ozarski, R. G. (1981). Beam Diagnostics on a Multi-Wavelength Fusion Laser. *Laser Focus.* 51-56.

Simmons, W. W., R. O. Godwin, C. A. Hurley, E. P. Wallerstein, K. Whitham, J. E. Murray, E. S. Bliss, R. G. Ozarski, M. A. Summers, F. Rienecker, D. G. Gritton, F. W. Holloway, G. J. Suski, J. R. Severyn, and the Nova Engineering Team (1982). Engineering Design of the Nova Laser Facility for Inertial Confinement Fusion. *University of California Conf-811040.* (Compendium of 11 papers from IEEE 9th Symposium on Engineering Problems of Fusion Research, 1981.)

Suski, G. J., and F. W. Holloway (1979). The Evolution of a Large Laser Control System - From Shiva to Nova. *IEEE Circuits and Systems Magazine.* 1, No. 3, 3-10.

Tanenbaum, A. S. (1981). Network Protocols. *ACM Computing Surveys.* 13, No. 4, 453-489.

DISCUSSION

Inamoto: Could you please describe the image processor? How do you share the tasks or jobs between three CPU's with one shared memory?

Suski: The image processor is a Floating Point System FPS-120B unit. It is connected directly to a Quantex video digitizer, which is connected only to the array processor that performs control functions for it. The array processor itself has to do some rather complex video programming. In Nova we develop our software on VAX which is a virtual array video address machine and therefore has the address space required to do image analysis in a high level language, such as PRAXIS. At that point we begin a migration - we remove some of the more commonly used functions inside the array processor. Eventually, when you have the situation where it is a function which must be performed very frequently, let's say, one every minute, then that function is totally recorded into the array processor in the same language. The FPS array processor comes with a FORTRAN complier. High speed requirements are such that, although this is a good compiler, it doesn't give us quite as optimized code as we need to perform our functions.

Your second question is on the task sharing in a multiport memory. We have two multiport memories in the system, one between the redundant VAX computers and one between the LSI-11's which control the power conditioning system. In the latter case, which is the simple one, the multiport memories are used largely as a data path. In the case of the VAX-11/780 we are also using the common memory to support common data for the synchronization of functions. It turns out that we are not going to actually use the full shared memory system for another year and a half yet. We have not solved all the problems that are required. One of the things that helps us significantly is the PRAXIS language with several constructs for controlling access to shared memory tables both for synchronizing this access and giving READ-WRITE. It does include sufficient structures in it to schedule common procedures. One of the more recent developments from DEC has been operating system support for multiple CPU's in the VAX. We are now investigating if that adds to our capability of solving that particular problem.

Joudu: What is the failure rate of your system? Have you any statistics on the failure rate of, say, a computer?

Suski: The computers in Nova tend to be much more reliable than the laser itself, which is fortunate. I have no hard statistical data, but I'll give a few pieces of information. When there are VAX central computers, we feel comfortable saying that we have a failure in a VAX which takes the machine down no more than twice a month. Each failure may, in fact, take a day to fix because we do not repair them ourselves - we wait for the manufacturer to come. In the Nova system we have redundant VAX's and we know how to restart the system, if we have to. In the preferable case you'd actually like to take the machine off line, the other two remain running. We are not convinced we can do that yet.

In Novette, which is the two arm system, we have only one VAX computer. That is because it is a rather small system. We were facing cost constraints, so that is going to be an interesting exercise. That system will be fully operational this fall, it is already partially operational. So far experience has been excellent.

It seems that the machines which are most prone to failing are those which are constantly used for diagnosis and for development and in that type of function which is not an on-line computer. Then, all sorts of microcomputers are extremely reliable; we have had no failures that I know of in different pulse power conditions.

In the fibre optic system that we use the fibres, of course, are reliable, but we have had some trouble with the transmitters which are very susceptible to the electromagnetic conditions. To summarize - the system itself has been extremely reliable. I know of no particular failures which have taken us down in the image processing system, for instance.

Tamm: Why do you use two nets, Novalink and Nova?

Suski: That's a very good question. As I said at the beginning, we wanted to allow the various subsystems to optimize control applications. It turns out that in the power conditioning system their job is, at the lowest possible cost, to control approximately a thousand devices distributed over 120 different nodes. They sometimes have to control these devices in sequences as close to each other as 20 to 100 microseconds. This implies an extremely low overhead network that you have total control over, no other traffic. That can be achieved with a single master if the rest of the system is fast enough.

The second aspect of the Novabus system is that it does not expect a separate reply which is circulating around the ring. There is also much lower error checking in the Novabus system. That is because, in that case if you have made an error, it is almost too late to recover, you've already fired the system. So in the Novabus system we had lower error requirements - we did not have the requirement to acknowledge each transfer. We did not want to wait for it - we wanted to compare and go on and do another function.

In the Novanet system we are trying to achieve a more general functionality. We do want to transfer data between computers and in fact, we needed a multimaster capability. In the Novanet system the principal function is to acquire data, so we do one good error checking on each piece of data transfer.

A MEMORY INTENSIVE FUNCTIONAL ARCHITECTURE FOR DISTRIBUTED COMPUTER CONTROL SYSTEMS*

D. G. Dimmler

Instrumentation Division, Brookhaven National Laboratory, Upton, New York 11973, USA

Abstract: A memory-intensive functional architecture suitable for distributed data acquisition, monitoring and control systems with large numbers of microprocessor based nodes has been conceptually developed and applied in several large scale and some smaller systems. This discussion concentrates on (a) the basic architecture; (b) expansions as they become feasible in view of continuing rapid developments in the microprocessor technology and in functional large-scale integration circuits; (c) the implementation of some key hardware and software structures; and (d) two system implementations, which are (1) a system for monitoring and managing of animal inhalation chambers and restricted access rooms as well as for experimental data acquisition and data analysis at an Inhalation Toxicology Research Facility and (2) a data acquisition and control system for a spectrometer with a 20 by 20 cm position-sensitive detector used in neutron scattering experiments at the Brookhaven High Flux Beam Reactor.

Keywords: Computer-aided system design; computer architecture; computer control; computer organization; data acquisition; distributed control; microprocessors.

INTRODUCTION

Continuing rapid decreases in the cost per function of microprocessors and functional large-scale integration circuits expose serious limitations in the traditional methods of conceptually outlining a computer control system application where one or several mainframe or host computer processors serve as a key focal point to which front-end electronics, sensors and peripherals are interfaced. Computer processors have become components which may be used in various places and levels of a system and, thus, become unsuitable as focal points.

BASIC CONCEPT

A system implemented in the distributed function architecture is geographically and logically organized along the paths of the data and control information of the application to be implemented. These paths now serve as the focal points. Numerous nodes, each roughly equivalent to a "computer" in the traditional terminology are placed and interconnected along these paths, geographically as close as possible to the source or destination of data and control information the particular node deals with. Each node is, then, assigned a well-defined function or a compatible set of functions. More detail on the basic concept is given by Dimmler (1974, 1978). Detailed studies on some key components of the architecture are reported in the Results of an R & D Effort on the Isabelle Accelerator Control System (Dimmler, 1979).

DESIGN OBJECTIVES

A user or observer of a system is unaware of many of those nodes placed along the paths in a system. Many nodes provide their functions without direct human interaction or attention. It is, therefore, a fundamental requirement that those nodes do not need the attention from support personnel, software analysts, operators and users as is customary for traditional computers; otherwise it would become logistically difficult or impossible to arrive at a reliable operation of a distributed system. The design objectives generated as a result of the functional orientation and above mentioned requirements are:

(a) Functional assignments to a node ought to be straightforward and well understood in every detail.
(b) The software operating in a node must be functionally complete and free of errors. This is probably the most ambitious objective.
(c) Emphasis should be placed on providing hardware and software for self-diagnosis of the nodes and of unsolicited problem reporting to appropriate personnel.
(d) A system is to be installed, tested and integrated into the application in an incremental fashion and along the

information paths. It should be avoided that, at any time, large portions of a system are under test or are in the process of being integrated.
(e) Manual handling of data and program media should be kept to an absolute minimum.
(f) It is vital that there is convenient, non-confusing and system-guided interaction between the system and all persons directly or indirectly involved in or affected by the system.

Extraordinary emphasis is placed on careful, selective adaption of communication speeds and timing of communication sequences between the system and people. A mismatch here is the traditional source of confusion, logistic complexity and dissatisfaction. For instance, in the connection of terminals to a system it is often overlooked that a person is able to interpret and act on written alphanumeric or graphic output information several orders of magnitude faster than one is able to insert information into the system via a keyboard or other input means. As part of the architecture elaborate use is made of high-speed, random-access raster-scan display monitors for alphanumeric and graphic information. The memories containing the display images are located either in the logic address space of a particular node, or a group of nodes, subsequently introduced as a cluster. An effective random-access update speed of approximately 100,000 bytes per second is achieved.

MODULES

For economic and practical reasons it is essential that the architecture is based on a well-developed, commercially available family of microprocessor equipment which is in widespread use. The LSI-11 family of modules which connect to the Q-bus has been chosen. This module family was introduced several years ago by the Digital Equipment Corporation. The instruction set and many of the structural features are compatible with earlier PDP-11 minicomputer systems. A fair amount of software and also some of the hardware modules could be adapted until more up-to-date equipment was available. Now a large variety of hardware modules and application software packages are offered by a large number of manufacturers. As a general objective, custom designed hardware modules and software packages are used only when commercial ones are not available.

NODES AND CLUSTERS

Each horizontal bus in Fig. 1 represents one node. The node components are interconnected via the Q- or node bus. They are, as a minimum, one LSI-11 processor, one serial diagnostic channel which allows for independent testing of the node, a random-access memory and a terminator module which also contains all the support hardware, such as time-out circuits, a real-time clock, a time-of-day clock, the node startup program and the node identification in read-only memory. Generally at least one display image memory is included. Up to eight functionally compatible nodes may form a cluster. Each node in the cluster is connected via an access port to the vertical M- or memory bus, as described by Dimmler (1977) and Alberi (1978), allowing access to the cluster memory of a potential size of 67 million bytes.

Figure 2 shows the photograph of a typical basic node. Figure 3 illustrates an example of an installed cluster. Presently available technology allows for up to 5 nodes and two megabytes of cluster memory in a 10-inch high and 19-inch wide drawer. Such a drawer includes a 750 W power supply serving up to 81 so-called option slots, i.e. connector space for a complete Q- or M-bus connection suitable to plug in a module which is 132 mm (5.2") high, 228 mm (8.9") long and allows for a spacing of 12 mm (.5"). Double height or triple height modules may connect to both busses simultaneously.

NODE BUS ADDRESS SPACE

The Q- or node bus, the standard bus for the LSI-11 as specified by the Digital Equipment Corporation, has a width of 16 bits allowing for addressing of 65,536 bytes or 32k words.

As shown in Fig. 4, the space between 0 and 24k words is populated by random-access memory. In general, 8k words each are reserved for the permanently loaded node specific operating system, 8k words contain the resident segment of the program which is in process of being executed and 8k words are reserved for one of up to 64 overlay segments which may be called in by the resident segment at any time. The cluster memory is partitioned into pages of 4096 words each. One page at a time may be selected into the page window between 24k and 28k allowing for random access of this page by the node.

The address space from 28k to 30k is reserved for up to 32 display image memory segments of 2048 words each. The processor may randomly address, through a program controlled segment address mechanism, one segment at a time. One raster-scan converter is associated with each segment; thus, all images can be displayed concurrently. Alphanumeric configurations of 16 lines by 32 characters, 16 lines by 64 characters and 24 lines by 80 characters are presently in use. There are plans to develop graphic modules with multiple layers for gray scale or color operation which are located in this space. Display image memory segments, which are potentially updated by several nodes, are located in the cluster memory address space and are accessed through the page window. The address space from 30k to 32k is assigned to input/output registers as specified for the standard Q-bus.

CLUSTER MEMORY

The M-bus access port of the node contains a program selectable page register as shown in Fig. 5. The bits 13 to 15 of the Q-bus are substituted, if they contain the pattern ११०, by the page register content and are used as bits 13 to 25 of the M-bus. The bits 0 to 12 stay unchanged. The M-bus, then, addresses the memory with the resulting 26-bit physical address. The M-bus resembles a PDP-11 modified UNIBUS where the bits 22 to 25 are added. Commercially available UNIBUS memory modules are used. A separate address decoder is used for cluster memories exceeding four million bytes.

A round-robin scanner provides arbitration between simultaneous accesses to the M-bus by several nodes. As shown in Fig. 6, the M-bus access control offers a port select option to a node access port for 80 nanoseconds. If there is an access request pending, the access port raises the scan stop allowing the node to execute one memory cycle. The photograph Fig. 2 shows that the 2 signals are distributed with point-to-point coaxial cables from the access control to each access port. The photograph of the access control module in Fig. 7 shows the connector terminals for 8 node connections.

Cluster Memory Protection

Each M-bus access port contains two UV-erasable read-only memory chips of the type 2716 which are used to protect the cluster memory against accidental write accesses. The memory is partitioned into 256 word protection segments. One bit is assigned to each segment. A "०" burned in the position prohibits write access to the segment from this node. The first four million words are protected this way. The first 16,384 words in the cluster memory are protected on a word-by-word basis by the second 2716 chip. Read access to the entire cluster memory is available to the node at all times.

CLOSELY COUPLED CLUSTERS

A very high-speed inter-cluster communication link can be established if the M-busses of two or more clusters are interconnected and a single access control is used. This mode of coupling is used where very close geographical and logistic proximity can be tolerated or is desired. The total M-bus length is limited to approximately 20 meters. A total of 8 nodes can participate in the coupling.

Figure 8 shows a block diagram of two closely coupled clusters.

COMMUNICATION PROTOCOLS

The communication protocols between closely coupled clusters are usually specific to the functions the clusters perform. The same is true for the communication between the various nodes in a cluster. Only general guide lines are established here. The control flags and data are communicated in an agreed-upon protocol and through pages in the cluster memory which are dedicated for this purpose. It was felt that a general purpose function-independent communication protocol for closely coupled clusters and for intracluster communication would introduce a substantial functionally unrelated overhead which would decrease the communication speed and increase the logistic complexity. In addition, it would have the tendency to divert attention from the functions the nodes are to perform and to prohibit some functionally desirable communication methods.

Clusters which are loosely coupled over a long-distance communication system as described below follow, on the other hand, a very strict function-independent communication protocol. One node in each cluster is dedicated to operate this communication protocol with a program stored in read-only memory of an approximate size of 4000 words. The functional independence at this point in the system allows coupling of functionally incompatible or unrelated clusters.

LONG-DISTANCE INTER-CLUSTER COMMUNICATION

Geographically distant and functionally incompatible clusters communicate among each other through a full-duplex, star shaped, serial communication system with an asynchronous rate of one megabit per second over a distance of up to one kilometer using two standard (RG 59) 75-ohm coaxial cables or fiberoptic cables. The distance may be extended to four kilometers by using low attenuation coaxial cables. Longer distances require active repeaters.

Modem

A record to be transmitted is, on the transmitting end of a cable, serialized and frequency modulated on an 8.4 MHz carrier by a three level voltage controlled oscillator. The resulting approximately $2V_{pp}$ signal at the receiving side is captured, amplified and demodulated using an RCA CA 3189E television FM-receiver chip. The automatic frequency control output which has a bandwidth of 1.5 MHz is then submitted to a serial-to-parallel converter. The upper left corner of Fig. 9 shows a photograph of the modem to which the coaxial cables are connected.

Communication Topography and Protocol

At any given time each uniquely identified node in the system is assigned the role of either an originator of a command for communication or of a server of commands. This role can be dynamically assigned, if functionally desirable, although in practical systems such a need has not yet arisen. In Fig. 10, for instance, a node in the control cluster A can direct, through a communication interchange switch, a WRITE, READ or REQUEST information command as detailed in Fig. 11, to a node in the Data Base cluster A. The arrows in the Figures 10 and 11 indicate the direction of control. The sequence of events is in the example of a WRITE command as follows: A command is executed by a series of transactions between (1) the originator and the switch and (2) concurrently between the server and the switch. In the idle state the server continuously interrogates the switch for commands to be serviced. The originator submits a PUT transaction as further detailed in Fig. 12 to the switch which routes as shown in Fig. 13 the data and the control information to the Wait Buffer of the server node desired to be contacted. The server node receives, via a GET instruction, the data and control information from the WAIT Buffer. The originator now interrogates its own Wait Buffer to await the return information.

It is noteworthy that the switch provides the functions of monitoring the logistics of the communication, of decoupling the nodes, of buffering, queuing and routing the information. It also monitors the performance of the communication hardware. A request block and an acknowledge block of a transaction as mentioned in Fig. 12 are each 32 words long; a data block has a variable length of up to 8192 words. In addition, each block is preceded by a hardware-recognized so-called Transfer Unit Start Word and is trailed by a cyclic redundancy check.

Multiple Path Interconnection

It is always a node in a cluster which controls a transaction or a command and never a communication interchange switch. Therefore it is possible to connect a cluster to several switches where the originating node in the cluster determines which way to send a command, and a server cluster determines (1) where to interrogate for commands to be served and (2) how to deal with the command. The originating cluster, of course, needs the information along which path a desired server can be found. In permanently interconnected clusters this topography description can be burned into read-only memories at each cluster. In dynamically interconnected systems that information is kept in a network description data base, which resides in one or several predetermined server nodes.

The specific function which is intended to be performed by using multiple paths is determined by the way the various clusters cooperate in the execution of commands and not by specific hardware or system software features. For instance, if path redundancy is intended as a function, a node in the cluster A sends the same information to both the switches A and B which are both interrogated by the data base A. The algorithm on how to deal with both communications, if they arrive, resides with the server node and is tailored to the specific service the node is intended to perform in the network.

In another example, a cluster may use the services of two otherwise independent networks. In Fig. 10 only cluster A would be interconnected with the switches A and B. The other crossconnections would not be used. Broadcasting will be accomplished if cluster A associates a command with several server nodes, causing the switch to stage copies of the information in several wait buffers. For such a function to be performed an agreement has to exist among the participating nodes on how to deal with multiple source return information.

A redundant data base scheme is established if the control cluster A names both data bases as command servers. Any one of the server nodes may, as mentioned before, temporarily assume the role of an originator and control an algorithm which determines data base integrity.

Complexity of Practical Systems

The achievable reliability and usefulness of a complex network, in general, depends on the degree with which the logistic complexity is understood and can be kept under control. Possible weaknesses in the software or hardware can usually be eliminated with available technology. It is very important that every feature introduced is functionally complete in the sense that there is a designed and complete response to every conceivable logical state of the system. This is particularly true for some of the communication features as described before. As a general rule, one should start in practical systems with simple and well understood configurations. Systems ought to be built up incrementally towards complex communication schemes, and this should be done only if (1) a demonstrated need exists and (2) the functional completeness of the new feature can be guaranteed. The potentially higher benefits of a more complex configuration are often accompanied by potentially more disastrous pitfalls.

Functionally distributed systems have been implemented at the Brookhaven National Laboratory since 1973 (Dimmler, 1976, 1976). The first such system, which controls ten neutron spectrometers, uses eleven single-node clusters with permanently assigned roles. PDP 11's have been used as processors. The communication interchange switch and all the server functions have been combined into one node. Serving of commands has been done in-line with the communication. Thus, a WRITE command has been reduced to one PUT transaction and a READ command consists of one GET transaction only. The most recent system which is in operation, the Inhalation Toxicology Data System as mentioned below, uses multi-node clusters based on the microprocessor based architecture described here. However, the roles of originators and servers are still permanently assigned and a single path interconnection is used. The communication interchange switch and all server clusters are geographically close and are closely coupled. In-line service of commands is still used. Thus, single transaction commands are still sufficient. For the next phase of this system, an urgent need for a geographically distant large data base with redundant recording has been demonstrated. This will be the first instance where the multi-transaction command scheme and a multi-path configuration will be put into practical operation.

SOFTWARE DEVELOPMENT

Software development is the process of creating, correcting, assembling, testing, confining and filing software of a wide variety of styles and linking that software into executable modules. In the case of distributed systems a reliable and automated distribution of these executable modules to the various nodes is a vital part of the process. In general, it is an effort by one or several persons to successively integrate information which is available in a highly dispersed fashion. In the present state-of-the-art this highly iterative effort is generally (a) tedious, (b) very time consuming, (c) very susceptible to human error, and (d) often confusing. The product is often not error free and not functionally complete.

As stated before, functionally and geographically distributed systems with many processors and several possible logical and physical communication paths demand considerable quantities of error-free and functionally complete software. That is a very different situation as compared with much more tolerant traditional main-frame or host computer-based systems. Successful application of distributed systems and advancements of the field will likely be limited in the foreseeable future by the degree to which the productivity of the development of such software can be improved. Several orders of magnitude in improvements are necessary here. An effort was undertaken in moving towards automation of the overwhelming procedural part associated with the development of large and integrated software systems.

A successful organization is highly disciplined, defined in every detail along the information flow from the creation by a person to the properly filed and loaded executable modules. The processes have to be repeatable with minimum chance for introducing new errors and confusion. Numerous tests and filters have to be placed along the paths in order to minimize the penetration of human errors into the executable modules.

One complete cluster is used for that task. As shown in Fig. 15 as Program Development Cluster 103, one node each is assigned the functions of (1) the source code manager, (2) the object code manager and (3) the communication terminal to the system. The source code to be edited, a segment of up to 1024 lines of 80 characters each, resides in 24 pages of the cluster memory together with the associated active directories. Inactive source code segments and directories are filed on removable disk cartridges dedicated to source codes and operated by the source code manager. They are archived off-line. Each cartridge is called a volume. The software for the source code manager resides on the cartridge. Thus, the editing process of a volume can be selectively parameterized to the system, the function, the language predominantly or exclusively used, the levels and qualifications of the people predominantly updating the volume, etc. In the implemented systems a distinction is made between the languages and the processor types used, e.g. the MACRO source editor behaves slightly differently than the FORTRAN editor. The editor for the RCA 1802 is different from the INTEL 8085 editor. Both are microprocessors which have been used for special functions.

Editing takes place using 3 keyboards and five high-speed, random-access raster-scan monitors. Figure 14 shows a photograph of a software development station. One screen displays selectable text available for comparison. The directory display shows the relevant selected directory information. Two screens are used for directory image copies which may be of interest to the user of the station. Very high emphasis is placed (1) on continuing improvements in convenience and relevance of the displayed directory information and (2) on increasingly complex linkages among the directories of software modules in order to allow for increasingly elaborate and automated sequences. It is at this time possible to start the process of compiling, integrating and describing one segment with one keyboard stroke. In another example the load module generation of a programming system consisting of 64 overlays and library segments organized in 36 libraries can be initiated by one keystroke after which no further attention is required.

The object code manager, the second node in the cluster, receives the edited source code which it compiles and integrates into one of several program systems. This is one of the very few places in the architecture where a traditional general purpose operating system is used as an integral part of the system. This has been done as a compromise, because almost all available compilers, linkers and other support software are intended for operation in a general purpose host computer and, thus, tied to one of those operating systems. It was considered essential to remove the continuous human attention such an operational system customarily demands. Therefore a command generator program has been developed which takes the necessary information from the previously described directories and compiles, links, interprets and updates object modules and provides appropriately labeled listings without human attention. The command generator has presently 10 overlays coded in FORTRAN with approximately 8000 words each. Functional completeness has been gradually achieved through an iterative process. Each problem or confusion experienced by a user has been carefully analyzed and the command generator has been appropriately updated. A terminal, as shown in Fig. 14 is still used to inform the user of the progress. The need for input at the keyboard has been eliminated.

A major limitation has been found in the awkwardly slow general purpose operating system which is representative of most commercially available systems. A complete compilation and integration can take up to ten minutes. Most of the time is used up in the functionally unrelated overhead caused by managing the utility processors and by the use of a disk based general-purpose file management. A substantial improvement would be gained (a) by elimination of the general-purpose system and (b) by designing the compilers and linkers as self-contained modules using function specific file management in random-access memory.

INHALATION TOXICOLOGY DATA SYSTEM

Starting in approximately July 1980 a Data System has been developed and put into operation in line with the construction schedule of the newly established Brookhaven Inhalation Toxicology Research Facility. The data system is used (1) to monitor and control a series of animal inhalation chambers and the chamber room environment, (2) to detect and enunciate alarm conditions and to schedule personnel concerned, and (3) to service experiments. In addition, a portable animal weighing station is part of the system.
Post analysis of collected data is performed to some degree. Elaborate analysis is referred to the Laboratory's computer center. Some of the features described here are scheduled to be completed in future phases. Figure 15 shows the configuration in operation or soon to be completed.

As indicated in the architectural objectives, the clusters are located at the most convenient places along the data paths. The clusters which service the controlled chamber rooms with 8 inhalation chambers each are located close to the chambers in an enclosure resistant to dirt, oil and water. This is done because the chamber rooms are routinely washed down and can be potentially contaminated with hazardous material. There is no air exchange between the closure and the room in order to avoid possible contamination of the electronics.

Three nodes combined to one cluster service one room. One node reads the sensors and serves control outputs on the inhalation chambers and the controlled room around the chamber. The chamber bottom temperature, the top temperature and the wet bulb temperature for humidity determination and some temperatures in the room are sensed by 40 solid-state temperature transducers of the type AD 590. The static pressure in each chamber and the flow rate of the intake air mixture at each chamber is measured by 40 piezoresistive pressure sensors, model 140 PC manufactured by the Microswitch Corporation. 120 digital inputs and outputs are used for sensing and controlling numerous on-off conditions.

Each sensor is described in a 128-word directory which includes information such as the mnemonic sensor name, a unique sensor identification, channel information, conversion factors, ranges for attention and alarm conditions, sampling time intervals, the last reading, etc. The directories are located in the cluster memory.

The second node continuously analyzes the directory information and forms several images which are displayed on monitors throughout the room and the facility. Fig. 16 shows a typical display image. The parameters of two of the eight chambers are displayed in four consecutive images which are shown for 15 seconds each. The images are kept updated during this 15-second interval.

Self-diagnosis is particularly important for above mentioned clusters as they are difficult to access for maintenance. Therefore, each cluster will soon be equipped with a 16-channel analog-to-digital converter and sensors with which all voltages of the cluster and the temperature in 6 selected places of the enclosure are continuously monitored. Problem reporting will be done through a low-speed, independent point-to-point RS 422 serial line to a maintenance node. Experience shows that early detection and correction of voltage and temperature problems will eliminate most failures of well designed electronic hardware.

A single path configuration with one communication interchange switch has been used. The presently implemented server clusters 100, 101 and 102 as shown in Fig. 15 are closely coupled. Thus, in-line serving of commands is sufficient. The chamber rooms and the digital equipment rooms are separated by a cable distance of approximately 100 meters.

SPECTROMETER CONTROL SYSTEM

A series of neutron spectrometers of various types are in operation or in development at the Brookhaven High Flux Beam Reactor. The spectrometers are used for neutron scattering experiments in basic research.

A neutron spectrometer control system has been recently installed which controls one of these spectrometers and takes data from a 20 by 20 cm position sensitive area detector with 256 x 128 resolution segments. As shown in Fig. 17, the system includes five clusters, which are (1) the spectrometer control, (2) a communication interchange switch, (3) a task library and data base storing the executable load modules and data files which are system-wide accessible, (4) a server for the computer peripherals, and (5) a software development station.

The spectrometer control, which is the only cluster located within restricted space close to the spectrometer, includes six nodes and one megabyte of cluster memory. The spectrometer control program is executed in one of the nodes. Such a program is typically specific to the experiment to be performed and is usually written by the experimenter concerned in FORTRAN at the software development station. Another node operates the interactive display terminal which is used for on-line monitoring and analysis of the data taken by the area detector.

Two of the nodes control the various motions and interconnect with the equipment at the spectrometer. Figure 18 shows clearly that the traditional notion of "interfacing equipment to a computer" appears inappropriate here. A better description of the situation is that microprocessor nodes surround the equipment to be serviced. The connections of a device to be serviced to the appropriate hardware module are made as directly and as function-specific as practical. A high overall bandwidth is achieved because the bottlenecks often caused directly and indirectly by a standard interface bus are avoided. The reasoning for some representative connections follows.

The area detector shown in Fig. 18 generates 15-bit addresses with a random rate of up to 10^6 events per second. These addresses serve to increment the appropriate variable of a data array which is located in the cluster memory. The data input module, a custom design, which extends over three option slots, is plugged into both the (vertical) M-bus and the (horizontal) Q-bus of the service node. The addresses are first captured in a FIFO buffer and then used for a read-increment-write operation which requires approximately 10^{-6} seconds. The data input is controlled by the service node through the Q-bus. The possible address range of the data input is larger than any presently conceivable detector would need. The achievable event rate is limited by the M-bus speed which in turn is limited by the speed of available mass-produced random access computer memory modules. It is unlikely that such memory speed will improve more than the factor of 2 or 3 in the foreseeable future. More extensive improvements would require rate reduction algorithms located between the detector and the M-bus connection.

Up to eight stepping motors are serviced by a motor control module which occupies 3 option slots in the Q-bus. The module includes 8 stepping motor control chips of the type PPMC-101B. The phased pulse sequences, which are the output of the control chips, operate low-resistance MOSFET power switches of the type IRF151 manufactured by International Rectifiers. They are in Fig. 18 identified as translators. The switches are supplied from a power supply which is common to all eight motors. The switches are mounted on a module requiring 3 option slots at the Q-bus. Nine more option slots around the power switches are kept unoccupied in order to allow for sufficient air circulation. The "interface to the experiment" consists here of the wires carrying the pulses for the stepping motors.

The temperature control shown in Fig. 18 has a standard RS-422 connection. Modules containing 4 such channels in one option slot are commercially available.

The various axes of the spectrometer are positioned with a precision of 10^{-2} angular degrees at a maximum speed of 10^3 steps per second. Incremental encoders are used to track the position during the motions and to determine the end position. The current positions of all axes are kept in directories located in the cluster memory. This allows for access from other nodes. The encoders may, in the worst case, generate a combined pulse rate of 10^4 pulses per second allowing only 100 microseconds to update an angular position. A single lost pulse would falsify such position information. In order to adhere to this strict timing requirement, a complete node has been dedicated for this purpose. The node also maintains a copy of the current position in a CMOS memory which is backed up by a battery. This feature assures that the position information is maintained if the power is turned off or if the cluster is disturbed otherwise.

An experimenter at the spectrometer receives essentially all the information from the system through several raster scan displays. A high bandwidth is achieved here by maintaining several concurrent images, each of which has a high potential update rate. In designing the images high emphasis has been placed (1) in assuring that the images are updated instantly within the frame of human perception and (2) that irrelevant or old information is removed. Fine tuning of these images will be a continuing process. Figure 19 shows a portion of the operating desk for the spectrometer.

ACKNOWLEDGEMENTS

I am deeply indebted to Mr. S. Rankowitz who developed most of the custom designed electronic hardware modules. He also was available for extensive discussions and advice on the hardware aspects of the architecture and the implemented systems. Dr. J. L. Alberi designed the data input module for the spectrometer control system.

Mr. J. Gannon's services in testing the modules, subassemblies and in installing the systems are greatly appreciated. Mr. R. Machnowski laid out several printed circuit modules. Mrs. M. A. Kelley designed several of the FORTRAN programs.

REFERENCES

Alberi, J. L. (1978). Multiport memory subsystem II. BNL Design note 1.1-032478.

Dimmler, D. G. (1974). Functional distribution - an architecture for multi-user computer networks in Instrumentation. IEEE Trans. Nucl. Sci., NS-21, No. 1, 838-850.

Dimmler, D. G., N. Greenlaw, M. A. Kelley, D. W. Potter, S. Rankowitz, and F. W. Stubblefield (1976). The Brookhaven Reactor experiment control facility - a distributed function computer network. IEEE Trans. Nucl. Sci., NS-23, No. 1 398-409.

Dimmler, D. G., D. W. Huszagh, S. Rankowitz, and J. Scott (1976). A controllable real-time data collection system for coastal oceanography. Proc. Ocean, 76, Sept. 1976, Washington, D.C. 14E1-14E13.

Dimmler, D. G. and W. H. Hardy, II (1977). A shared random access memory resource for multiprocessor real-time systems. IEEE Trans. Nucl. Sci., NS-24, No. 1, 469-475.

Dimmler, D. G. (1978). Computer networks in future accelerator control systems. IEEE Trans. Nucl. Sci., NS-25, No. 2, 974-988.

Dimmler, D. G. (1979). Results of an R&D effort on the ISABELLE control system. Brookhaven National Lab., Upton, N.Y. BNL 25528.

* This research was supported by the U. S. Department of Energy: Contract No. DE-AC02-76CH00016.

By acceptance of this article, the publisher and/or recipient acknowledges the U. S. Government's right to retain a nonexclusive, royalty-free license in and to any copyright covering this paper.

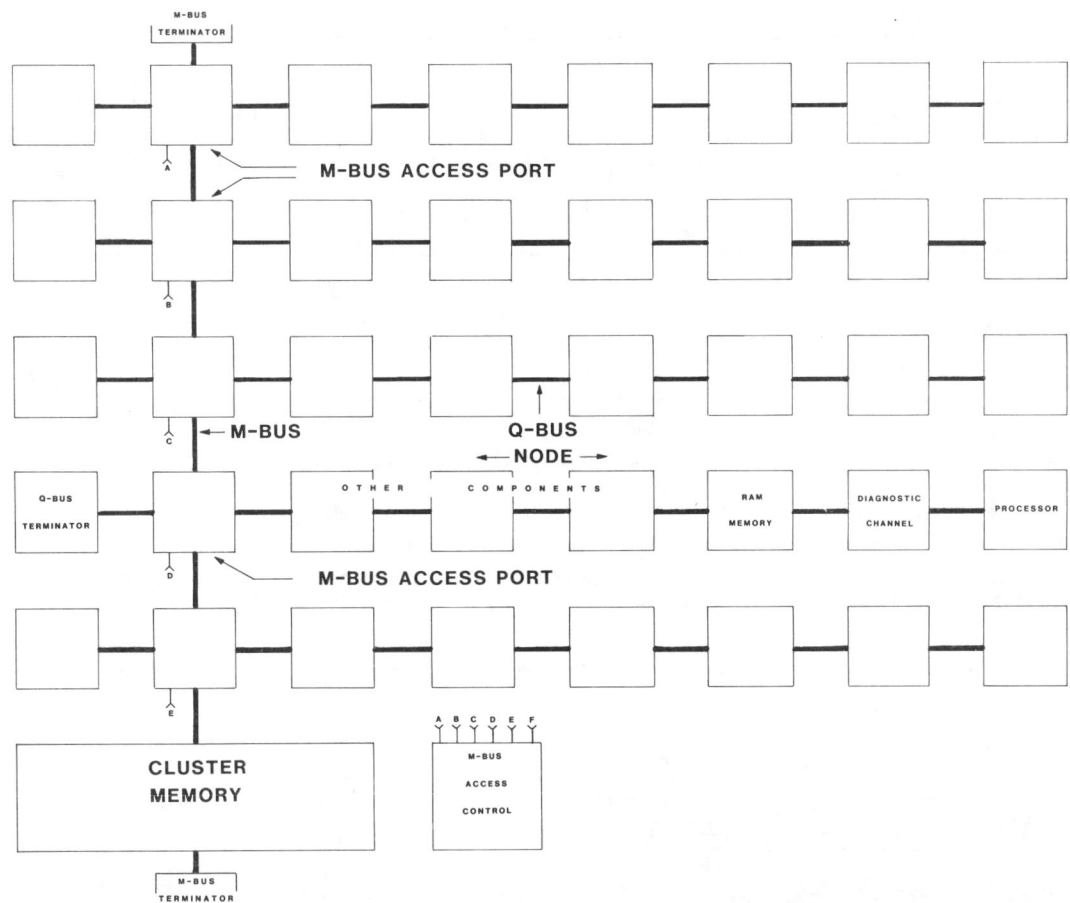

Fig. 1. Cluster organization

Fig. 2. A typical basic cluster node

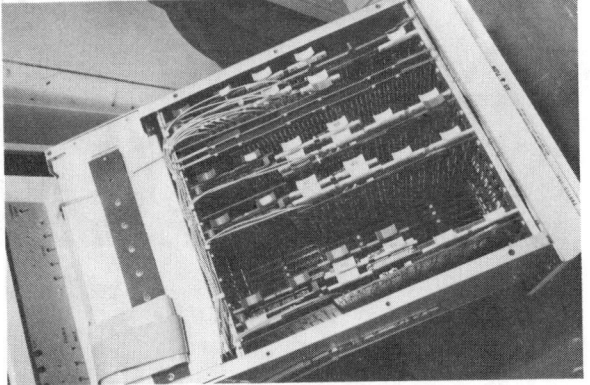

Fig. 3. An example of an installed cluster

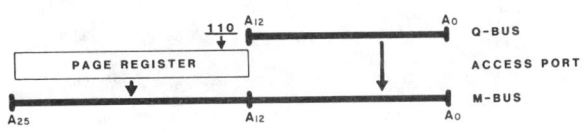

Fig. 4. Memory allocation of a node

Fig. 5. Memory address calculations between Q-bus and M-bus

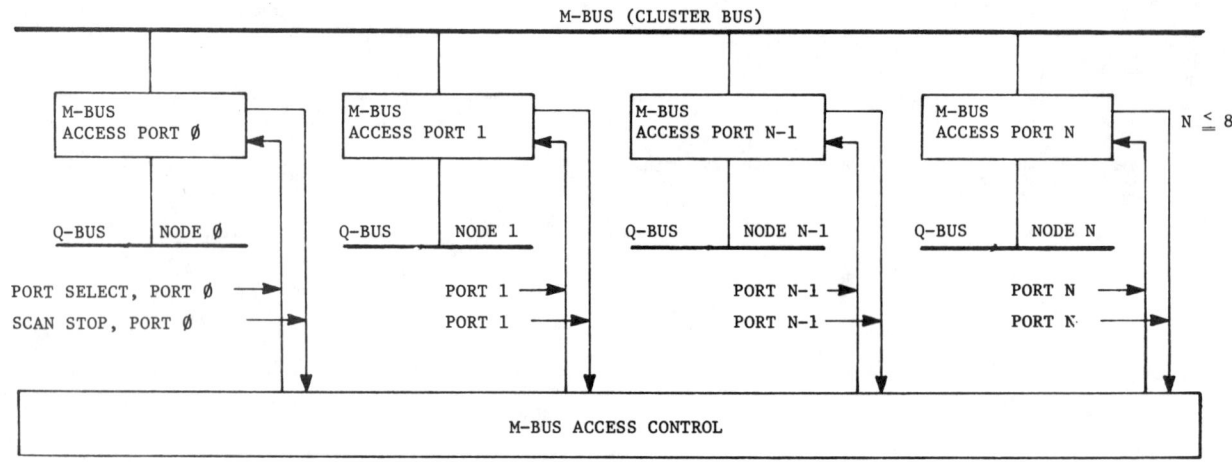

PORT SELECT OFFERED TO PORT: ~80 nanoseconds
SCAN STOP TIMED OUT AFTER: ~100 microseconds

PORT SELECT AND SCAN STOP ARE ROUTED AS POINT-TO-POINT COAXIAL CABLES FROM THE ACCESS CONTROL TO EACH PORT. ONE PAIR FOR EACH PORT IS USED.

Fig. 6. Interconnections between the cluster memory access control and the memory ports

Fig. 7. Cluster memory access control module

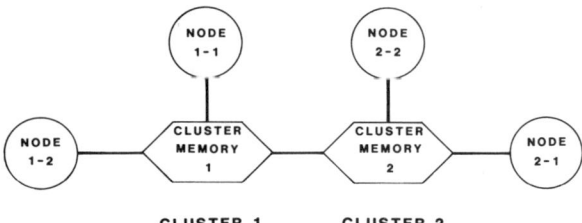

Fig. 8. Closely coupled clusters

Fig. 9. Long-distance communication module

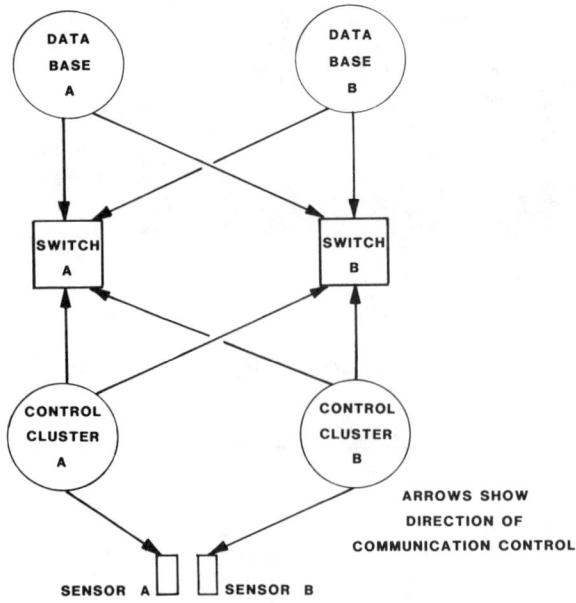

Fig. 10. Network with multiple paths

A Memory Intensive Functional Architecture

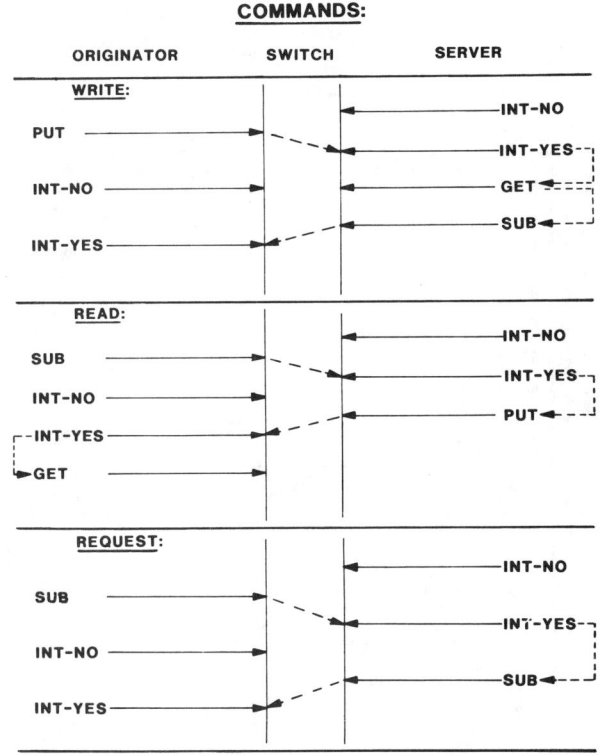

Fig. 11. The three types of commands

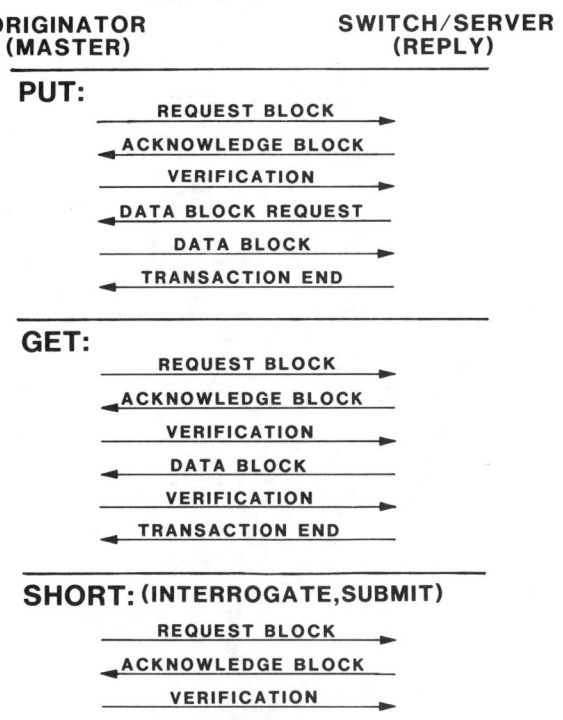

Fig. 12. The three types of transactions

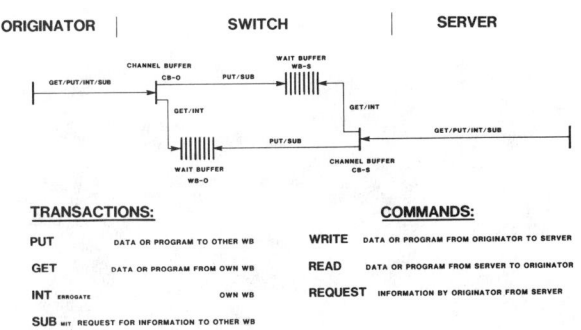

Fig. 13. The paths of a command between the originator and a server

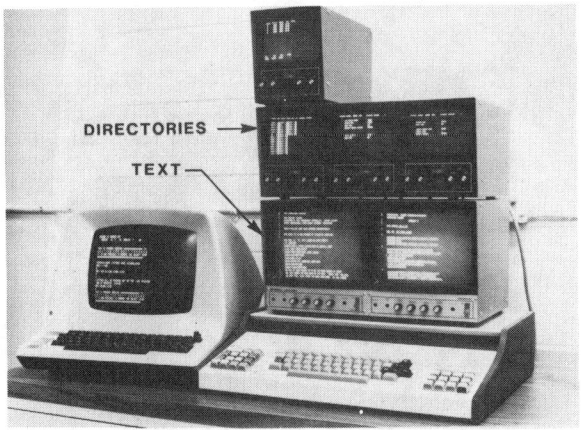

Fig. 14. A software development station

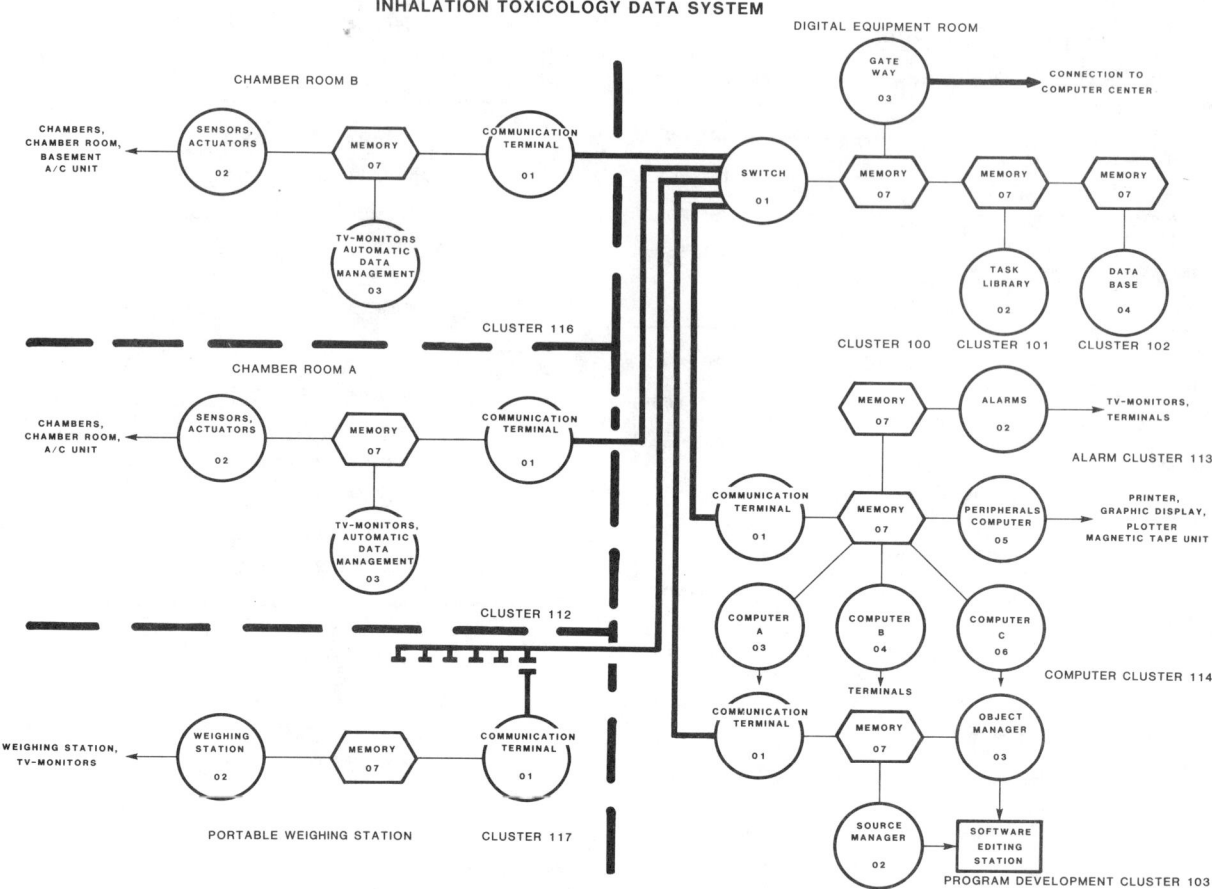

Fig. 15. Interconnection of the nodes and cluster of the Inhalation Toxicology data system

Fig. 16. Typical display image in chamber room

Fig. 19. Partial view of the Neutron Spectrometer operating desk

A Memory Intensive Functional Architecture

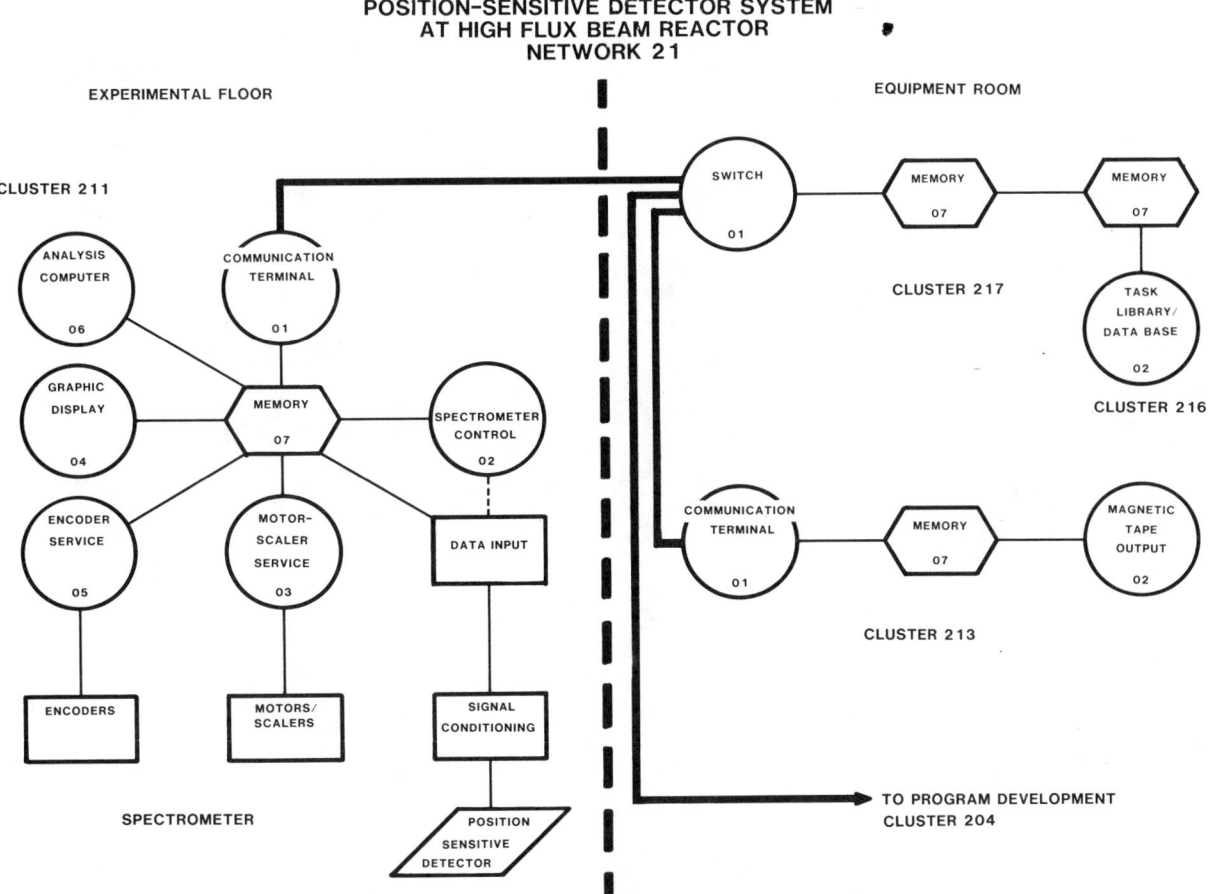

Fig. 17. Interconnection of the nodes and clusters of the Spectrometer control system

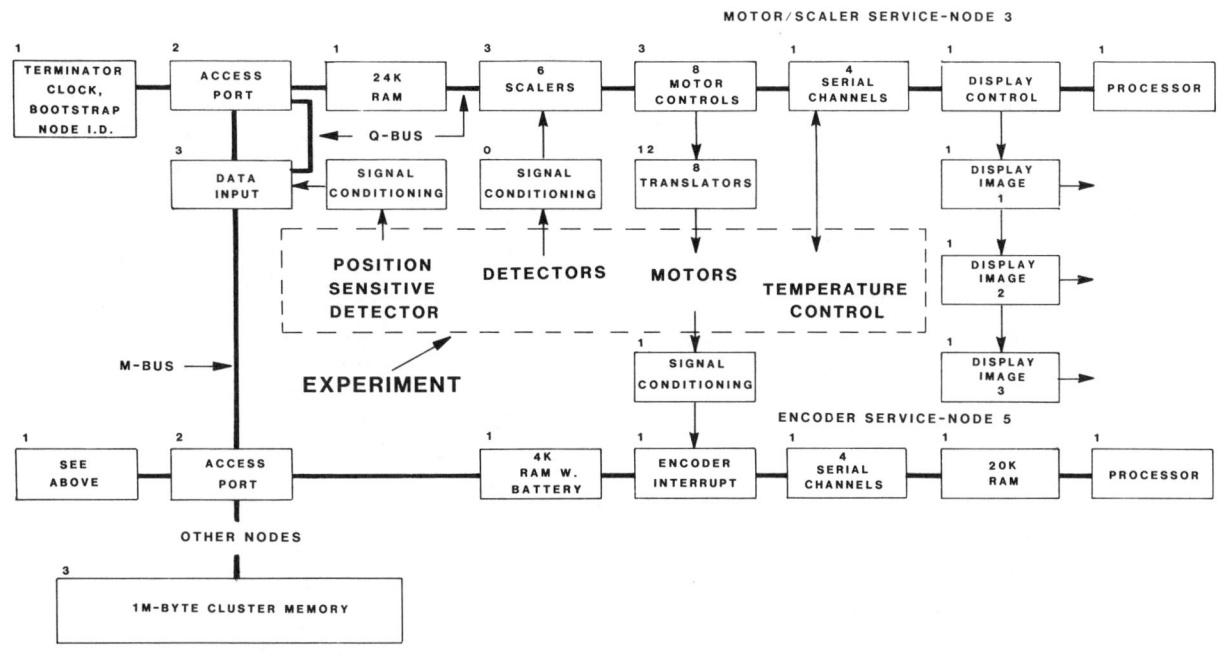

Fig. 18. The connection of the control system to the experimental setup

DISCUSSION

Rodd: Mr. Dimmler, I would be interested to hear more about the M-bus structure in your cluster. Is the M-bus also a standard parallel bus?

Dimmler: Shared memories are very useful devices to communicate with. The M-bus is a PDP-11 modified UNIBUS where the bits 22 to 25 and an interrupt scheme are added. By means of this bus the nodes are grouped around the shared memory. A separate address decorder is used for cluster memories exceeding 4 M bytes.

Inamoto: I'm curious about the configuration of your software development station, especially since you have three keyboards and several CRT's. Usually a programming station has only one keyboard and, maybe, one hardcopy device. Have you conducted any studies on the productivity of equipment of this type?

Dimmler: Let's assume that we have a very large system, for instance a FORTRAN system. These spectro-meters are highly automatic devices and generally their programs have about 40 overlays of FORTRAN and many library programs. If you want to make one change in such a system, you'll have to be able to do that within a minute, hit a button, and have everything come together in a reliable way. That automation process is actually the key to why there are three keyboards. There is one keyboard in the middle where you have your text, one keyboard on the right which deals with the initialization or termination of the session and of the text or previous lines; and the left keyboard is for editing. In that way you can make farily complete functions, you do not have to type a command with a sequence of alphanumerics. You just hit the button and you know that the thing is inserted.

The five screens come as follows: You have two main screens; one is the directory which at any given time describes exactly the text you have on the bottom screen. That text is automatically updated if anything is attached, then is stored with a segment. On the two other screens - left and right - you can make copies of the directories because usually you want to compare directories so that you do not have to copy the things down on a piece of paper. On the bottom there is a copy of the text, so that you can compare two texts with the hit of a button. The listing output is a function on the other side that happens to be dealing only with the source text.

Dealing with an object code is completely automated and there are just several sequences possible. If a new sequence is desired, you automate it rather than explain to a programmer how to do it. A part of this process is a line printer output which is completely labelled. There is no way to avoid line printing output because a few screens don't solve the problem. You want to compare, you want to have that stuff on the table, and there is an awful lot of information on line printers. You would need high resolution monitors to avoid line printer output.

I think the system couldn't have been built without automation program development. The total number of people involved in the software is two - I'm personally one of them.

Inamoto: Are very sophisticated programs necessary?

Dimmler: No, quite on the contrary - a casual programmer should be able to handle the thing and make complex systems. We never managed to write a manual for it.

I should like to clarify something. When I say programmers, I mean system programmers. The objective is to get these people to write complex systems because they will never read a programming manual. They will read the very simplified FORTRAN manual - that's the only thing they have, and everthing is self-explanatory.

IMPLEMENTATION AND PERFORMANCE EVALUATION OF A DISTRIBUTED SPLITTED-BUS MULTIPLE COMPUTER SYSTEM

Lan Jin, Wei-min Zheng, Ding-xing Wang and Mei-ming Sheng

Department of Computer Engineering and Science, Tsinghua University, Beijing, China

Abstract. A simple scheme of implementation of one-dimensional splitted-bus structure is proposed. Insertion of routing switches into the bus enables a distributed multiple-computer system to transmit variable-length messages simultaneously between several distinct processor-node pairs so long as the data links used for them do not conflict with one another. The interconnection topology and the mode of control for such an experimental system are described. Its characteristic performance values, such as data-line utilization, mean total message transmission time, etc. are given as results of simulation. These values are compared with similar results for other well-known loop networks found in the literature.

Keywords. Computer architecture; parallel processing; distributed computing; distributed computer systems; computer networks; performance evaluation of computer systems.

INTRODUCTION

Bus structure is widely used to interconnect computer modules to organize distributed multiple computer systems (Jin, 1981). It has the advantages of simplicity, reliability, high bandwidth, incremental expandability, etc. Special consideration should be taken, however, to reduce the possibility of bus-contention. For this purpose, splitted-bus structures have been proposed (Jafari, 1977, 1980; Jin, 1980), for which routing switches were inserted to dynamically partition a bus into segments, so that separate data links can be available for simultaneous transmission of variable-length messages between several distinct pairs of processor nodes so long as the communication paths for them do not conflict with one another. Consequently, the characteristic performance of the system, such as bus utilization, throughput, response time, as well as flexibility and reconfigurability, can be considerably enhanced.

In the first part of this paper, a scheme of implementation of one-dimensional splitted-bus structure will be described. Then the ways of realization of various data-communication patterns and processing functions will be explained in order to reveal the advantages of the proposed system.

An experimental distributed splitted-bus multiple computer system is being implemented on the basis of PDP-11/03's as computer nodes, a microcomputer system with INTEL-8080 as control processor, and various IC chips as interfaces and routing switches.

The interconnection topology and the control algorithm of this system will be described subsequently.

For measuring characteristic performance values of the proposed splitted-bus network, a series of computer simulation experiments has been conducted. The third part of this paper is devoted to brief description of the simulation method, followed by typical simulation results obtained. Comparison of these results with similar values for other existing loop networks would show a considerable effect of splitted bus on performance enhancement of distributed multiple computer systems.

DESCRIPTION OF THE SPLITTED-BUS DATA LOOP

A great variety of architectural forms has been developed for implementation of distributed computer loop networks. They differ from one another not only by the ways in which processor nodes are interconnected, but also by the mode of message transmission between them.

As to the ways of processor interconnection, there are two major approaches: by a common bus in the form of a ring, on which all processor nodes are attached; or by a sequence of processor-to-processor links closed in on itself. The system proposed in this paper belongs to the first category, but has some special properties, which make it competitive with systems of the second category.

A shared bus serves as the common data path for communications between all pairs of processor nodes attached to it. Compared with the point-to-point links, bus technique makes easier to implement distributed systems composed of mini- and microcomputers which have limited processing power and resources individually. Since messages transmitted along the bus between two processors need not be routed via any intermediate processor node, it is more reasonable to allow bi-directional data transfer around the loop.

Although the proposed system is built on the basis of bus structure, it is totally different from a simple integrated-bus network, which may suffer from serious problems of bus-contention and thus of limitations both in the number of processor nodes attached and in the distance of message transmission. Owing to the special features of splitted busses, the system behaves, in fact, as a point-to-point communication loop network, which is characterizad by small transmission delay, high throughput, simplicity of routing control, and possibility of implementation of multiple-destination message transmission.

The one-dimensional structure of the splitted-bus system proposed in this paper looks like a wheel, as shown in Fig. 1. Each processor node P is connected to the ring by a set of three switches: one central switch S_c for segmentation of the data loop, and two side switches S_p for connecting the two ports of the processor to the bus segment across the central switch. Each processor node has a third port which is used for communication with the centralized control processor CP, thus forming a star configuration of control links.

This arrangement provides an efficient means for realizing different modes of data transmission, peculiar to the proposed system. Like all other types of loop networks, broadcast operation can be used to implement one-to-all or any multiple-destination transmission. Besides this, two other modes of data transmission are possible. One is shifting, and the other is multiple random access. By shifting we mean that every node sends messages to its neighbour at one side and, at the same time, receives messages from its neighbour at the other side. Shifting may take place between neighbours by any modulo count. By multiple random access we mean to split the common bus into segments by means of routing switches for establishing separate non-conflict communication paths at once for several distinct node pairs to transmit variable-length messages.

All these modes of processor interconnections are very useful for distributed computing on the proposed system. They lead also to a number of important features, characterizing it. At first, the shifting operation is due to the fact that each processor is connected to the data loop via two ports. It can be shown that the performance of a two-port network is much better than that of a single-port one, nearly by a factor of 2. Secondly, the average value of the multiplicity of random access to a splitted bus lies between 3 and 4, depending upon the number of nodes of the system. That means, the communication power of a splitted-bus loop is equivalent, in average, to 3 or 4 point-to-point links, available for simultaneous transmission of variable-length messages around it. Thirdly, the installation of three switches with each processor node permits any number of processors to be isolated or grouped without affecting the operation of the rest of the system. This property, in turn, improves flexibility, reconfigurability, partitionability and fault-tolerance of the system.

DATA LOOP CONTROL ALGORITHM

Control of the data loop is not completely distributed. The centralized control processor, however, is much limited in its functions. In fact, it is not concerned with individual data transfers, but performs only routing control by handling communication requests from each node, setting corresponding switches before start of the data transmission, and releasing them after its completion. Star configuration of control links enables to derive a simple control algorithm and give fast responses to communication requests delivered to and queued up in the control processor. For relatively small systems, microcomputers may fulfill this duty easily and satisfactorily. They cost so cheap that duplicated control processors may be used, if necessary, for ensuring high reliability.

In the experimental system being implemented on the basis of PDP-11/03's as computer modules, the microcomputer system of the type INTEL 8080 is used as control processor. The routing switches are implemented by IC chips of bus buffer gates with three-state outputs connected back-to-back to form symmetrical, bi-directional bus drives. The devices are controlled by bus control signals, generated by the control processor through LSI chips of general-purpose programmable I/O interface, such as INTEL 8255A, programmed to operate in mode 0 to provide 24 output lines per chip. To establish or delete a route, all the related switches should be set or released as required by the routing algorithm, and, at the same time, the table of switch status should be updated accordingly. Communication between PDP-11/03's and the control processor is accomplished through ordinary serial I/O interfaces.

The control processor accepts communication requests from all nodes on the basis of interrupts. The flowchart of the interrupt-handling program is shown in Fig. 2. If the request is to establish a route, then it is put into a waiting queue A, from which the jobs can be delivered subsequently one by one to seek for the data communication paths around the loop according to the First-Come-First-Serve (FCFS) discipline. If the request is to delete a route, then the switches en route should be released as soon as possible, so that more free paths may be available for the jobs waiting for communication.

The process of seeking for data communication routes is performed by the program with the flowchart shown in Fig. 3. The control processor holds a table of switch status which represents whether each routing switch is open or closed. Each time when a route is to be established, the status of the related switches are taken from the table and compared with those required by the routing algorithm. If they coincide with each other, the switches are set in proper positions with corresponding status in the table updated. The source and destination nodes are then informed about the success of route-searching. Since the data loop is bi-directional, there are two alternative communication paths between every pair of source and destination nodes. The short route is searched at first. If any one of the switches en route happens to be busy, the long route is to be searched next. If in both cases the control processor fails to find paths for some job in queue A, this job enters the second waiting queue B.

Higher priorities are assigned to the jobs in queue B, than in queue A. Every time when any job completes its data transmission and the switches used by it have been released, all waiting jobs in queue B are examined one by one in order to determine whether any free path has become available for any job. If some job in queue B fails to find path again during this scanning process, it is to be returned into the same queue. Thus the process of scanning and recirculation of jobs in queue B may be performed repeatedly. In order to distinquish the jobs returned after unsuccessful route-searching from those jobs which were originally in queue B and have not yet been examined in current scanning, two flags, n_1 and n_2, are introduced in the program. The former is used to count the jobs currently existing in queue B, and the latter is used to count the jobs having been returned into it. Obviously, the condition $n_1 = n_2$ means that all the jobs in queue B have been examined in current scanning, and therefore this scanning process can be ceased until a new deletion of communication route takes place. Each time when any communication process is completed and such deletion of route occurs, n_2 must be reset to zero in order that a new scanning process of queue B can be originated once more. Only in the interval when queue B is waiting for switch-releasing, the queue A can be searched to handle new communication requests.

SIMULATION RESULTS

A series of simulation experiments have been conducted on PDP-11/03 for measuring the characteristic performance values of the proposed system. For formulation of the simulation problem, the following assumptions about the queuing model of the system were taken:

1. All N processor nodes of the system are assumed to be identical with one another, so that they are delivering message-communication requests with equal probabilities, and the destination nodes to be addressed by them assume a uniform distribution.

2. The arrival of jobs delivered from each node to queue A of the control processor assumes a Poisson distribution with mean arrival rate λ packets per second, or, in other words, the interval of arriving of jobs from each node assumes a negative exponential distribution with mean value $1/\lambda$ sec.

3. The message length of each packet assumes a truncated negative exponential distribution with mean value $1/u$ bytes per packet.

These are the same assumptions which have been generally accepted in analysis and simulation of the performance of many networks discussed in the literature (Reames and Liu, 1976). Under these assumptions, the queuing model of the system can be considered as to be composed of three parts.

The first part is queue A in the control processor, which accepts input jobs of Poisson distribution with mean total arrival rate $N\lambda$ packets per second and serves on the basis of FCFS discipline with constant service time equal to the time of performing the above route-searching program.

The second part is the data loop of the system, which provides a multiplicity of concurrent non-conflicting data paths between separate distinct node pairs for their intercommunication.

As the third part of the model we have queue B, which must be searched once after each input job has completed its communication.

In correspondence with the actions of these three parts, the whole communication process originated by an input job can be modeled by the state-transition diagram of its source node as shown in Fig. 4. For this purpose, three possible states are assigned to each node.

At state 0, the communication request delivered by the processor node is served through queue A, and then it changes to state 1.

At state 1, the processor node is seeking for the communication route. As the result, there may be two alternative outlets from state 1: either changing to state 2 after the required route has been found and communication is performed; or being suspended in queue B and remaining at state 1 to represent a case of failure in the route-searching.

At state 2, the communication process is completed and returns to state 0 for handling new communication requests. But prior to this, all jobs in queue B should be awakened so that they can be examined in a new route-searching process.

The main program simulating the above state-transition diagram consists of three modules, which correspond just to the actions taken at the above three states respectively.

For convenience of comparison of the proposed system with such loop networks as Newhall, Pierce, DLCN (Reames and Liu, 1975, 1976), and Jafari (1977, 1980), not only the same assumptions which have been stated above, but also the same parameters of the statistical model of the system were taken. They are as follows:
  number of nodes N = 6;
  mean packet length $1/u$ = 50 bytes
    with minimum of 10, and maximum
    of 512 bytes;
  transfer rate V = 125 Kbytes/sec.

The results of simulation performed under these conditions are shown in Fig. 5 and 6 as curves of data line utilization and total mean transmission time (including queuing time and message transfer time) versus mean arrival rate of communication jobs. The simulation results for other four loop networks are taken directly from the paper of Jafari (1977), which can be referred to for further explanation. To make the results compatible with one another, the time of transmitting one character is taken as the time unit for plotting the curves of Fig. 6. Comparison of these results shows a considerable effect of splitted-bus structure on the performance enhancement of a distributed multiple computer system.

CONCLUSION

The great advantages of the splitted-bus structure in organization of distributed multiple mini- and microcomputer systems have been shown and explained from a viewpoint of multiplicity of communication paths available for simultaneous execution of multiple jobs of transmitting variable-length messages. Using the conceptual classification scheme of parallel processing computer architecture given by Jin (1981, 1982), the above explanation can be pushed further. If we say the work of Newhall network is based on resource sharing (time-multiplexing), and the parallelism in Pierce and DLCN networks is due to time-overlapping (pipelining), then the parallelism realized in the proposed system, together with Jafari network, can be regarded as resembling resource replication (Multiple Instruction stream Multiple Data stream system). Owing to the use of two ports for processor interconnections, the degree of parallelism is increased even further.

REFERENCES

Jafari, H., and T. Lewis (1977). A new loop structure for distributed microcomputing systems. 1st Ann. Rocky Mountain Symp. on Microcomputers: Systems, Software, Architecture, 121-141.

Jafari, H., and J. Spragins (1980). Simulation of a class of ring-structured networks. IEEE Trans. Comput. C-29, 385-392.

Jin, L. (1980). A new general-purpose distributed multiprocessor system structure. Proc. of the 1980 Int'l Conf. on Parallel Processing, 153-154.

Jin, L. (1981). Architectural considerations of distributed computer control systems. Paper presented on the 3rd IFAC Workshop on DCCS, Beijing, China.

Jin, L., D. Wang, and M. Sheng (1982). Parallel Processing Computer Architecture (in Chinese). Guofang Gongye Publisher, Beijing, China.

Reames, C.C., and M.T. Liu (1975). A loop network for simultaneous transmission of variable length messages. Proc. 2nd Ann. Symp. on Comput. Arch., 7-12.

Reames, C.C., and M.T. Liu (1976). Design and simulation of the Distributed Loop Computer Network (DLCN). Proc. 3rd Ann. Symp. on Comput. Arch., 124-129.

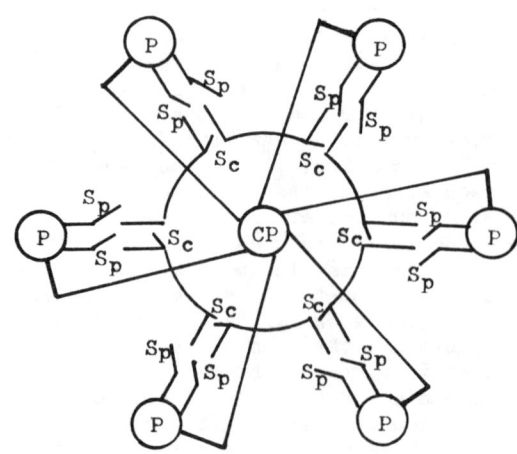

Fig. 1. An one-dimensional splitted-bus wheel structure

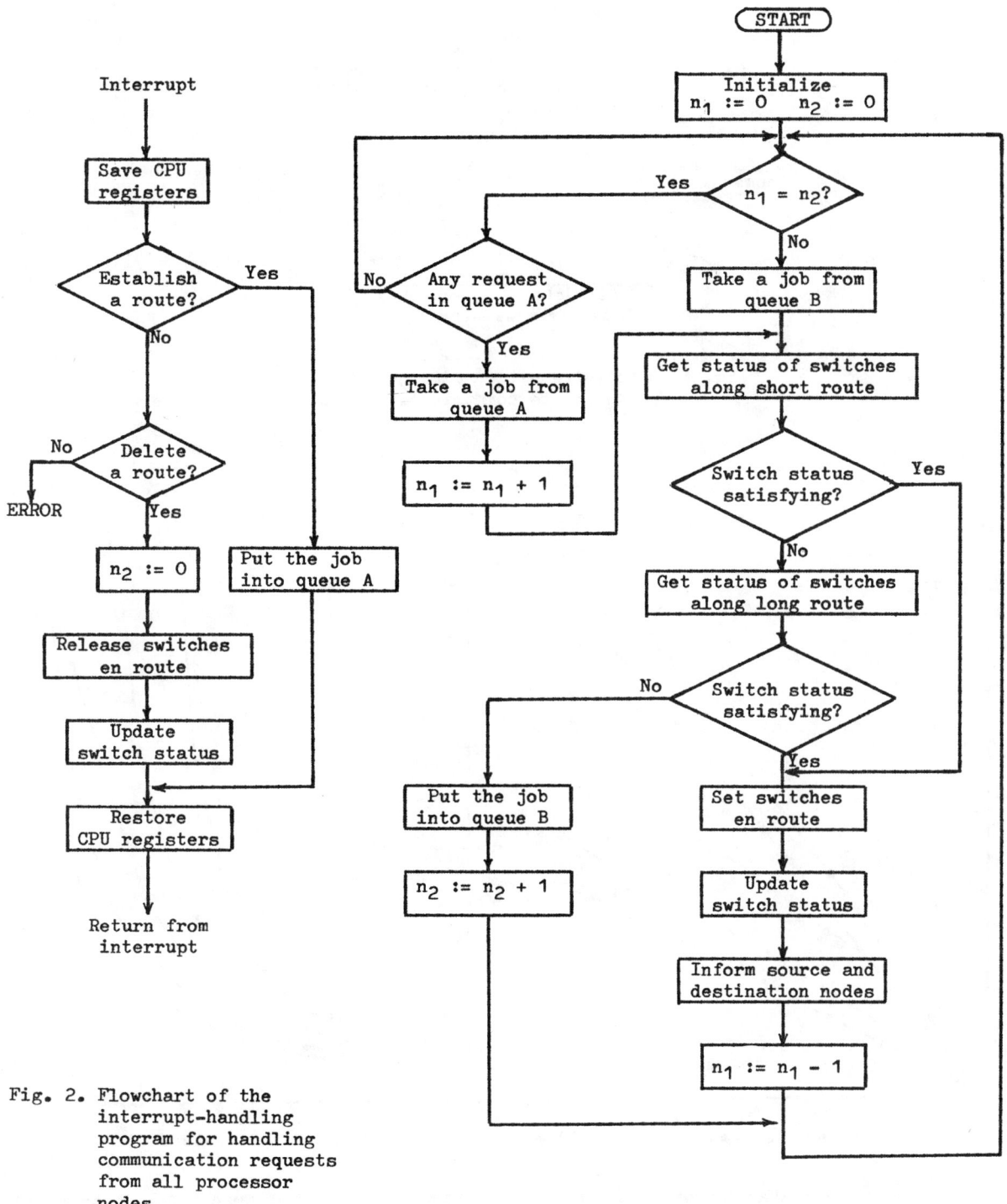

Fig. 2. Flowchart of the interrupt-handling program for handling communication requests from all processor nodes

Fig. 3. Flowchart of the main program for route-searching and monitoring queues A and B in the control processor

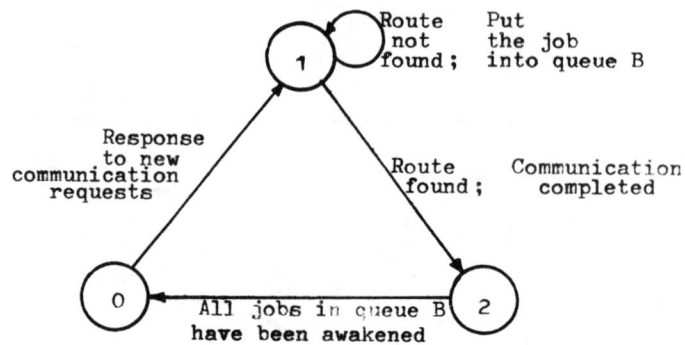

Fig. 4. State-transition diagram of a node in its communication process

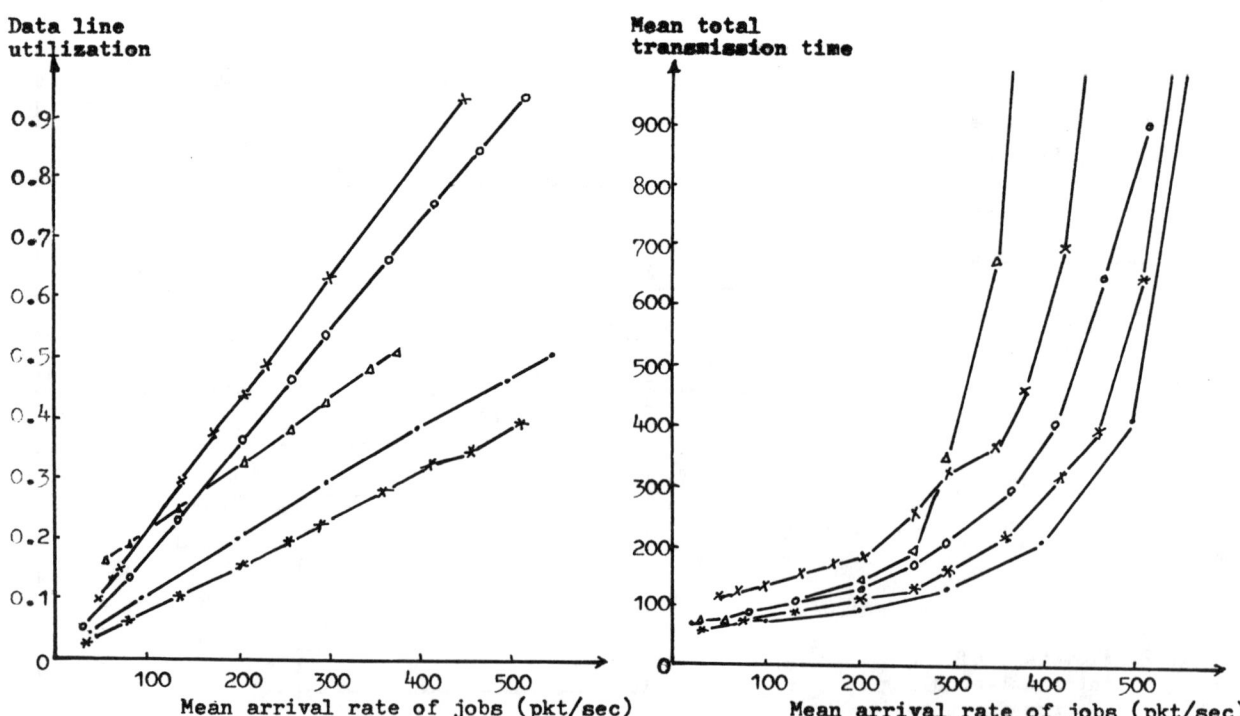

Fig. 5. Curves of data line utilization versus mean arrival rate of jobs for Newhall network (△), Pierce network (×), DLCN network (○), Jafari network (*), and the proposed system (•).

Fig. 6. Curves of mean total transmission time versus mean arrival rate of jobs for Newhall network (△), Pierce network (×), DLCN network (○), Jafari network (*), and the proposed system (•).

DISCUSSION

Work: Your splitted bus seems very similar to a wheel with circuit switching. If a station sends a message to another station, it uses a certain route which cannot be used by other stations at the same time. Therefore, you can't claim that any pair can work simultaneously. In your wheel a message doesn't have any address-fields, the destination address (or addresses in broadcasting mode) is known only to the control switch. Let me point out that the so called splitted bus is a circuit switching network with specialized topology and with a central switch with its well known reliability problems.

Lan Jin: Yes, I quite agree with both your comments.

Rodd: You use a 16-bit parallel bus. Are the processes syncronized to each other? Is the whole system synchronized?

Lan Jin: Yes, it is.

ON THE DESIGN OF HIERARCHICAL PROCESS CONTROL COMPUTER SYSTEMS

J. Davidson and J. L. Houle

Ecole Polytechnique, Montreal, Canada

Abstract. The definite turn of process control computer systems towards distributed configurations represents one of the most challenging aspects of computer control engineering for the 80s. The distributed computer control configurations should be designed to achieve a balance between cost and performance while at the same time matching the specific requirements of each application. This can be done by detailed evaluation of a number of configurations for a given physical process and control strategy through computer simulation. In this paper, we present software tools for designing hierarchical process control computer configurations in close relation to the mathematical model of the physical process and the control system itself.

Keywords. Computer control, distributed hierarchical systems, simulation, software tools, computer graphics.

INTRODUCTION

Distributed mini and microcomputer systems today represent the fastest growing segment of process control systems. Distributed systems for data acquisition and control have many attractive features: low cost, graceful degradation, modularity and simple real-time software requirements. Centralized, computer-based data acquisition and control systems, on the other hand, lack these advantages, but do provide very fast response times. Modern multiprogramming event driven real-time operating systems provide simple, straightforward communications among software modules in centralized systems. Such computer hardware provides fast and easy control over interrupts, and input/output devices. For comparable response times, distributed systems are much more costly than centralized systems because of the need to reliably communicate between modules which are physically separated. And because of the inherent communication lags, overhead due to communication protocols, and the need to insure data base integrity, response times can be a problem, especially in an emergency situation. Analysis of performance of distributed systems requires the design of their configurations, the distribution of their functions, and the planning of their capacities, so that the response is adequate under all specified conditions. In order to design or specify a distributed system it is necessary to be able to predict response times under a variety of conditions. To achieve an acceptable balance between cost and performance in designing distributed systems, architectures must be selected to match the specific requirements of each application. This can be done by using a number of building blocks as necessary components of a representation from which system characteristics can be derived. A design methodology for distributed systems can be built through derivation and generalization from techniques established for centralized systems and computer networks. Fault tolerance is also important in real-time control environments due to the possible catastrophic consequences of faults. Distributed systems may provide greater fault tolerance than centralized systems owing to their ability to disperse hardware elements and perform redundant processing. The lack of a general theory for system design is a cause of projects often plagued with difficulties, e.g. delays, inability to meet objectives, escalating costs. The situation could be worse with distributed systems, where new options continuously become available, unless a methodology is developed to provide designers with sufficient understanding of system properties to predict performance and to generate more accurate specifications. There is a need to create a formal description, including all necessary and sufficient components from which system characteristics can be derived. The critical nature of many applications has led to tightly coupled computer systems, most specifically, to hierarchical configurations in which multiple processors can be switched into various configurations for backup purposes.

Software tools can be used for designing hierarchical process control computer configurations in close relation with the mathematical model of the physical process and of the control system itself. The performance of a hierarchical process control computer system can be defined in terms of impulse response and steady-state services. The proposed simulation technique is a tool which will evaluate in great detail a number of different alternatives of configurations for a given physical process and a given control strategy by taking into consideration the required impulse response and steady-state service characteristics. The simulation will allow a much better understanding of the interrelation of the process and the control system, and will also determine the type and the number of components interconnected in the hierarchical configuration. At the same time, this simulation may be used to bring modifications and improvements of the control strategies and algorithms leading towards more efficient and economical plant operations. By designing, through simulation, different hierarchical computer control models, combined with the physical process model and control algorithms, the designers will be able to choose the best-suited, technologically-available computer configurations.

HIERARCHICAL COMPUTER CONTROL SYSTEMS

The current approach to the majority of computer control systems involves the use of a single, general-purpose, centralized computer, Fig.1. Often a dual centralized computer system is installed, with one of the computers serving as a back-up in case of failure, Fig.2. Usually, the centralized computer initially handles tasks involved in control functions, and secondly, some plant supervisory and management tasks.

All input/output signals are individually hard-wired to the central control room, which results in a massive amount of cabling between the control area and the rest of the plant. To minimize the amount of cabling in some installations, the actual computer/process interface is accomplished with multiplexors installed remotely on site, as close as possible to the equipment, Fig.3. One major limitation of this configuration is the overhead, a result of the concentration of all activities in a single processor. Since the operating system has to supervise the sharing of the central processor in terms of task scheduling, interrupt processing, memory contention, etc., an important percentage of the processor power is lost in non control activities, reducing the number of jobs which could be handled simultaneously. Furthermore, the application software for the entire project must reside in a single computer, which greatly increases the cost of the system in terms of software development, manpower and time. A central computer makes planning for growth very difficult. Today, users are often forced to choose between upgrading their central computer or using a distributed configuration by linking their central computer to distributed mini or microcomputers.

The most used distributed configuration in industrial control plants is the hierarchical computer system which usually consists of a tree structure of mini and microcomputers, Fig.4. The host processor is charged with supervising and controlling the configuration, while taking care of the management-related functions and providing facilities for program development. The host often maintains a data base in order to provide up-to-date information on plant performance and to help optimize the production cycle through rescheduling decisions based on overall plant performance. The host processor also often down-loads tasks to satellite computers through communication lines, thus assuring a complete recovery of a remote system after a hardware or software failure. The communication method and control strategy are dependent on the type of function the computer configuration has to perform, as well as the number of connected computers, distance between computers, average message length, peak load throughput requirements, acceptable error rate and recovery method, required response time to critical alarms, and software overhead. Satellite computers are typically small mini or microcomputers without disk storage, executing tedious functions such as data acquisition, analog to digital and digital to analog conversions, local algorithms and functions for control, alarm gathering and transmission, etc. In this way, remote computers greatly reduce the number of functions which otherwise would have to be handled by the host processor. The host processor typically performs functions such as data processing, control distribution and data base management. The most important benefit of a minicomputer network is the distributing of processing power to the right locations in the plant, thus improving the response time to the process without increasing the workload of the host.

The choice of a hierarchical configuration is justifiable because of the high degree of reliability attainable by spreading the supervising and control functions over several distributed mini and microcomputers, and by providing one or more levels of back-up for each mode in the multilevel configuration. For example, in Fig.4., the host at level 2 could provide the back-up for the mincomputers at level 1; and each one of the processors at level 1 could replace in back-up mode the minis at level 0.

PERFORMANCE CRITERIA FOR A HIERARCHICAL PROCESS CONTROL COMPUTER SYSTEM

Among the criteria for defining the performance of a hierarchical computer control system (Weitzman, 1980) are the impulse response and steady-state service. These are defined as follows:

The impulse response (T_{in}) or settling time represents the average input processing time from its initiation on computers at various levels to the generation of the command by host computers.

The steady-state service measures the efficiency of the hierarchical configuration for a number of tasks, as the ratio of their acutal execution time.

$S = (\sum_{i=1}^{n} ITi) / (\sum_{i=1}^{n} ATi)$ where:
ITi = Ideal turnaround time for job i,
ATi = Actual turnaround time for job i,
S varies between 0 and 1.

Supposing that in a given heirarchical computer system only one job at a time is activated, it will then be very easy to determine its total execution time. It becomes much more difficult to predict the execution time of a particular job when several jobs are overlapping in time. The factors involved in maintaining a job active for a certain period of time, in excess of the ideal turnaround time, are: the processors waiting queues, the interprocessor communication links contention, the variations in interrupt arrival rates, etc. Nevertheless, the impulse response and steady-state service criteria are extremely useful in pre-establishing a hierarchical computer control configuration for given T_{in} and S values.

Other major technical criteria for a hierarchical process control computer configuration are the overall system reliability, availability and fault tolerance (Schoeffler, 1979). Reliability is a measure of the success with which the system conforms to the specification of its behavior. It is expressed as MTBF (Mean Time Between Failure) of the hardware components of a system.

Availability, A = (MTBF)/(MTBF+MTTR), where MTTR is Mean Time To Repair. The MTTR includes the actual time to repair a failure and also a part of preventive maintenance time.

The mini and microcomputers are usually associated with a reliability figure established by the manufacturer. This figure helps determine the availability of the system as long as the MTTR is a known entity provided by the local representative of the manufacturer.

Fault tolerance architectures should be used in order to insure that sub-systems failures do not result in total system failures. These types of configurations are achieved through redundancy of processors, communication links, memory, etc., Fig.5.

Perhaps one of the greatest advantages of hierarchical computer control configurations is that a high degree of reliability can be achieved through partial redundancy as compared with a centralized system where the same reliability can be achieved only through a complete duplication of the central computer.

Other performance criteria, less important for the design of the hierarchical configuration for process control, are: the form factor, the growth factor, the ease of development factor and the life-cycle cost (Weitzman, 1980).

PERFORMANCE REQUIREMENTS FOR A HIERARCHICAL PROCESS CONTROL COMPUTER SYSTEM

In order to design a tailored hierarchical computer control system for a given application, it is necessary to clearly establish its performance requirements. Performance requirements are generally quantitative measures against which the system can be tested. For example, the system must be able to process information from 200 digital inputs and 100 analog inputs, and return the commands with a maximum delay of 2 seconds (in automatic control) to a defined output point. Performance requirements can also include the number of cathode ray tubes (CRT), the number of asynchronous or synchronous communication lines, or the number of processors to be included in the hierarchical configuration. Some of these requirements will have to be established before beginning the design of the hierarchical configuration, and some will be determined as a result of the design itself.

The control hierarchy in a process control system can usually be divided into the following three functional levels corresponding to three levels of a hierarchical computer control configuration:

- data acquisition and control (level 0)
- production control and process optimization (level 1)
- plant management (level 2)

A number of known or unknown elements must be taken into account at each level before beginning the design of the hierarchical configuration (Golemanov, 1981). For level 0, data acquisition and control, the elements to be taken into consideration are:

- number of processors - NPØ
- type of processor - TYPRØ
- type of operating system - TYPOSØ
- total memory - MMØ
- number of analog inputs - NAI
- number of digital inputs - NDI
- number of analog outputs - NAO
- number of digital outputs - NDO
- number of analog loops - NAL
- number of digital loops - NDL
- number of alarms - NALA
- number of local operator interfaces - NLOIØ
- number of supervised control loops - NSCL
- number of Direct Digital Control loops - NDDCL
- number of multiplexors connected to the processors - NMUX
- scanning speed in points per second - SS
- number of synchronous communication channels - NSCØ

- number of asynchronous communication channels – NAC0
- communication speed on synchronous channels – CSS0
- communication speed on asynchronous channels – CAS0
- standard used in communication – STANC0

If there are any redundant processors at level 0, the following elements should be specified:
- number of redundant processors – NRP0
- number of redundant analog loops – NRAL0
- number of redundant digital loops – NRDL0
- type of transfer to the redundant processor – TTR0

At level 1, the elements to be specified are:
- number of processors – NP1
- type of processor – TYPR1
- type of operating system – TYPOS1
- total memory – MM1
- number of local operator interfaces – NLOI1
- number of synchronous communication channels – NSC1
- number of asynchronous communication channels – NAC1
- communication speed on synchronous channels – CSS1
- communication speed on asynchronous channels – CAS1
- number of units on the synchronous channels – NUSS1
- number of units on the asynchronous channels – NUAS1
- standard used in communication – STANC1
- type of auxiliary storage device – TYASD1

If there are any redundant processors at level 1, the following elements should be specified:
- number of redundant processors – NRP1
- number of redundant synchronous communication channels – NRSC1
- number of redundant asynchronous communication channels – NRAC1
- number of redundant units on the synchronous channel – NRUSS1
- number of redundant units on the asynchronous channel – NRUAS1
- type of transfer to the redundant processor – TTR1

The elements to be specified at level 2 are:
- type of processor – TYPR2
- type of operating system – TYPOS2
- total memory – MM2
- number of plant operator interfaces – NPOI
- number of synchronous communication channels – NSC2
- number of asynchronous communication channels – NAC2
- communication speed on synchronous channels – CSS2
- communication speed on asynchronous channels – CAS2
- number of units on the synchronous channels – NUSS2
- number of units on the asynchronous channels – NUAS2
- standard used in communication – STANC2
- number of controlled loops per CRT – NCLCRT
- number of alarms per CRT – NALCRT
- number of process schematics displayed graphically – NPSDG
- type of auxiliary storage devices – TYASD2
- programming capabilites and languages – PROCL
- number of special and advanced control algorithms – NSACA
- process optimization capabilities – POC
- advanced control strategies – ACS
- adaptive control capability – ADC
- management information system functions – MISF

The selection of factors chosen among the elements presented above, for a particular hierarchical computer control system, should allow a maximal possible coverage of different configurations.

DESIGN OF HIERARCHICAL PROCESS CONTROL COMPUTER SYSTEMS USING AN INTERACTIVE SIMULATOR

Large industrial plants present many complex problems for which computer simulation offers an ideal means for assessing the controlability of a given plant. An important aspect of simulation work is the development of mathematical models to describe the dynamics of various components of a plant. These models provide simulation blocks for each of the components, so that in order to study the behavior of a given plant, the control engineer selects appropriate simulation blocks, assembles them on the computer, and subjects them to disturbances in order to obtain system responses (Tanuma, 1981).

The complete simulator package for a distributed computer system is divided into two parts: the control system simulation and the plant simulation. In this way, the function of the control system can be completely simulated for almost every situation which occurs in the plant, be it normal or emergency.

Usually the total production activity of a chemical plant can be described as a stochastic, continuous and dynamic process. Therefore, simplification is needed in order to be able to use simulation for the control system and the plant itself. Consequently, assuming the controlled process is a deterministic one, with pre-established performance requirements, we can find, through interactive simulation, the hierarchical computer control configuration best suited to it. An interactive simulator and a graphic terminal can be used for designing a hierarchical computer control system based on its performance requirements and performance criteria. By using computer simulation, the designer has the capability of verifying multiple configurations before continuing to the actual implementation.

The interactive simulation system should consist of software for interactive graphics, data transfer, and analysis. By entering control system data into the simulator, we will draw a block diagram on the graphic terminal according to the information previously stored in a data base. A block diagram should consist of process input/output, control processors, intermediate processors, a host processor and communications lines between each block, Fig.4. In order to support interactive analysis the following functions should be prepared:

1. Display and modification of parameters on block diagrams.
   This function should make it possible to display a list of parameters related to the control system, and to change them interactively.

2. Display and modification of process input/output.
   This function will allow the display of the list of process inputs and outputs, and the exchange of their type and parameters.

3. Diagnosis of block diagram.
   The paramters'values and the communcation links are examined by this function. If inadequate parameters, or blocks not linked with others are found, the name of imperfect blocks and their errors will be displayed.

4. Guidance for determining details of impulse response time and steady-state service calculation.
   At the demand of the user, a "help" file should be used to indicate the method and the calculations needed.

5. Graphic display of the results of calculations and of final block diagrams.
   Calculations results concerning the performance criteria of the hierarchical computer system should be shown on the terminal in the form of tables or trend graphs.

The analysis of hierarchical computer control configurations may be summarized in the following steps:

1. Consider the model of each processor in the configuration as being a block having a number of parameters associated with it.

2. Model the configuration as an interconnection of hierarchically connected blocks.

3. Formualte and state the average steady-state process input/output rates and the transactions among processors, such as:
   - average process input arrival rate
   - average service time - control processors
   - average service time - intermediate processors
   - average service time - host processor
   - average communication time among processors

4. Pre-establish the values associated with the performance criteria of the hierarchical system.

5. Calculate individual processor parameters starting from an assumed configuration.

6. Calculate and display overall system performance parameters of this configuration.

7. Compare the calculated system performance with the pre-established performance criteria for the assumed configuration. If significantly different, estimate new values and go back to step 5.

The final results of the interactive simulation will be the hierarchical configuration with the closest system performance values to the chosen performance criteria.

CONCLUSIONS

This paper has described an approach that can be used to design a hierarchical process control computer system. It is proposed that an interactive simulator together with a graphic terminal be used to design a hierarchical configuration. The designer can communicate with the simulator using a block diagram. The analysis of the control system with respect to the impulse response and steady-state service can be computed interactively by using simulator guidance. The simulation package will be defined in view of its implementation in the near future on a "Digital Equipment Corporation - VAX 780" computer environment, using a "Tektronix 4112", as a graphic terminal, and a "Tektronix 4663" as a plotter. It is obvious that more details are needed for the design of the interactive simulator, nevertheless, the approach presented in this paper is considered feasible and appropriate for the design of a hierarchical computer system. Moreover the analysis which will be performed on different configurations during the interactive simulation will allow a better understanding of hierarchical systems characteristics, and consequently, will accelerate the detailed design of hierarchical computer configurations.

REFERENCES

Binder, Z., Perret, R. and Rey. D. (1981). A modular multi-microprocessor distributed control system. Proc. of the 3rd IFAC Distributed Computer Control System Workshop. Beijing, China.

Dobrowolski, M. (1981). Guide to selecting digital control systems. InTech. June, pp.45-56.

Golemanov, L.A. et al. (1981). A united approach in the analysis and design of integrated production management systems. Proc. of the 8th IFAC Triennial World Congress. Kyoto, Japan. Vol.22, pp.35-41.

Greenwood, J.R. et al. (1980). Hierarchically structured distributed microprocessor network for control. Proc. of the 3rd Rocky Mountain Symposium on Microcomputers Systems, Software, Architecture. pp.50-60.

Kezunovic, M. (1981). A system approach to the design of an integrated microprocessor based substation control and protection system. Proc. of the 8th Triennial World Congress. Kyoto, Japan. Vol.20, pp.60-65.

Larsen, P.M. (1981). Distributed microcomputer control systems for electrical power plants. Proc. of the 3rd IFAC Distributed Computer Control System Workshop. Beijing, China.

Malik, O.P. and Hope, G.S. (1981). Design concepts for a distributed microprocessor-based transmission line control and monitoring system. Proc. of the 8th IFAC Triennial World Congress. Kyoto, Japan. Vol.11, pp.143-167.

Reinig, G. (1981). Application of a distributed microcomputer based control system in a new petrochemical plant. Proc. of the 8th IFAC Triennial World Congress. Kyoto, Japan. Vol.22, pp.48-54.

Schoeffler, J.D. (1979). Reliable error recovery in distributed computer control systems. Power Industry Computer Applications Conference - PICA - 79. Pp.245-249.

Shaw, W.T. (1980). Now to choose a distributed system architecture. InTech. December, pp.41-43.

Tanuma, M. et al. (1981). Interactive simulator using a graphic terminal for linear control system. Proc. of the 8th IFAC Triennial World Congress. Kyoto, Japan. Vol.11, pp.35-40.

Weitzman, C. (1980). Distributed Micro/Minicomputer Systems. New Jersey: Prentice-Hall.

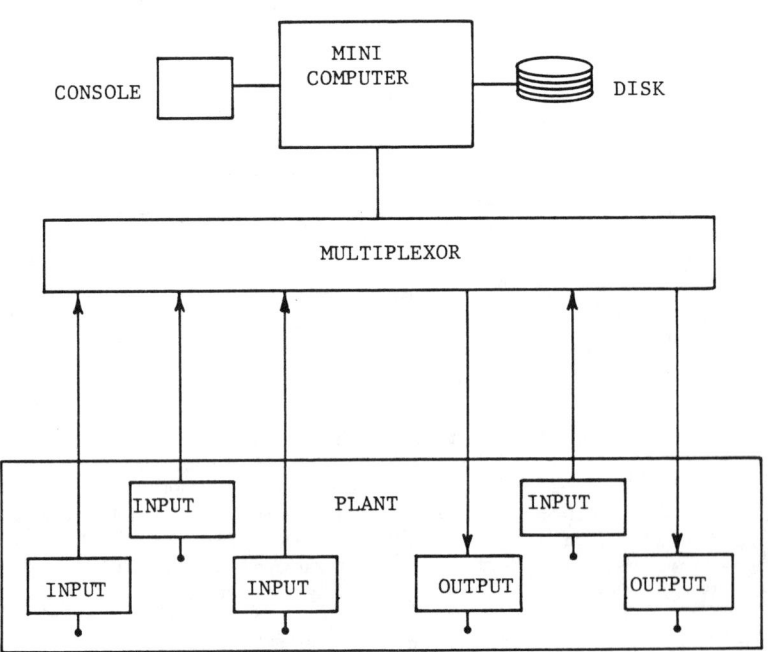

FIG. 1 CENTRALIZED COMPUTER CONTROL SYSTEM

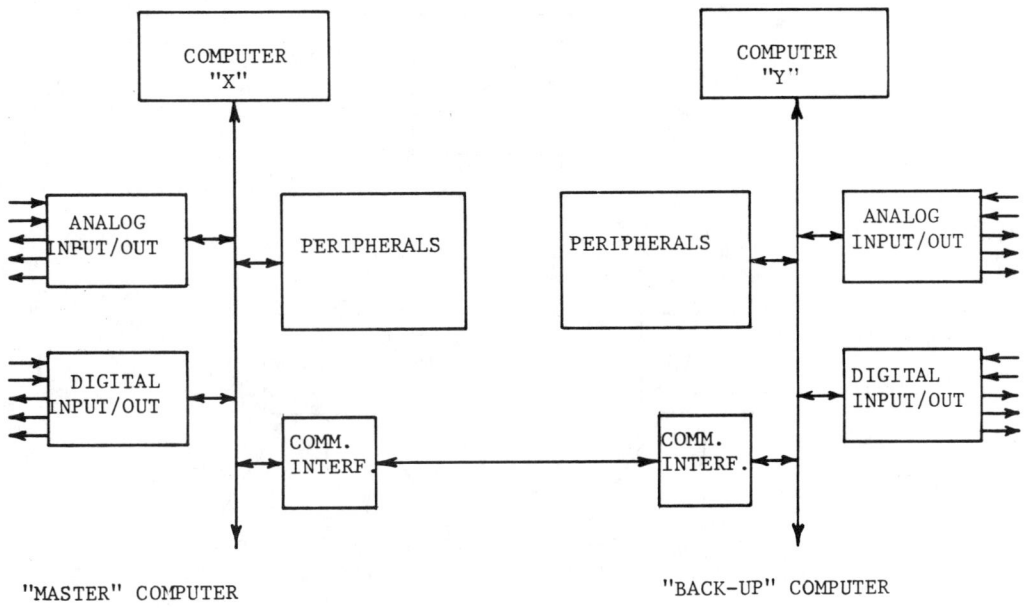

FIG.2 DUAL CENTRALIZED COMPUTER SYSTEM

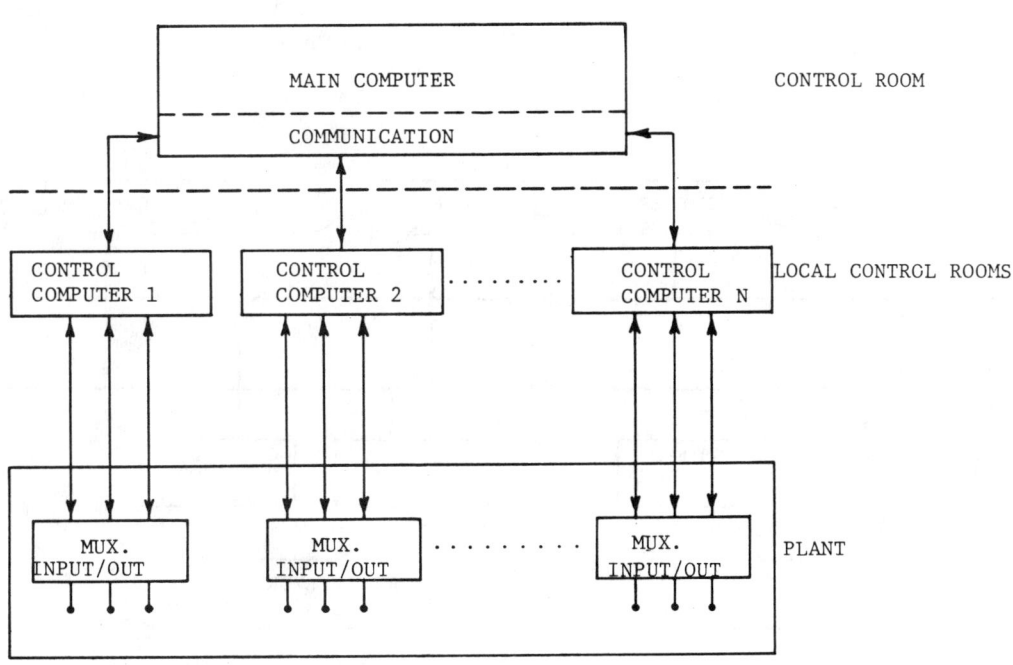

FIG. 3 CENTRALIZED-DISTRIBUTED CONTROL SYSTEM

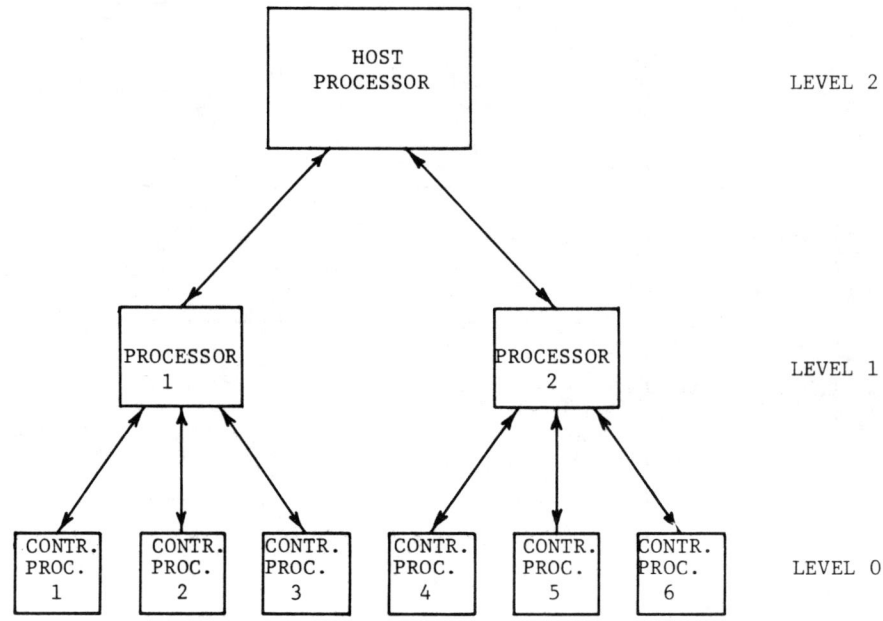

FIG. 4 HIERARCHICAL COMPUTER CONTROL CONFIGURATION

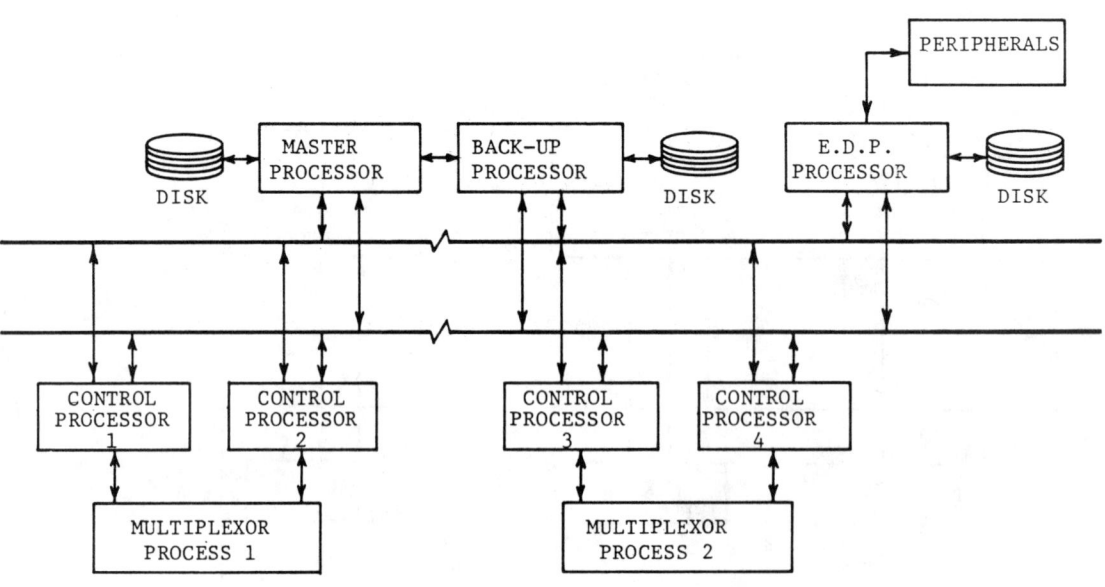

FIG. 5 DUAL-REDUNDANT HIERARCHICAL COMPUTER CONTROL SYSTEM

DISCUSSION

Narita: What kind of an index or measure do you use for the evaluation of the overall reliability of DCCS? It may consist of many processors and peripherals connected by communication links. Maybe you need that single index to compare two alternative configurations? You mentioned on your paper that you use MTBF for component realibility. I think we need something major for the overall system reliability. How do you calculate that reliability.

Davidson: There are formulas to calculate a distributed system. An overall reliability - that's a mathematical formula - I don't have it here. It depends on whether the processors are linked in parallel or in serial - the formula is different in each case. For processors linked in a serial fashion the total reliability is lower than in the case of parallel linkage.

Narita: This approach is very effective for a simple configuration, but DCCS's have many processors in a diverse area and that approach might fail in some cases of large applications.

Davidson: As I said, to my knowledge there are not many methods around for designing distributed computer control systems. What we tried to do, and maybe it's a long story but I'll try to make it short -, we tried to do something practical. I personally participated in a few conferences and almost nothing practical came out. I've personally been involved in two major distributed computer control systems in the Montreal area. One was a waste water treating plant, for which I was responsible, with one PDP-11/70 computer as a host, three PDP-11/44 at the intermediate level and twenty LSI-11 microcomputers at the low level. We designed it by taken into consideration only one principle: the number of points per processor at the level 0 should be equal. That is, each of them should have the same number of points. It was a complete mistake. The mistake was that we had bottlenecks all over the system. Why? We tended to concentrate all the alarms on only one processor, saying that all the analog inputs should be in only one multiplexor because they need special power supplies etc. It surely was a mistake and we've learned from it. Designing the second system - it was the automation of a postal plant in Montreal, with the PDP-11/70 again as a host processor and 21 so-called subset control stations, each one of them consisting of 2 back-to-back PDP-11/34 computers. In this case we tried to be a little bit smarter. We divided the input/output points in such a fashion as to reduce the bottlenecks. We simply calculated by hand the input response time and the steady-state service for few alarms and few points. It surely is not enough, the system was not perfect and takes a lot of effort and a lot or work. But taking the same approach with a simulation package where the computer is doing the calculation for you is feasible, workable and necessary. We can do that at the moment when communication links are already established. I mean, we know what kind of communication we are going to use, and we know fairly well the computer system; PDP-11 computers linked point-to-point by DECNET. It is an isolated case, you might say, but from my practical experience I know at least five plants where this type of configuration is used. I'm saying, let's go ahead with this type of configuration, let's try to do it, to simulate it, and after that we can extrapolate to more complicated systems or to other types of computers.

Inamoto: I am interested in your interactive simulation system. In order to simulate the computer system we have to simulate the operating system. Dynamic simulation is especially difficult in the distributed computer control system where we have many processors as well as many operating systems operating in parallel. We also have to take into account the data transmission dynamics. Have you developed any algorithm to simulate such a multi-operating system and data transmission mechanism?

Davidson: That is a very good question and I'll try to talk about it. We didn't have to simulate fully the operating system for the simple reason that by choosing PDP-11 for the configuration I spoke about a few minutes ago, we use compatible operating systems. They are running under the RSX - 11M plus operating system, the intermediate level PDP-11/44 and one other system were running under RSX-11M, which is completely compatible with the M-plus. The operating system for the level 0 processors is RSX-11S which is also a subset of M, and at three levels they communicated through DECNET protocol. Clearly, we needed data in order to be able to simulate all this. Fortunately, in the literature we found all the necessary data. To give you an idea, we have graphs, provided by DEC for the CPU percentage use per type of computer, PDP-11/34 or PDP-11/70 dependent on the message size transmitted between computers. Similarly, we have CPU percentage useage dependent on the number of messages per second, also for different processors PDP-11/44 or PDP-11/70, etc. We are not interested in simulating the entire operating system, but just the part of it which can create bottlenecks.

Inamoto: Is this dynamic or some kind of static simulation?

Davidson: You have a mixture because, by the data we have available, first of all we have to carry out something like a continuous process simulation. At the same time we do have a static type of simulation because, for example, in the calculation of the impulse response you have to take into consideration some static values - we cannot do it otherwise.

Gong Bingzhen: What type of a mathematical model do you use in your simulation, and, secondly, how do you determine the control loop parameters?

Davidson: As I said previously, this is a bottom-up approach. We did simulate the processes on both projects I talked about on the plant itself. The first one being a continuous process, we used the language called CSMP (Continuous Simulation Modelling Program). The second one was a discrete process, a postal plant automation system, and we used the SIMSCRIPT language which is specifically designed for discrete simulation. Concerning the second question, the algorithm tuning is a problem most of the people here know. If you are talking about tuning process control algorithms they are very well defined and documented in the literature. Usually, the instrumentation engineer will tell us in advance that for this point he needs a proportional-integral type of controller and he know how to implement it. The models are existing in the literature, it is no problem whatsoever.

Maxwell: It seems possible that even if the process engineer specifies the PI-algorithm the simulation might show that PID-algorithm was needed. Have you run across this situation?

Davidson: Sure, very many times. The simulation of the process itself will establish, if we need a PI, or a PD, or a PID. By using a simulation package we are capable of fine tuning the algorithm, either the PI or PID. I think in every company involved in process control they need such a simulation package. This problem should be solved before designing a distributed computer system. We need this information in order to be able to establish the amount of memory and CPU time needed for solving these control problems.

A MODEL BASED DESIGN OF DISTRIBUTED CONTROL SYSTEM SOFTWARE

L. Motus and K. Kääramees

Institute of Cybernetics, Computer Division, Tallinn 200105, USSR

Abstract. In this paper we suggest a formal model for specifying control system behaviour with special attention to timing and general software structure. The model gives an implementation independent description of the problem. Several timing and structural properties can be proved using the model, it can also be useful for comparing alternative designs. A trivial example demonstrates the use of the model. The paper ends with discussion of related papers and some future problems.

Keywords. Distributed computer control systems software; system behaviour specification; formal model; timing; deadlock avoidance.

INTRODUCTION

Distributed computer control systems can in many aspects be considered as a more advanced alternative to centralized computer control system. Admitting that, we can still notice a number of similarities in basic software features of both systems. The reason is that the underlying technological process will not change because of changing the computer control system. The application oriented functional tasks to be solved are the same in both types of control systems.

In the case of distributed control systems there certainly appear specific problems, such as alloting functional tasks to processors (nodes of network), physically organizing communication between tasks executing in different nodes, etc. However, the essence of control system software -- parallel execution of functional tasks in the presence of constraints generated by physical properties of the technological process -- remains unchanged.

On the other hand, control systems software differs substantially from parallel programming problems arising for example when solving complicated mathematical problems, such as a system of partial differential equations or large dimensional matrix calculus. The main difference is that a system to be programmed is normally ill-defined in the case of control systems, whereas that of a mathematical problem is usually well-defined.

An ill-defined system appears due to complicated and non-formal nature of a technological process and its environment. It is extremely hard to get a complete and noncontradicting specification of a control system without using a CAD system. It is not a rare case that a number of errors are discovered in the system after it has successfully passed testing and has been operating for some time already. Many of these errors are due to incorrect timing of parallelly executing tasks, or are otherwise connected to an ill-defined specification.

Many articles and a book (Yeh, 1977) have been published on software specification. Unfortunately, timing problems have not been too popular in the literature. Nevertheless, in control applications correct timing is often of great importance.

The aim of this paper is to assist in developing tools for specification and design phase of distributed control systems software. Two ideas form the starting point. First, we are convinced that a formal mathematical model will help to obtain a complete and noncontradicting specification and hence facilitate the software design. Second, we feel that software design should be decomposed into two phases -- the design of software structure and the design of algorithms for the structural units of the software system.

We have adopted a model (Quirk and Gilbert, 1977) that supports our approach. The model specifies a software systems structure as a system of communicating sequential processes. It also provides constraints to selecting algorithms for each process and a good foundation for practical design of the software. Since the model concentrates on timing and system structure, some facets, such as an exact specification of algorithms and data structures which are traditionally considered of main importance, are neglected. However these facets may be cleared out in the second design phase using traditional methods (see for example (Yeh, 1977)).

In this paper we present a slightly modified model, originally developed in (Quirk and Gilbert, 1977), discuss its main properties and demonstrate its use in avoiding potential deadlock. We also compare the model with well known net models (Petri nets) for analysing parallel systems. The paper ends with a discussion of presented results and some future problems.

A MODEL

The description of control system software as a set of interacting parallel processes is rather generally accepted by practitioners, although it is not always unanimously agreed on what is a process. Because of the great variety of interaction conditions it is difficult to design and to implement such systems, even if we had an ideal parallel programming language. It is still more difficult to test all possible cases of interactions.

A model should fix the structure of process interactions and prove some of its properties. It means that using the model we introduce partial ordering on a set of processes -- the list of necessary processes together with their interactions comes from analysing the application. Partial ordering determines a sequence of process executions. In control systems it means that some of the processes should be connected to certain events from the technological environment, some others to time instants and some processes depend on the result of execution of another process. The time instants may be relative (with respect to an event in the system) or absolute (connected to astronomical time). A characteristic feature of control systems software is that most of the processes are cyclic -- some of them are executed with fixed periods, some are aperiodic.

Among properties to be proved by the model is timing on the first place. Incorrect timing produces dynamic errors in system's behaviour -- they occur at random and are therefore very hard to find. A deadlock danger can also be discovered already in the specification phase, the same applies to the completeness of the system description. The model thus provides us with restrictions and constraints needed to specify algorithms that will fit in the designed structure so that the demands of the application are met.

In parallel programming the problem of determinacy is also connected to the structure of the system (Coffman and Denning, 1973). However, due to cyclic execution of processes and dependence of the system behaviour on on-line input data (so-called inner memory of algorithms), a special investigation of the determinacy problem in control systems software is needed. This problem is not handled in the present paper.

Since distributed computer control systems are in the most cases dedicated, dynamic resource sharing is not used too often. In the most cases hardware resources are allocated permanently to processes, reallocation is considered as an exception. The model neglects sharing in reusable resources. It considers only consumable resources, i.e. the information that processes exchange while interacting. Communication should be organized using message exchange since it has many advantages compared with shared variable method (Zave, 1976), especially in the case of distributed computer control systems.

The model that seems to suit the best for solving the above listed problems was originally developed in (Quirk and Gilbert, 1977).

This model supports hierarchical organization of a system. The philosophy is as follows: a system consists of interacting subsystems; each subsystem may be considered as a system. This division into subsystems, subsubsystems, etc. goes on until we reach a reasonably detailed description. A system on the lowest level is called a process -- it usually fulfils some elementary functional task and produces values for variables that characterize the process. These variables form the state of a process, which is a part of the state of the system. The process here incorporates code and data, which is good for specification of timing and interactions. The process produces its state values only at the end of its execution, but it can consume input data (state values of other processes) while it is executing. To avoid reusable resource sharing at the specification phase, each process is supposed to have a private processor. A more realistic approach is taken in the design phase.

The Description of Cyclic Processes

Let $P = \{p_1, \ldots, p_n\}$ be a set of processes that should form a system. Let us denote the state of the i-th process p_i by s_i, then the state of the system is determined by a set $S = \{s_1, \ldots, s_n\}$.

For any process $p_i \in P$, the state of the process may take one of the following values: a real number ($s_i \in S_R$), a set of real numbers ($s_i \in S_B$) and a predicate ($s_i \in S_L$). Thus we have that

$$S = S_R \cup S_B \cup S_L .$$

A state of a process gets new value at the end of process execution. The processes are executed cyclically, recalculation of a process $p_i \in P$ starts at certain time instants that form the process timeset $T(p_i)$.

Now we can better formulate a process concept. The set P of processes can be divided into three subsets, each process is a mapping. Depending on state values, we have for

$$s \in S_R \Rightarrow p: T(p) \times \text{dom } p \to \text{val } p \subset R$$

$$s \in S_B \Rightarrow p: T(p) \times \text{dom } p \to \text{val } p \subset 2^R$$

$$s \in S_L \Rightarrow p: T(p) \times \text{dom } p \to \text{val } p = \{\text{true}, \text{false}\},$$

where dom $p \subset S$ is domain of definition of process p in S
val $p \subset S$ is the range of states of process p in S
R is the set of real numbers
2^R is the set of all subsets of R.

For the sake of simplicity we shall further omit system states from the argument list of a process and use a notion p(t) instead of p(t,s).

For a timeset T(p) of any process $p \in P$ we can formulate some intuitive constraints:

- T(p) is a strictly ordered set

- there exists a common starting point of the system, i.e. for any $p \in P \Rightarrow \min (T(p)) = 0$

- in any finite time interval a process can be initiated only a finite number of times, i.e. power of the set $[0,t] \cap T(p)$ is finite.

The last constraint is too abstract for practical use. It could be substituted by a demand that the distance between any two consecutive elements $t,t' \in T(p)$ must be in externally (i.e. from the application) supplied bounds:

for any $p \in P$ and $t,t' \in T(p) \Rightarrow$

$$\Rightarrow t_{min}(p) \leq t' - t \leq t_{max}(p),$$

where $t_{min}(p)$ and $t_{max}(p)$ are given functions and

$$t_{min}(p) \leq t_{max}(p).$$

The timeset T(p) of a process p allows us to describe the cyclic behaviour of the system by describing separately that of every process.

To **guarantee** cyclic execution of a process $p \in P$, initiated at $t \in T(p)$, in practice, we must ensure that its execution time d(p,t) is strictly bounded. It is a random variable, but a rough estimate can usually be given

$$d(p,t) \in [a(p), b(p)]],$$

where [.,.] denotes an interval
a(p) and b(p) are given functions and
$a(p) \leq b(p)$.

The Description of Interactions

So far we have been dealing with independent cyclic processes. To have a software system it is necessary to introduce process interactions. From the designer's point of view it would be advisable that processes are not aware of each other, each process is supposed to know only its necessary input data. Input data can only be state values of some other process. It should be entered to a process at necessary time instants while the process is executing.

Consequently, we must have a set of logical input devices that provides a process with proper data at necessary time instants and thus implements the wanted interaction scheme. We will not distinguish between communication (exchange of data) and synchronization (exchange of control information).

Hence, a system is formed by a set of processes together with their logical input devices, called channels for short. A channel is used for one way transmission of specified data, it is a special process connecting a data source (a producer) to a data sink (a consumer) and checking that only earlier specified data pass from producer to consumer.

Let us denote a channel connecting the state of process p_i (the producer) to the process p_j (the consumer) by an ordered pair

$$\sigma_{ij} = \langle p_i, p_j \rangle \in \Sigma ,$$

where $\Sigma \subset P \times P$ is the set of channels needed to describe the interactions between processes of the system. For different types of interactions we need different **types of** channels, but with four basic types it is possible to describe a wide class of applications. These types are as follows:

- the null channels (Σ_n) carry no information; they only demand that two processes work on the same timeset

- the synchronous channels (Σ_s) transmit a specified number of consecutive producer process state values to a consumer process and demand that the processes work on the same timeset

- the semisynchronous or Petri channels (Σ_p) transmit information as synchronous channels but the connected processes execute cofrequently, i.e. the producer process generates the timeset of a consumer process, the initiation instances of the processes are biased although processes are activated with the same frequency

- the asynchronous channels (Σ_a) transmit information between two processes that are executing on different timesets.

The union $\Sigma_n \cup \Sigma_s \cup \Sigma_p \cup \Sigma_a$ need not in general coincide with Σ. It should be possible, if necessary, to add new types of channels.

A channel σ_{ij} is actually a mapping

$$\sigma_{ij}: \text{val } p_i \times T(p_i) \times T(p_j) \to \text{dom } p_j,$$

where val p_i is range of process p_i

dom p_j is domain of definition of process p_j in S.

In principle a channel creates a permanent connection between producer and consumer processes. To restrict the connection to certain time instants only, a channel function is introduced.

The channel function $K(\sigma_{ij},t)$ determines for any $\sigma_{ij} \in \Sigma$ and for any $t \in T(p_j)$ a subset of $T(p_i)$:

$$K(\sigma_{ij},t) \subset T(p_i), \quad t \in T(p_j).$$

This means that only those producer process states, calculation of which has been activated at the moments $t' \in K(\sigma_{ij},t)$, are available to the consumer process at the time $t \in T(p_j)$.

In the case of synchronous and Petri channels, the channel function $K(\sigma_{ij},t)$ is usually given as a pair $[\mu,\nu]$ determining the interval on the strictly ordered set $T(p_i)$. The interval is given relative to the present moment, thus 0 means the present execution, 1 means immediately previous to the present, etc. For example, the function $[2,1]$ means that state values of two executions of the producer process are accessible through this channel -- the immediately previous to the present moment and the one before that.

Since we allow processes to input data not only in the beginning of the execution but at any time during the execution, still another parameter of the process should be estimated. For any channel $\sigma_{ij} \in \Sigma$, a delay between initiating the recalculation of the consumer process p_j at $t_j \in T(p_j)$ and its needing the state of the producer process p_i, should be determined.

This delay is denoted by $d_1(\sigma_{ij},t)$, $t \in T(p_j)$ and is in general a random variable, an estimate in the form of an interval could usually be provided

$$d_1(\sigma_{ij},t) \in [a_1(\sigma_{ij}), b_1(\sigma_{ij})],$$

where $a_1(\sigma_{ij})$ and $b_1(\sigma_{ij})$ are given functions and

$$a_1(\sigma_{ij}) \leq b_1(\sigma_{ij}).$$

Now it is possible to describe which states are in principle accessible through an asynchronous channel. All the states of the producer process (p_i) calculation of which has started at $t' \in T(p_i)$, where t' satisfies

$$t' + d(p_i,t') < t + d_1(\sigma_{ij},t),$$

with $t \in T(p_j)$, are accessible to the consumer process p_j at the time instant $t \in T(p_j)$. It is possible to select among accessible states by defining a channel function in the same way as for synchronous channels.

PROPERTIES OF THE MODEL

The procedure of describing an application in terms of any model has by itself a positive effect on increasing correctness of problem statement and analysis. In the case of the Quirk's model this procedure is not difficult to computerize, adding at the same time some checks. For example, it is straightforward to check correspondence of processes and channels ensuring thus practical completeness of the specification -- each consumer must have a producer and vice versa.

However, the Quirk's model enables also more sophisticated tests of system properties. In this paper some aspects of correct timing and deadlock avoidance are considered.

Timing of Processes in the System

The problem of timing in software systems has got surprisingly little attention in the literature. The example of producer--consumer system functioning at natural rate, studied in (Sifakis, 1979), is closest to timing problems of control systems software. A Petri net model is used in (Sifakis, 1979).

Nevertheless, incorrect timing causes errors which are extremely difficult to find by testing because of their dynamic nature. It is important to ensure correct timing at the specification and design phase already. A significant step towards the solution of this problem has been made in (Quirk and Gilbert, 1977). Some of their results are listed below.

Proposition 1. (Quirk and Gilbert, 1977). The constraint $b(p) - a(p) < t_{min}(p)$ ensures that for any $t_1, t_2 \in T(p)$, $t_1 < t_2$, $p(t_1)$ is available before $p(t_2)$; i.e. the state of the earlier initiated copy of process p is available earlier.

Proposition 2. (Quirk and Gilbert, 1977). The constraint $b_1(\sigma_{ij}) - a_1(\sigma_{ij}) < t_{min}(p_j)$ ensures that for any $t_1, t_2 \in T(p_j)$, $t_1 < t_2$, the request from $p_j(t_1)$ arrives the channel $\sigma_{ij} \in \Sigma$ before the request from $p_j(t_2)$.

Proposition 3. (Quirk and Gilbert, 1977). A process which needs its own values from its previous execution must never have to wait for this or else its wait time and completion time will increase without limits.

These results do not solve all the timing problems of interacting processes, nevertheless it seems to be a good first step.

Deadlock Avoidance

We have a set P of processes, each of which executes on its own processor. We also have a set Σ of channels realizing process interactions by transferring messages (the states of producer process) from their source to their destination (to the consumer process).

We do not consider problems connected with sharing reusable resources at this phase. Thus, the only reason for potential deadlock may be a circular wait for messages. The cases where circular wait is caused by absence of channels will be discovered when analysing practical completeness of the system -- every consumer must have a producer. Our intention is to prevent deadlock by constraining the system structure, i.e. by suitable selection of channels and channel functions.

In the case of null channels there is nothing to wait for. In the case of asynchronous channels we get the state value from the last execution (prior to the request) of the producer process. The asynchronous channel may result in incorrect timing -- the data may not be produced at the designed moment, it may be either too old or too new, but it does not result in permanent wait condition.

The circular wait may occur when using synchronous and/or Petri channels. Let us consider a group of processes connected with each other by synchronous and/or Petri channels.

Definition 1.
A sequence of parallel processes $p_1, p_2, \ldots, p_n$ is called a synchronous chain, if
$\sigma_{i,i+1} \in \Sigma_s \cup \Sigma_p$, where $i = 1, 2, \ldots, n-1$,
Σ_s is a set of synchronous channels and
Σ_p is a set of Petri channels.

Definition 2.
A synchronous chain is a synchronous loop, if $p_n = p_1$ and the channel function is given as an interval $[\mu, 0]$ on their common or generating timeset $T(p)$ and μ may change from channel to channel.

Proposition 4.
The existence of synchronous loop in the system's structure is sufficient condition for deadlock.

Proof: To prove the proposition it suffices to demonstrate that the system containing synchronous loop satisfies definition 2.3 from (Coffman and Denning, 1973) of a deadlock situation. The paraphrased definition is as follows: a deadlock situation in executing a group of processes of a system arises when these processes request more resources than is the overall capacity of the requested resources in the system.

Let us consider a process as a chain (Coffman and Denning, 1973) consisting of two tasks -- the first task reads in a message (messages) and releases no messages, the second task releases messages (produces a state value of the process which is formed into messages by channels). Before the initiation of a synchronous loop the resource capacity vector is a zero vector, consequently the definition 2.3 of (Coffman and Denning, 1973) holds and we have a deadlock situation. QED.

AN EXAMPLE

In order to demonstrate the possibilities of the Quirk's model let us analyse a description of a software structure for a direct digital control (DDC) regulator. In the following we describe the regulator's software by using Petri nets and by using Quirk's model. For those not familiar with Petri nets we recommend (Nader, 1980). The comparison will be limited to the analysis of software execution time diagrams (firing diagrams for Petri nets).

A DDC Regulator

A regulator takes in measurements from the technological process, it computes control actions according to the given algorithm. The algorithm is not specified at this phase -- it may be PI, PID or some of the multivariable control algorithms. The computed values should be output to an actuator, for safety reasons we demand a feedback signal from the actuator.

A Petri Net Model for the DDC Regulator Software

In Fig. 1 a net description of the DDC regulator software structure is given. A similar net has been thoroughly analysed in (Sifakis, 1979) by using timed Petri nets for evaluating the natural rate of functioning the producer--consumer system. However, in our case we cannot count on natural rate because the producer is activated by arrival of measurements and the consumer depends on feedback signal from the actuator.

The behaviour of the model in time is described by the net firing diagram in Fig. 1a. Although the software parts operate asynchronously, attention should be paid that the measurements input rate is not too high and that the actuator is not too slow.

A Quirk's Model for the DDC Regulator Software

Making use of basic features of the Quirk's model we can neglect two processes (SYNALG and DUMMY) which were necessary in the Petri net model. The Quirk's model is presented in

Fig. 2. The type of channels and channel functions is not specified since by varying them system designer can have substantially different behaviour of the software system.

In this example the aim is not to specify processes (timesets and other parameters), but rather to demonstrate several design possibilities. From the same set of processes it is possible to get different systems by modifying the type of channels and channel functions.

We consider three sets of channels in this example:

- set A, where ch1 = ch2 = ch3 = ch4 = Petri channels with the function [0,0] and ch5 = null channel, ch6 = asynchronous channel, ch7 = synchronous channel with function [1,1]

- set B, where ch1 = ch2 = ch3 = ch4 = ch6 = ch7 = synchronous channels with function [1,1] and ch5 = null channel

- set C, where ch1 = ch2 = ch3 = ch4 = ch6 = = synchronous channels with function [0,0] and ch5 = null channel, ch7 = synchronous channel with function [1,1].

The corresponding time diagrams are presented in Fig. 2a, 2b, and 2c, respectively.

From comparing Fig. 1a, 2a, 2b, and 2c the following conclusions are evident:

- the Quirk's model with Petri channels imitates rather well the behaviour of the net model (Fig. 1a and Fig. 2a); the design can be implemented on one-processor system, on multiprocessor system, or on a network since no strict synchronization is needed; process SYNOUT acts explicitly as a watch-dog timer

- the Quirk's model with synchronous channels and functions [1,1] (Fig. 2b) suits for implementing on a multiprocessor computer; it is too complicated to synchronize separate nodes on a network and the resulting system on one-processor system will work ineffectively; this system enables to reach high accuracy in time for inputs and outputs

- synchronous channels with functions [0,0] (Fig. 2c) describes multitask work of a one-processor system, i.e. software operation in centralized control systems.

In many aspects there seems to be close connection between Petri nets and the Quirk's model. However, the Quirk's model is more flexible and allows us to compare different implementation alternatives. The comparison of the Quirk's model with other parallel programming models needs further study.

DISCUSSIONS AND CONCLUSIONS

A number of papers have recently been written on the problem of specifying and designing the structure of a software system separately from detailed data organization and processing algorithms (see for example (Riddle and others, 1979; Laventhal, 1979; Ludewig, 1980)). Paying much attention to specifying detailed data types and processing algorithms in a too early phase of a project is considered confusing to the designer, especially in the case of computer control systems (Ludewig, 1980; Boebert, 1980; Motus, 1980).

The Quirk's model, described in this paper, is also oriented to specifying and designing software structure only. From the model follow restrictions for a more detailed design and also demands to the computer configuration for implementing the software system.

Principles of good specification system were developed in (Balzer and Goldman, 1979). We claim that the Quirk's model meets all the eight principles, and can thus be considered as a good basis for building a control system's software specification and design system. However, by Quirk's model it is possible to specify the structure and behaviour of the software system and solve its timing problems. For detailed specification of algorithms other means should be used.

The Quirk's model fully separates logical structure and description of the software from the problem of its physical implementation. The mapping of abstract network of processes (description of application in terms of Quirk's model) to the physical computer network is done separately. The model provides necessary data for designing/selecting suitable computer network for the specified control system (Motus and Vain, 1982).

An interesting approach is developed in (Mott, 1979). Starting from the functional decomposition -- similarly to Quirk's model -- a dedicated computer system architecture is developed. Among other analogies with the approach developed in our paper, a hardware implementation of a communication mechanism (which in principle is similar to our channel concept) is suggested.

The necessity of subscription for data on a periodic basis in real-time control system has been pointed out in (Wilhelm, 1980). The Quirk's model includes the idea of data subscription -- a consumer process does not need to consume all the data produced by a producer process. The opposite assertion is also true -- the results of every execution of a producer process must not be consumed by some other process. The selection (subscription) can be made by a suitable channel function at the system specification phase.

Many specification systems for real-time application are based on PSL/PSA or RSL/REVS languages (Ludewig, 1980; Furia, 1979). There exist other systems (for example (Biewald and others, 1980; Goldsack and others, 1980; Ludewig and Sandmayr, 1981)), but as a rule, timing problems are not considered explicitly enough. The timing problem has one of the central places in the Quirk's model. No doubt, it needs further investigations, but already now quite subtle potential timing errors can be discovered in a specification phase.

Provided that the abstract network of processes (resulting from the control system description in terms of Quirk's model) corresponds to the logical structure of the underlying technological process, the Quirk's model facilitates maintenance and modification of already operating computer control system. It is necessary to create a suitable interface between the end-user and the model. The same interface is used for system specification and design.

A number of other questions remain unsolved in connection with the Quirk's model. For example, the implementation of channels by using:

- the existing operating system primitives for synchronization and communication together with necessary extensions to them

- communication protocols in the underlying physical computer network

We hope to handle this problem in a future paper.

Another interesting topic is the usage of the control system specification in terms of the Quirk's model for well-founded design of dedicated real-time operating system.

General properties of the process interaction and many specific timing problems need a careful study in the future.

REFERENCES

Balzer, R., and N. Goldman (1979). Principles of good software specification and their implications for specification languages. Proc. of the IEEE Conference on Specification of Reliable Software, 58-67.

Biewald, J., E. Joho. S. Jovalekic, and H. Shelling (1980). Application of the specification and design technique EPOS to a process control problem. Proc. of the 6th IFAC/IFIP Conference on Digital Computer Applications to Process Control, Düsseldorf, 517-522.

Boebert, W.E. (1980). Formal verification of embedded software. ACM SIGSOFT Software Engineering Notes, 5, No. 3, 41-43.

Coffman, E. G., and P.J. Denning, Jr. (1973). Operating systems theory. Prentice Hall, Inc., 331 pp.

Furia, N.J. (1979). A comparative evaluation of RSL/REVS and PSL/PSA applied to digital flight control system. AIAA 2nd Computers in Aerospace Conference, Los Angeles, 330-337.

Goldsack, S., V. Haase, and H. Halling (1980). A step towards application oriented specification. In H. Meyer (Ed.) Real-time data handling and process control. North-Holland Publishing Co., 525-533.

Laventhal, M.S. (1979). Synchronization specifications for data abstractions. IEEE Conf. on Specification of Reliable Software, 119-125.

Ludewig, J. (1980). PCSL -- A process control software specification language. Institut für Datenverarbeitung der Technik, Kernforschungszentrum Karlsruhe, KfK 2874, 45 pp.

Ludewig, J., and H. Sandmayr (1981). On the specification of distributed computer control systems. The 3rd IFAC Workshop on Distributed Computer Control Systems, Beijing, China.

Mott, D.R. (1979). A distributed computing architecture for real-time system control and information processing. Proc. of the 1st Intern. Conf. on Distributed Computing Systems, Huntsville, 204-211.

Motus, L. (1980). Introduction to the specification of computer control software. Res. Report SÄH 39/80, VTT Electrical Engineering Lab. 35 pp.

Motus, L., and J. Vain (1982). A set of tools for designing and evaluating communication protocols in industrial computer networks. Submitted to the 3rd IFAC/IFIP Symposium for Computer Control, Madrid, Spain.

Nader, A. (1980). Petri nets for real-time control algorithms decomposition. In T.J. Harrison (Ed.) Distributed Computer Control Systems. Pergamon Press, 197-210.

Quirk, W.J., and R. Gilbert (1977). The formal specification of the requirements of complex real-time systems. Harwell, AERE, Report No. 8602. 57 pp.

Riddle, W. E., A. M. Stavely, J. H. Sayler, A. R. Segal, and J. S. Wileden (1979). Abstract monitor types. IEE Conference on Specifications of Reliable Software, 126-138.

Sifakis, J. (1979). Use of Petri nets for performance evaluation. Acta Cybernetica, 4, No. 2, 185-202.

Wilhelm, R.G. Jr. (1980). Transaction processing in distributed control systems. In T.J.Harrison (Ed.) Distributed Computer Control Systems. Pergamon Press, 133-142.

Yeh, R.T. (Ed.) (1977). Current trends in programming methodology. Vol. I. Prentice-Hall Inc. 276 pp.

Zave, P. (1976). On the formal definition of processes. Proc. of the International Conference on Parallel Processing, 35-42.

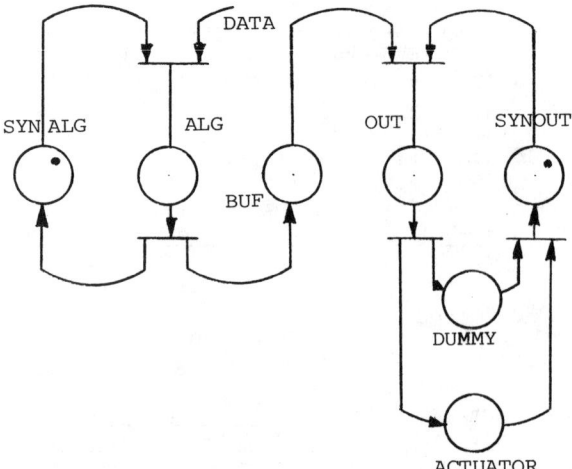

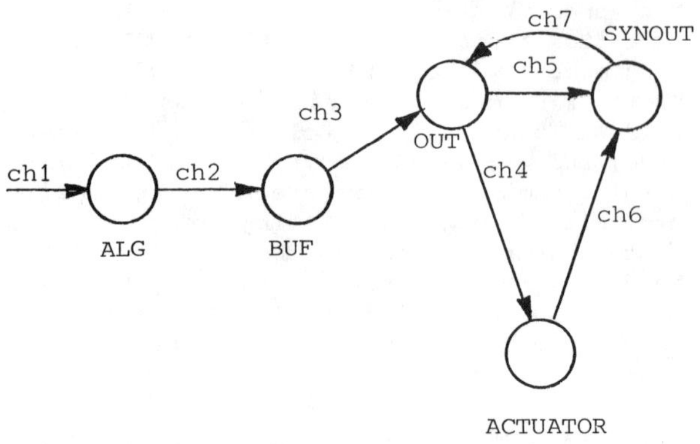

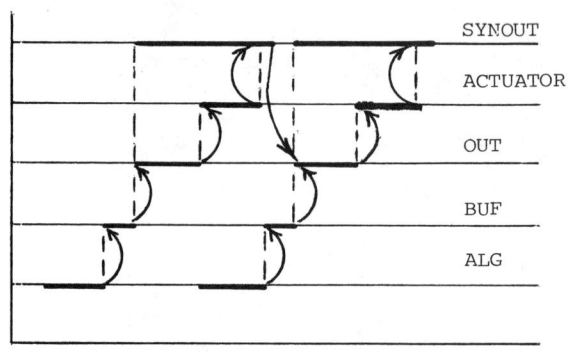

Fig. 2. A structure of the DDC regulator software using Quirk's model.

ALG - implements control algorithms
SYNALG, SYNOUT - input synchronization
BUF - simulates buffer or communication line
OUT - prepares signals of ACTUATOR
ACTUATOR - simulates the work of real actuator

Fig. 1. A structure of the DDC regulator software using Petri nets.

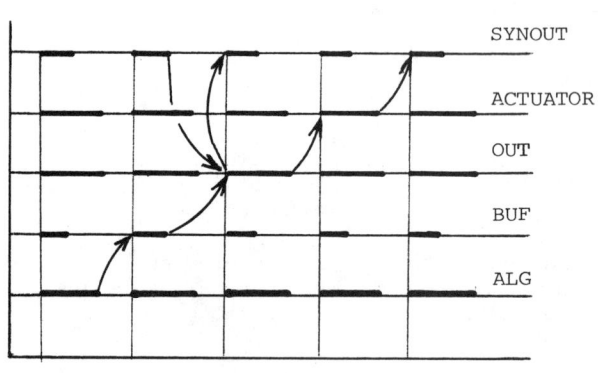

Fig. 2a. Time diagram of the Quirk's model using set A of channels.

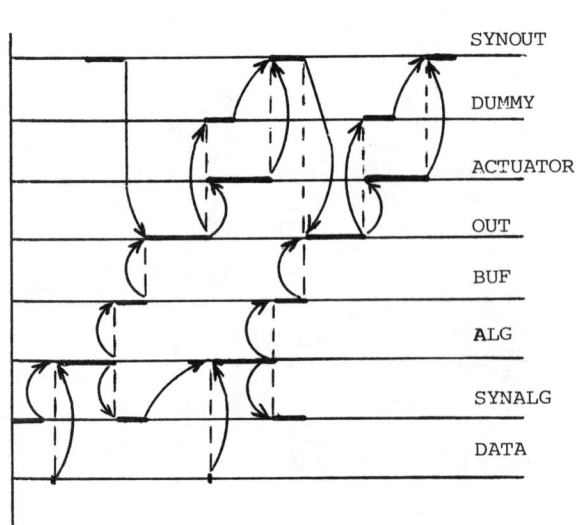

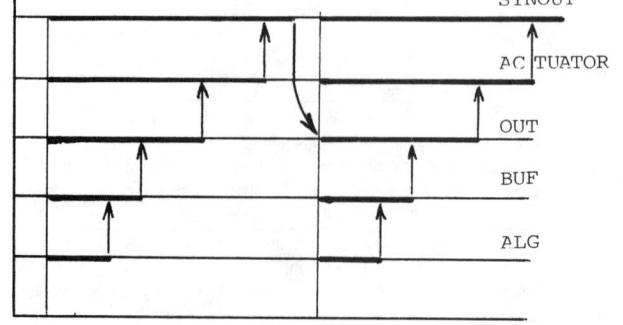

Fig. 2b. Time diagram of the Quirk's model using set B of channels.

Fig. 1a. A fragment of the DDC regulator software Petri net model firing diagram.

Fig. 2c. Time diagram of the Quirk's model using set C of channels.

ENSV TA rotaprint. Tell.128 Tir.70 MB-03792

DISCUSSION

Vamos: We listened to a paper (Davidson's) just before yours on a more practical approach where the author claimed that for their purposes the use of the well-known simulation languages was quite practical and adequate. You've tackled similar problems which of course need to be simulated. Have you some new methodology which is more efficient and more far-reaching than those that are used regularly?

Motus: At the moment it is quite difficult to say whether our methodology is more efficient and more far-reaching. Actually we must also do some simulation, not on the level of a process algorithm but on the level of system structure and its dynamical behaviour. I believe that this way we can quite remarkably economize on the time of simulation.

Vamos: In programming or running of simulation?

Motus: We don't need to program the simulation for different applications because we have a rather formal model and we specify each application in terms of this model. Thus, simulation is practically reduced to executing a certain set of tests on the specified parameters of the model. The user himself should not write simulation programs. Instead, he has to specify his problem according to the brainwork of this particular model. We can use the same simulation program for a number of applications. That is what we hope.

Rotanov: I think the deadlock avoidance is really a serious problem, but could you say a few words about the infinite loop avoidance problem?

Motus: Perhaps a simple drawing (Figure 1) would be helpful. Let us suppose that we have three processes that are connected by synchronous channels, with the channel functions given by intervals. The last number determining the interval is 0, the first one can be different for different channels. As all these processes are working synchronously, they are activated at the same moment. Before process A can finish its work, it needs the state of the present execution of process C because we have zero in the channel function. Since process A can't come to and end, its state from the present execution is not accessible to process B. Process C is waiting for the end of the present execution of process B. So it is a circular wait and it is a rather trivial one if we draw it separately like that. Anyway, it occurs quite often when we specify a practical system - we don't even pay any attention to it. The circular wait can be discovered by our model.

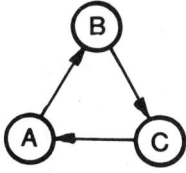

Figure 1

PARALLEL PROCESSING FOR REAL-TIME CONTROL AND SIMULATION OF DCCS

H. Kasahara[1] and S. Narita

Department of Electrical Engineering, Waseda University, Tokyo, Japan

Abstract. This paper concerns itself with the resource allocation and task scheduling problems for real-time control and simulation of distributed computer control systems. The problems are formulated and solved at two levels. The Interblock Problem deals with the allocation of computing resources to a set of interrelated blocks (program modules). A solution algorithm is presented to minimize the total processing time, with due consideration given to interprocessor communication times. The Intrablock Problem essentially is an optimum resource allocation and task scheduling problem in a parallel processing environment. Heuristic algorithms are combined with a rigorous solution algorithm to reduce the overall solution time. The proposed solution algorithms are applied to a few real-time control and simulation problems to demonstrate their usefulness in the design, implementation, and operation stages of DCCS.

Keywords. Parallel processing; distributed control; resource allocation; task scheduling; real-time control; simulation; interprocessor communication.

INTRODUCTION

Along with the development of LSI technology and that of high-end microcomputers, in particular, decetralized system architecture with multiple processors distributed spacially or functionally has become common practice in system control applications. The superiority of distributed system architecture to centralized one has been discussed from a wide variety of points of view such as reliability (RASIS), expandability and maneuverability. In order for those advantageous features of distributed system architecture to be fully utilized, there still remain a number of technical problems to be resolved which are inherent to distributed control systems, although a wealth of technical know-hows accumulated so far in the course of designing, manufacturing and operating conventional centralized systems may serve as a basis on which to implement advanced design philosophy.

The well-balanced combination of interprocessor communication, computer and control technology is essential for successful implementation of DCCS. Among other things, functional distribution, resource allocation and task scheduling are the most important key problems. This stems from the fact that a plurality of processors are distributed spacially or functionally and the tasks processed on respective processors require, to a greater or lesser extent, some data exchanges, which is not the case in centralized system architecture.

In the case of centralized system architecture, primary efforts to enhance the processing efficiency or the system throughput have been directed toward the minimization of the so-called operating system overhead through judicious job scheduling. In the case of decentralized system architecture, the problems of task partitioning and resource allocation must be duly taken into consideration at an early stage of system design and implementation. In software engineering terminology, these problems are closely related to the problem of

---

[1] Presently with the Department of Electronics Engineering, Saitama Institute of Technology.

structurization which, despite a number of pioneering works, is a kind of problem most difficult to apply quantitative methods. For this reason, actual design procedures of DCCS have almost always been dependent on empirical rules or rules of thumb based on past experience of system design, implementation and operation.

This paper is aimed at the establishment of a quantitative approach to the resource allocation and task scheduling problems in a distributed computing environment. Although the contents of the present paper are centered around the real-time simulation and control of distributed computer control systems that can be described by well-defined mathematical models, the proposed approach is equally applicable to general distributed computer control systems in charge of a variety of control, monitoring and logging functions if necessary input data are all provided.

THE INTERBLOCK PROBLEM

A set of tasks or processes is generally referred to as a job. In this paper, an assembly of tasks or jobs grouped by some means is referred to as a block. In DCCS, a block may represent the logical or mathematical operations to perform a variety of control functions for an area or subsystem of the target (controlled) system. The Interblock Problem formulated here is an optimizing problem to minimize the sum of the execution termination time and the total interprocessor communication time, where the execution termination time is defined to be the longest processing time of a plurality of processors running concurrently. This minimization problem is subject to a wide variety of soft and hard constraints such as real-time execution time limits for on-line control operations and available storage capacities.

The interblock problem assumes that the following conditions hold.

1. The processing time of a task assigned to a computational resource is given as a deterministic value.
2. No interprocessor communication time is required between a pair of tasks if the two separate blocks containing these two tasks are allocated to one and the same processor.
3. When a pair of tasks having some interactions are processed on separate processors, a given time is required for interprocessor communications. The communication time can be assumed to be almost determinstic with small random variations around some average value, although it may be data-dependent.
4. The number of blocks within a study system is greater than or equal to that of processors available. Two or more processors cannot be assigned to a block. The case where a block requires a plurality of processors because of real-time execution limits and/or storage limits will be dealt with separately in the succeeding section as the Intrablock Problem.
5. The total execution time of a block on a given processor does not depend on the order or sequence of tasks to be processed; it is simply the sum of the respective task execution times.

A wide variety of distributed computer control applications may be formulated as the interblock problem or its modifications. An air defence monitoring system (Ma, Lee, and Tsuchiya, 1982), a patient monitoring system (Weitzman, 1980), and a data acquisition/processing system using a multiprocessor configuration for high-level performance (Weitzman, 1980) are typical examples. In industrial real-time control applications, the interblock problem may play an important role especially in the design stage to determine an appropriate and hopefully optimal computing resource allocation scheme.

A wealth of relevant works have so far been reported. Gylys and Edwards (1976) studied an allocation scheme of logical resources (programs and data files) to computers which minimizes the communication traffic. Buckles and Harden (1979) employed the notion of proximity or similarity for the solution of the partitioning/allocation problem to break down an integrated logical resource into several nonoverlapping subsets and, at the same time, showed an allocation scheme that minimizes the sum of storage and access costs using mathematical programming.

In this paper, particular considerations are given to the effects of communication requirements in the formulation of the interblock problem. Otherwise, the results obtained can give rise to an excessive high interprocessor traffic, which, in turn, may cause intolerable delays in data transmission or data losses due to intermittent communication channel overloading.

In what follows, the interblock problem is formulated in mathematical terms.

Objective Function

$$\text{Min } f = \sum_{j=1}^{n} a_{\alpha j} x_{\alpha j}$$
$$+ \sum_{i=1}^{m} \sum_{j=1}^{m} \sum_{k=1}^{n} \sum_{l=1}^{n} p_{ij} c_{kl} x_{ik} x_{jl} \quad (1)$$

where

x_{ij} (assignment variable)
= 1 : if block j is assigned to processor i.
= 0 : otherwise.

a_{ij} = execution time of block j on processor i.

p_{ij} = interprocessor communication time per unit amount of data including communication overhead (from processor i to procossor j).

c_{kl} = amount of data to be transmitted from block k to block l.

Notice that the first term of Eq. (1) stands for the total execution time of the blocks assigned to processor α while the second term for the sum of interprocessor communication times. In case the processing capability (in terms of MIPS, for example) is equal for all m processors, we can set, with no loss of generality, $\alpha = 1$ in Eq. (1) to account for the longest execution time of the n processors. In the general case where the throughputs of the processors are different, we need to solve Eq. (1) repeatedly for α = 1, 2, .., m to find out the optimum assignment of processors.

Constraints

Several constraints (equalities or inequalities) need to be augmented to Eq. (1) to account for the conditions listed before as well as for storage limits and on-line execution time requirements. In addition, a set of artificial variables and associated inequalities are introduced to reduce the original nonlinear integer programming problem (due to the product terms $x_{ik} x_{jl}$ in Eq. (1)) to an equivalent linear one. Constraints incurred from practical considerations such as the preference of a particular processor for a particular block, the non-uniform spacial distribution of processors, and the provision of special hardware in a processor can be accounted for by assigning appropriate large or small values to the relevant parameters. A slight modification of Eq. (1) can solve the so-called work-load balancing problem to level the execution times of the respective processors.

Solution Method

The nonlinear integer programming problem of Eq. (1) may be solved by the use of a suitable mathematical programming solution package. This paper employs the so-called implicit enumeration method. The only problem to be resolved is the dimensionality problem due to the introduction of the artificial variables. If the null elements in the parameter matrices are duly accounted for, the overall computing time may be reduced by a large margin (approximately 75 percent reduction in the numerical examples to follow). Another solution algorithm on the basis of the branch-and-bound method was recently developed by Ma, Lee, and Tsuchiya (1982) and applied to an air defence control system. This approach seems to be advantageous for problems with many constraint conditions.

Examples

The multicomputer control of a continuous production plant represented schematically by the block connection diagram of Fig. 1 was considered. Each rectangular box of the diagram represents a block, which mainly consists of an assembly of application programs for control and monitoring of a subsystem (or a stage) of the total plant. An arrow between a pair of blocks represents the existence and orientation of interblock data communication.

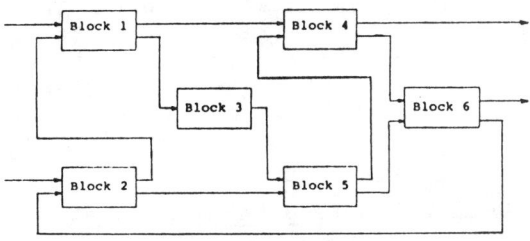

Fig. 1. The block connection diagram.

In an attempt to examine the effects of interprocessor communication, four cases were tested with different parameter sets. Table 1 lists the reference-case parameters for processing times. The interprocessor communication times and amounts of data to be transmitted are also given Table 1.

- Case A : Reference case.
- Case B : Interprocessor communication times are all disregarded.
- Case C : The communication data amounts are partially changed. (c_{13} and c_{35} are doubled.)
- Case D : The processing time of a block is dependent on the processor; the interprocessor communication times vary with the processor pair. This is the most general case.

TABLE 1 The Reference-case Parameters

a_{ij}

j i	1	2	3	4	5	6
1,2 and 3	132	336	280	157	210	79

p_{ij}
 =0 : if i=j.
 =15 : otherwise.

c_{kl}
 =1 : if there exists an arrow from block k to block l in the block connection diagram.
 =0 : otherwise.

The results of computing resource allocation are summarized in Table 2 for the respective cases. Comparison of Case A with Case B reveals that, despite the relatively small communication time requirements, the resource allocation is greatly affected by interprocessor communication. It is also seen from Case C that even partial changes in the amount of communication data need reallocation of computing resources. In Case D, Eq. (1) was solved repeatedly for α = 1, 2 and 3 to find out the best allocation. Table 3 shows the minimum execution time for the parameters a_{ij} and p_{ij} given in Tables 4 and 5, respectively.

TABLE 2 The results of Allocation

processor	Case A	Case B	Case C
1	2 5 6	2 3	1 3 5
2	1 4	1 4 6	4 6
3	3	5	2

TABLE 3 The Minimum Execution Times in Case D

α	the minimum execution time	value of objective function
1	928	928
2	928	978
3	904	904

TABLE 4 a_{ij} for Case D

a_{ij}

j i	1	2	3	4	5	6
1	132	336	280	157	210	79
2	132	357	220	321	240	92
3	100	393	165	206	164	120

TABLE 5 p_{ij} for Case D

p_{ij}

j i	1	2	3
1	0	50	50
2	75	0	75
3	100	100	0

The next problem of practical importance is: "How many processors are needed to be able to meet the real-time execution time limit specified by the response time requirements for on-line, real-time control?" Table 6 shows the optimum (shortest) execution time obtained as the results of the proposed allocation scheme for m = 1, 2, and 3, m being the total number of processors available. For this particular example, the optimum number of processors is two if the real-time execution limit lies in the range between 1194 u.t. (unit time) and 685 u.t.

TABLE 6 The Optimum Execution Times for Each Number of Processors

number of processors	execution time
1	1194
2	685
3	685

THE INTRABLOCK PROBLEM

When a block cannot be processed on a single processor within a specified real-time execution time limit because of the lack in throughput or when the program size exceeds the storage limit, a plurality of processors running concurrently must be allocated to that block. The Intrablock Problem considered in this section is how to allocate the multiple processors to the tasks which constitute the block in such a manner that the resultant execution time be minimized while the precedence relations existing among the tasks be satisfied. This problem also includes the optimum ordering of the tasks allocated to the respective processors. The solution to the intrablock problem gives the minimum number of processors that can process a given block within a specified real-time execution limit. In case the number of processors available is specified, one can determine the minimum execution time of a given block.

It is assumed that the following conditions are met in the intrablock problem.

1. The computing capability (e.g., MIPS) is the same for all processors.
2. The processing time of each task in the block is given deterministically.
3. The data transfer time between a pair of interrelated tasks is disregarded if they are allocated and processed on the same processors because the data transfer may be realized very quickly by means of memory reference operations and the like.
4. If a pair of interrelated tasks are allocated to separate processors, a predetermined deterministic communication time is required for data transfer.
5. Each task must be executed once and only once within a specified time span. In other words, no conditional branches are involed in the block.
6. No task preemption is assummed. No task may be terminated at its intermediate stage once it has been initiated.
7. Each task must meet the precedence conditions. That is, a task must wait for execution until all preceding tasks have been completed.

There exist a number of pioneering works on the allocation and scheduling of a set of partially ordered tasks. However, few papers take into consideration the communication time requirements between processors, which are found to be very sensitive to the final task allocation and scheduling scheme obtained. Since solution of the intrablock problem is very time-consuming and rigorous solutions are sometimes difficult to obtain within a reasonable time span except for fairly small problems, a two phase solution algorithm is employed in this paper. In the first step or Phase 1 of the algorithm, heuristic methods are utilized to find out suboptimal (hopefully very close to global optimum) solutions. These suboptimal solutions are then used as the bounding values in Phase 2 which employs a modified branch-and-bound method to find the optimum solution.

In order to represent the precedence relations and the communication time requirements between tasks, use is made of Petri net like graphs in place of the conventional task graph. The graph illustrated in Fig.2 shows the Petri net representation of an example block.

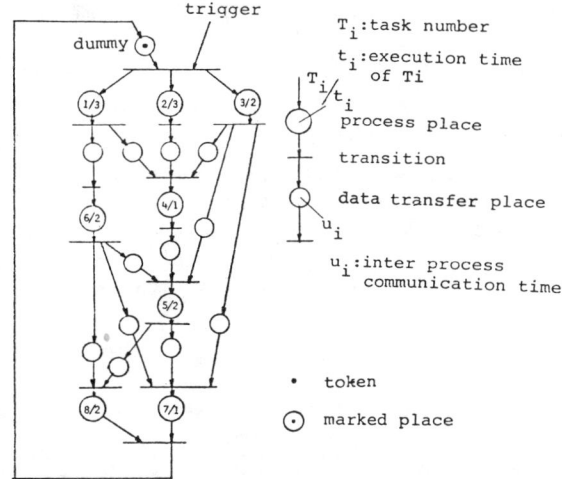

Fig. 2. The Petri net representation for an example.

The Petri net N consists of four basic elements, i.e., N={P,T,Z,Q}. P stands for a "place" or a state. For the sake of convenience, two types of places are distinguished. A "process place" depicted by a large circle represents the execution of a task or a process while a "data transfer place" shown by a small circle stands for data transfer. The figure T_i attached to a process place stands for the task number and t_i for the execution time of the task. The figure u_i inside a data transfer place (a small circle) represents the interprocess communication time. Should the process places (tasks) immediately before and after a data transfer place be assigned to the same processor, u_i is automaticallay set to zero. A short horizontal line (bar) represents transition T. An arrow Z directed from P to T represents a forward incidence function while an arrow Q from T to P a backward incidence function. These functions specify the precedence relations among tasks. A dot or a token inside a circle represents the current position of processing. As soon as all the places preceding a transition are "marked" or dotted, the transition is activated and fired and the token is passed to the succeeding place(s).

Before we can go into the description of the heuristic algorithms in Phase 1, we need to define the earliest possible execution time $\underline{\tau}_j$ and the latest possible execution time $\overline{\tau}_j$ for the task T_j, where, for the time being, the interprocess communication times u_i are ignored.

Earliest possible execution time :

$$\underline{\tau}_j = \max_k \sum_{u \in \pi_k} t_u \qquad (2)$$

where π_k is the k-th path from the triggered transition to the process place T_j in the direction of the arrows.

Latest possible execution time :

$$\overline{\tau}_j = \min_k [t_{ub} - \sum_{u \in \hat{\pi}_k} t_u] = \min_k [t_{ub} - l_j] \qquad (3)$$

where $\hat{\pi}_k$ is the k-th path from the triggered transition to the process place T_j in the direction opposite to the arrows. t_{ub} is the real-time execution time limit of the block, which cannot be shorter than the critical path time of the Petri net.

In short, $\underline{\tau}_j$ stands for the earliest possible time when the execution of T_j can be initiated while $\overline{\tau}_j$ for the latest time when the execution of T_j must be started in order for the execution of all the tasks in the block to be terminated at no later than t_{ub}. In other words, should the processing of the block be completed no later than t_{ub}, the task T_j must be started at some time in the interval $[\underline{\tau}_j, \overline{\tau}_j]$.

We are now ready for the explanation of two heuristic algorithms for Phase 1 of the intrablock problem.

Heuristic Algorithm 1

Step 1. Initialization.
Step 2. Select those task(s) which have no preceding task(s) or those whose preceding task(s) have all been executed.
Step 3. Choose the currently executable task having the largest l_j. Allocate the selected task to the processor which gives rise to the minimum interprocessor communication time.
Step 4. Find out the processors which first finish processing.
Step 5. Check if all the tasks have finished scheduling. If not, go to Step 2.

Heuristic Algorithm 2

Step 1. Initialization.
Step 2. Choose the task t_j which has not yet been allocated to any processors and has the largest l_j. If there are more than one tasks which have the largest l_j, choose the task T_j which can start earlier than the other tasks, taking into account the interprocessor communication time.
Step 3. Allocate the selected task to the processor on which the task can start earliest.
Step 4. Check if all the tasks have finished scheduling. If not, go to Step 2.

The distinction between the two heuristic algorithms lies in the fact that the heuristic algorithm 1 employs $\underline{\tau}_j$ as the principal index of scheduling and $\overline{\tau}_j$ as the subsidiary index while the situation is reversed for the heuristic algorithm 2. Although the relative advantages of the two heuristic algorithms vary with the problem at hand, the heuristic algorithm 2 gives better results when some processors involve "forced idle times" (introduction of intentional delays for currently executable tasks.)

Optimal Scheduling by a Modified Branch-and-Bound Method

The second step or Phase 2 of the intrablock problem is based on a "modified" branch-and-bound method, using the result obtained by the use of the heuristic algorithms for initial scheduling. Different from the conventional branch-and-bound method which extends the branch having the smallest value at each node (see Fig. 3(a)), the modified method finds out, at first, a feasible scheduling by means of forward search and then tries to find a shorter scheduling by repeating backward and forward searches (see Fig. 3(b)). This approach has the following advantages.

1. A proper bounding value can be found quickly.
2. Storage requirements can be small.
3. In case the search procedure to find the optimum solution is interrupted because of time limit, the intermediate result may be used as a suboptimum scheduling.

Since the modified branch-and-bound algorithm is rather involved, the details are given in Appendix.

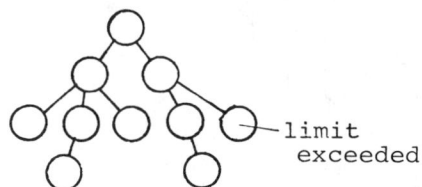

(a) the conventional b-a-b method

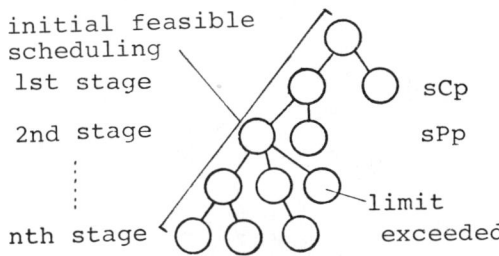

(b) the modified b-a-b method

Fig. 3. The branching schemes for two types of b-a-b methods.

Example 1

For the purpose of demonstrating the effectiveness of the proposed approach, the set of partially ordered tasks (a block) represented by the Petri net of Fig. 2 is considered. Figures 4 (a), (b) and (c) show the resource allocation and task scheduling results obtained by the use of the heuristic algorithms 1 and 2, and the modified branch-and-bound method, respectively, where the number of processors available is two. In this case, both heuristic algorithms 1 and 2 provided satisfactory results, although not optimal.

Figures 5 (a), (b) and (c) show the case where the number of processors is three. In this case, the two heuristic algorithms gave the optimum schedule. Since the heuristic algorithm 1 gave the optimum schedule for the case where the communication times were ignored, this optimum schedule remains unchanged and applies for all cases with the communication time less than 1 [u.t.].

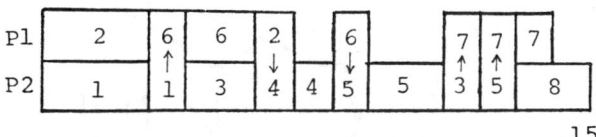

(a) heuristic algorithm 1

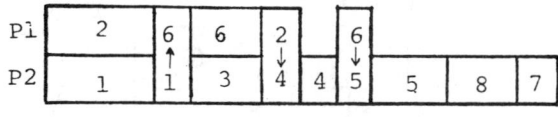

(b) heuristic algorithm 2

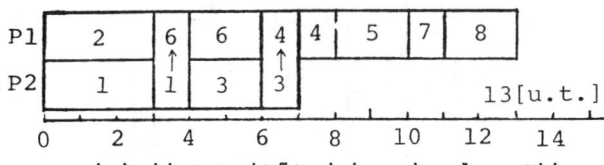

(c) the modified b-a-b algorithm

Fig. 4. The results of scheduling (for two processors).

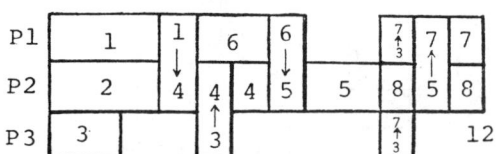

(a) heuristic algorithm 1

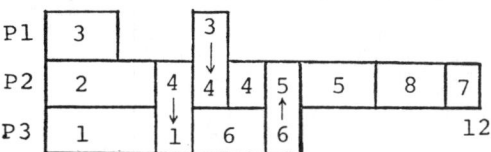

(b) heuristic algorithm 2

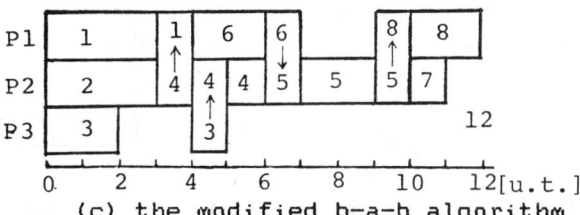

(c) the modified b-a-b algorithm

Fig. 5. The results of scheduling (for three processors).

Example 2

Fig. 6 shows the Petri net representing the tasks (mathematical operations such as additions, subtractions, multiplications, divisions, numerical integrations, square roots and trigonometric functions) necessary to perform the digital simulation of a landing aircraft on a multiprocessor digital dynamic simulator. The intrablock problem here is to find out the allocation of a plurality of processors to the tasks

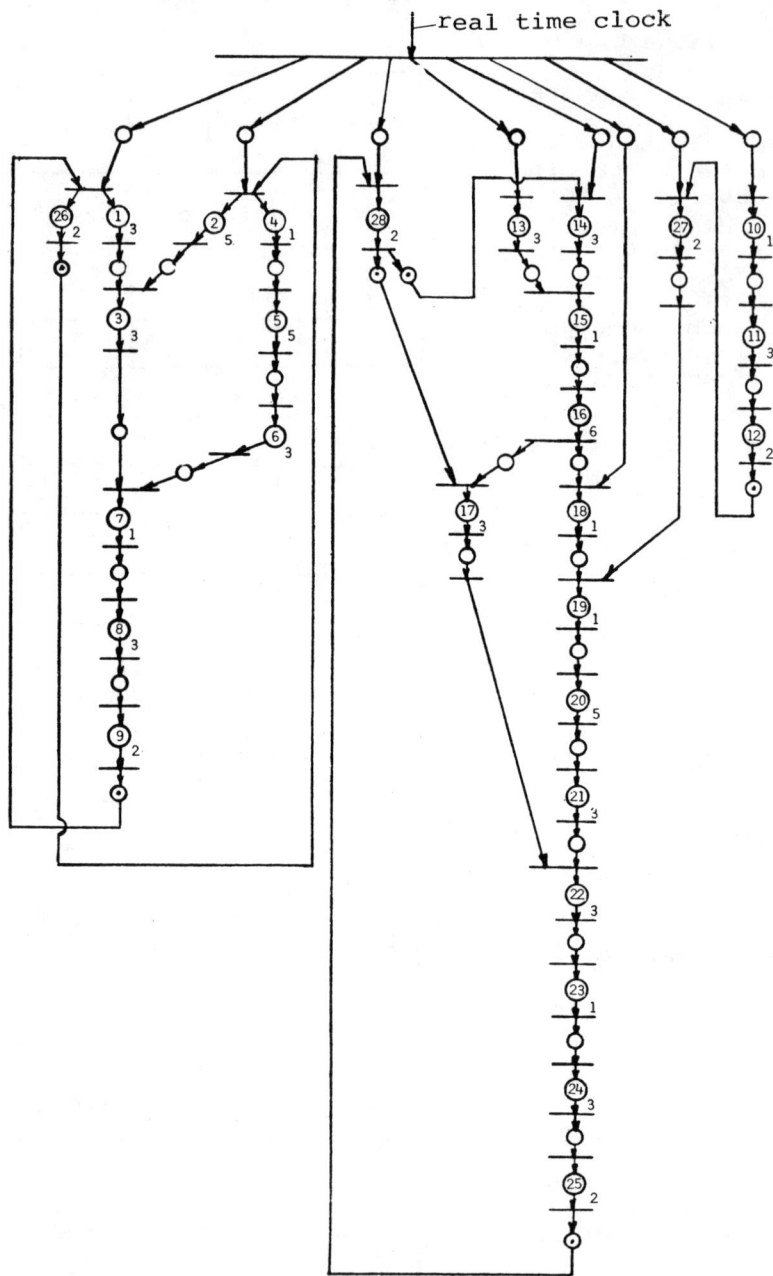

Fig. 6. The Petri net representation for the simulation of a landing aircraft.

and the optimum ordering of the tasks allocated to the respective processors including the insertion of "forced idle times".

Problems of this type seem to be of practical use in many computer control applications. For example, the operator of a DCCS may want to have some knowledge of the dynamic behaviours of the controlled plant for a set of assumed contingencies, disturbances or malfunctions of the process components. This can be realized by performing high-speed digital simulation using tightly or loosely coupled multiple computers concurrently. This type of simulation for contingency evaluation may be very helpful for "preventive control" of DCCS.

Another example occurs when the hierarchical control theory is to be applied to DCCS, where dynamical equations for process dynamics and optimal control algorithms need to be solved repeatedly for each subsystem.

The resource allocation and task scheduling for an industrial robot controlled by multiple microcomputers to compensate for the lack of computing power of monoprocessor configuration also belongs to this category of problem (Luh, 1981).

The resource allocation and task scheduling of Example 2 consisting of a total of 28 tasks are quite involved and application of the modified branch-and-bound method was prohibited because the computing time to obtain the optimum solution was found to be over half an hour on a large-scale mainframe computer (a HITAC 200, Model H). For this reason, we had to resort to the heuristic algorithms.

Fig. 7 plots the speed-up factor (the relative computing speed of multiprocessor configuration to that of monoprocessor configuration) against the number of processors m when the heuristic algorithms 1 and 2 are employed for resource allocation and task scheduling. Since optimum scheduling was not available for this particular example, the quality of the suboptimal schedules obtained by the use of the heuristic approach was checked by comparing the execution termination time with the critical path time length of the Petri net of Fig. 6. Different from the conventional task graph, some arrangements are needed to find the critical path of a Petri net because it varies with the existence or non-existence of data transfer places. Table 7 lists the critical path times for different data transfer speeds.

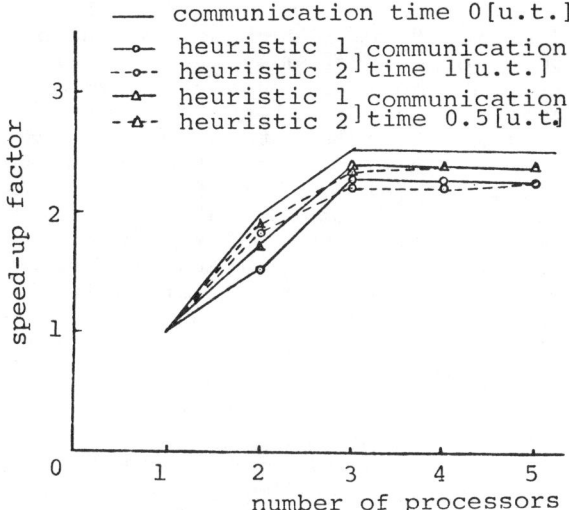

Fig. 7. The speed-up factor.

Fig. 7 shows that the speed-up factor exhibits saturation when the number of processors is three. This stems from the fact that the execution termination time has reached the critical-path time. Hence the optimum number of processors for this particular case turns out to be three.

As for the choice of the number of processors to be employed, Hu(1961), Ramamoorthy(1973) and Fernadez (1973) developed formulas to determine the lower bound of the number of processors for the case where interprocessor communication times can be ignored. The ignorance of interprocessor communication times cannot be justified in DCCS except for the case of very tightly connected multiprocessor configurations.

DISCUSSION AND CONCLUSION

The resource allocation and task scheduling problems for real-time

TABLE 7 The Critical Path Times for Different Data Transfer Speeds

data transfer time [u.t.]	critical path time [u.t.]	execution termination time [u.t]	
		heuristic 1	heuristic 2
0	29.0	29.0	29.0
0.5	31.5	31.5	32.0
1.0	34.0	34.0	35.0

control and simulation of DCCS have been studied at the inter- and intra-block levels, and solution algorithms to obtain global optimum or suboptimum solutions presented, together with illustrative numerical examples. These algorithms may be applied to advantage to the preliminary design stage of actual DCCS in conjunction with existing empirical approaches. The authors strongly feel that success or failure of quantitative approaches to the design of DCCS such as presented in this paper depends, to a large margin, on the availability of a proper system model of the controlled system with relevant numerical data. They also want to add that the heursitic algorithms for the intra-block problem have been successfully implemented on a multiprocessor digital dynamic simulator assembled by the authors and their colleagues for high-speed (real-time) simulation of a wide variety of dynamical systems.

ACKNOWLEDGEMENT

The authors wish to thank Professors Nobuyoshi Tanaka and Hitoshi Kobayashi of Saitama Institute of Technology for constant encouragement and support. Their thanks also go to Messrs. Naoshi Wakatsuki, Akira Chugo, and Hirosi Saito (presently with Toshiba Electric) for numerous suggestions and comments.

REFERENCES

Buckles, B. P. (1979). Partitioning and allocation of logical resources in a distributed computing environment. General Research Corporation Report.

Fernandez, E. and B. Bussel (1973). Bounds in the number of processors and time for multiprocessor optimal schedules. IEEE Trans. Comput., Vol. C-22, Aug., 745-751.

Gylys, V. B. and J. A. Edwards (1976). Optimal partitioning of workload for distributed systems. COMPCON 76 Fall, IEEE, 353-357.

Hu, T. C. (1961). Parallel Sequencing and assembly line problems. Oper. Res., Vol. 9, Nov., 841-848.

Luh, J. Y. S. (1981). Scheduling of distributed computer control systems for industrial robots. 3rd IFAC Workshop on Distributed Computer Control Systems. Edited by R. W. Gellie.

Ma, P. R., E. Y. S. Lee, and M. Tsuchiya (1982). A task allocation model for distributed computing systems. IEEE Trans. Comput., Vol. C-31, Jan., 41-47

Nader, A. (1979). Petri nets for real time control algorithms decomposition. IFAC Distributed Computer Control Systems, 197-210.

Ramamoorthy, C. V., K. M. Chandy, and M. J. Gonzalez (1972). Optimal Scheduling Strategies in a multiprocessor system. IEEE Trans. Comput., Vol. C-21, Feb., 137-146.

Weitzman, C. (1980). Distributed Micro/Minicomputer Systems. Prentice-Hall, Inc., Englewood Cliffs, N. J.

APPENDIX

MODIFIED BRANCH-AND-BOUND ALGORITHM

Step 1. Initialization.

Step 2. Select those tasks which have no preceding task(s) or those whose preceding tasks have all been executed. Furthermore, add to the above task group a total of p-1 "idle tasks" to account for "forced idle times" where p is the number of processors currently available. Store the number of such tasks as s and task identification numbers.

Step 3. Choose a total of p tasks from the above task group including idle tasks. The number of combinations of choosing p tasks from the task group is eqaual to sCp at the initial stage of search. From the second stage on, it is equal to the permutations sPp. Select one combination of tasks and processors and go to Step 4. When all combinations sCp are checked at the initial stage of search, the optimization algorithm is completed. (Set GOEND=1). For the second stage on, set SBACK=1 and go to Step 9 when all permutations sPp are scheduled.

Step 4. Check the possibility that the selected combination can yield a schedule having an execution termination time shorter than the current lower bound. If there exists the possibility, go to Step 6. Otherwise, go to Step 5.

Step 5. Choose a new combination and go to Step 4. When all combinations are checked, GOEND=1 and go to Step 10 or all permutations are checked, SBACK=1 and go to Step 9.

Step 6. Calculate the communication times for the allocated

tasks. In this case, the order of interprocessor communications are duly taken into account Calculate the execution times of the tasks.

Step 7. Check if allocation of all tasks has been completed. If so, go to Step 8. Otherwise, increment the stage number and find out the processors which first finish processing, then go to Step 2.

Step 8. Compare the execution termination time of the schedule just obtained and the minimum time (lower bound) schedule obtained so far. If the former is greater than or equal to the latter, go to Step 9. Otherwise the old lower bound is taken over by the new schedule just obtained and go to Step 9.

Step 9. Decrement the stage number. Go to Step 2.

Step 10. Print out the optimum schedule.

DISCUSSION

Motus: Since you insisted, Mr. Chairman, that I should have a comment, I'd say that I liked professor Narita's talk very much. I have also tried to use Petri nets for my problems and I found that they have a drawback. Namely, with Petri nets it is rather difficult to describe the situation where interacting processes, occurring in parallel, should be executed in different periods or at different frequencies.

Narita: I quite agree with you. I tried to find some suitable way for representing parallel processing procedures. There are many other ways that graphically represent the parallel processing system and I don't think that Petri net is the best solution. I purposefully included the data transfer place in addition to the process place. My primary concern is just to solve the problem, not its graphical representation.

Joudu: Can you tell us how to compute the necessary number of processors for a given task?

Narita: There have been many works on how to determine the lower bound of the number of processors but all of them have ignored interprocess communication requirements. The point here is - I included the interprocess communication time, essentially multiplying their approach. If the system is a very closely coupled one, then you can ignore the interprocess communication time and you can apply rather simple rules.

Baranov: I can't really see the difference between this problem and its solution and the classical operations research methods, for instance for task distribution.

Narita: Most of the research of this kind is centred around the monoprocessor in connection with operating systems. The problem I've just described is for DCCS. Actually I dealt with two cases: one is a loosely - coupled multiprocessor system, the computer located at a long distance. In such a case we have to take into account the communication time very exactly. Most of the work that appears in conventional handbooks on computer science deals with the so-called monoprocessor task scheduling in the case of very closely coupled systems.

Lan Jin: How do you estimate the interprocessor communications time in scheduling, as it depends on many factors which are difficult to estimate. Can you include this time in the execution time of the related task?

Narita: The interprocessor communications time has no fixed value, it has stochastic value. If you want to treat the problem exactly you'll probably have to choose some stochastic approach, maybe you'll have to solve some stochastic non-linear equations. We employ the mean values for the interprocessor communications time and for control applications and we can estimate it quite precisely, with small variations. So, this is the reason why I assume that the interprocessor communications time is deterministic.

Lan Jin: Is it possible to obtain automatically the Petri net representation diagram of the multitask system which you have used as the starting point of your scheduling algorithm?

Narita: In this particular application of the landing aircraft I said it was the block diagram type of representation and we assigned the processing time, for example, ten time units for integration, three for addition, etc. We had a one-to-one correspondence between the block diagram representation and Petri net. But, strictly speaking, we have to take into account some other factors to get the truly optimum results. The results I just obtained are sub-optimum.

Rodd: I think that was a question we'd all like an answer to. That is how do you start giving a problem? How do you represent it in a form you can feed into a Petri net-type representation? I think we are a long way off that problem. At the moment I think it is a purely manual procedure.

Narita: That is right. We need some kind of a human intervention.

ON THE DISTRIBUTION OF TASKS IN AUTOMATION SYSTEMS

M. Ollus and B. Wahlström

Electrical Engineering Laboratory, Technical Research Centre of Finland, SF-02150 Espoo 15, Finland

Abstract. One of the most important phases in the development of automation systems is the decision of how to distribute the functions within the system. This has to be decided at an early stage of the development project, because changes of the main principles are difficult and tedious. The intended functions and the distribution principle should be written down in detail in a specification, which is used to guide the development of the system. The specifications should rely on a thorough analysis of the needs for all users of the system. Starting from these needs a functional specification of the system and its components can be made. The functions can be divided into subfunctions or modules performing well defined tasks with simple and clear communication between the modules. The module realization can be done by different types of hardware and software and the decision about the realization should rely on the functional needs. The use of this type of top-down approach has been used in the development of two different automation systems.

Keywords. Computer organization; direct digital control; industrial control; man-machine systems; microprocessors; mining; multiprocessing systems; optical variables control; process control.

INTRODUCTION

During the recent years the use of distributed systems for supervision and control of technical processes has increased rapidly especially in the process industry, but distributed systems are also coming in the manufacturing industry. This trend can be seen as a result of the combination of advances in the following fields
- process technology
- electronics
- methods

In the process technology the growing requirements on the economy and product quality usually lead to larger processes with decreased product tolerances. At the same time there is an increasing need for better safety and reliability usually due to new regulations.

Hence the amount of information to be processed in the operation of the process is rapidly increasing and the operation must be more exact and safe. A consequence of these requirements is an attempt to distribute the function of the growing automation systems to several independent data processing units in order to enable local use and processing of data. As a result of this distribution the need of data transmission can be reduced and usually the availability of the process will increase. The great amount of information, which has to be presented to the operators, has effects on the size of the control room, which can be reduced using CRTs to present the information on a time shared basis.

The well known development in microelectronics has enabled the economical utilization of microprocessors for rather complicated calculations. Hence, the implementation of new control methods, which have been developed using system theory, has been possible.

In the development of distributed automation systems the decision of how to distribute the functions within the system is the most important phase, because the function of the system will depend on how well this division has succeeded. Changes of the main principles after the specification phase, are later on both difficult and tedious.

The Electrical Engineering Laboratory at the Technical Research Centre of Finland has participated in the development of automation systems for different purposes. A systematic, hierarchical approach to the specification procedure has appeared to be very useful in these development works. This paper describes this design procedure, where the main functional criteria for the systems are coming from the problem to be solved and the users' needs. The principles of the distribution with these functional needs as basis are also discussed. As examples of developed distributed systems two different microprocessor based systems are presented.

DESIGN BASIS

Main requirements

The design of any automation system starts with a thorough analysis of the intended use of the system. In this analysis the main interest is given to the wanted functional behaviour of the system. In this stage the technical solutions are not essential although available technology to some extent will give limits to the available functions.

An automation system is an operation tool or an interface between the users and the process to which it is applied. Hence, the requirement on the system is mainly defined by the users and the application process.

Different persons, such as operators, maintenance personnel, system planners, managers, etc. are in contact with the automation system and are some kind of users. When defining the main functions of the system all users should be taken into account and the interface to the users should be planned in accordance with the users' needs, which will affect on many levels of the system. As general needs of all users may be an easy use and profitability. Because the use of the system varies widely with the users these needs lead to different requirements on the system.

The operator wants clear information about the state of the process and ability to easily transfer the process to new states without any need to understand the complex algorithms of the system.

The system planner uses the system mainly when adapting it to new processes or when there is a need to implement new features in the application. He has a need to understand the available algorithms and to easily adapt them to his application. Usually he wants to do this without need to understand computer programming.

The profitability of the system usually means that it should be tailored just for the application, because in industrial applications generality usually causes extra costs. In a general purpose instrumentation system this question has to be considered very carefully, because the system has to be applied to very different processes which means a rather great amount of generality. Also from the maintenance point of view the use of general modules may be preferable.

The process to which the system should be applied defines many of the functions of the system. For example, the control principles are defined by the wanted use and the dynamics of the process. From the control principles many of the functions of the automation system can be defined. Such are connection to other information processing systems, interlocking needs and wanted information processing capacity.

Also the environment of the application causes some requirements on the design of the system. One usual requirement is that the system should work reliably 24 h a day in industrial circumstances.

Functional specifications

The functional specifications can be written using the requirements of the system. In these the intended functions are written down in detail. Hence the functional specifications can be used as a guide in the development of the system.

In the specifications the system can be broken down into functional modules. When defining the modules the main criteria are
- well defined tasks
- well defined internal communication
- simplicity.

The division into modules should be done in a "natural" way so that the tasks of the modules are easily understood. The communication between the modules is then usually well defined and the need of communication can then be minimized. When this approach is used the structure is usually also simple.

Although the structure of the system is built into the functional specifications the modules are still only functionally given. The modularization can, however, serve the administrative needs of the system design. The modules are also basic building blocks of the system.

SYSTEM REALIZATION

System structure

The geographical distribution of the system to local processing units is done taking into account the needs of interactions in the system. A natural approach is to try to transmit only necessary data through the system and to process as much information as possible locally where the information is generated.

The tasks of the modules can be performed by hardware and software. The decision between hardware and software implementation of the functions is not straight-forward. The decision about the realization has to be done taking into account the ability to perform the wanted tasks and the costs. The maintainability and the ease of use are also very important factors for the decision.

A hardware solution usually gives a higher throughput while a software implementation is easier to maintain and change. Many tasks may be more efficiently performed by hardware than by software especially when real-time function is needed.

Hardware realization

The use of hardware realization in automation systems is usually feasible in modules, which are connected to the process. These modules get usually much processinformation and the necessary processing speed is defined by the speed and dynamics of the process. On this level there is also a need to tailor the module for the application, which may require special hardware arrangements. This is however no general rule and the decision of the division between hardware and software has to be done separately for every module.

A hardware module may be performed be general hardware or special hardware. General hardware consists of modules which can be used for different applications, to which they are tailored through some kind of manipulation, which can be done by, for example, switches, jumpers, registers, etc. The use of general hardware modules is preferable when the same system or module is used in different applications or in systems which are not produces in large series. A general purpose hardware module can then be used for different purposes within the system. The advantages of a general module are that maintenance is easier and the production series are larger. The module contains, however, facilities which are not needed in a certain application. This may affect the costs.

A special purpose hardware module is tailored for the application and contains only necessary functions. This approach is suitable if a tailored approach is necessary of if the module is used to perform a common task in the system, when maintenance will be easier and the costs can be cut down due to larger series.

Software realization

Software realization is usually suitable for modules performing data processing tasks, for example, mathematical algorithms. The realization could be done either by the system software or by the application software. The system software consists of standard modules which are available in different applications and are some kind of basic software elements of the system. The system modules are fitted to a specific application by the application software. When defining what type of software is to be used, the function of the system and the users' needs has to be considered.

The system software offers the tools which are available for the system planner when he does the application. This is done by connection of the modules to a working application. In a large automation system the connection may contain many different phases and an easy use means that the planner can make it convienantly. The functions and the structure of the modules define very much of the planning routines. These are easier if all modules are used in a similar way and the module structure is simple. Some kind of computer aided application planning and use of data bases for the planning information are useful in many applications.

Although some software module performs a general task, for example, a mathematical algorithm, tailoring of the method to serve the needs of the automation system is usually necessary. Such solutions are realizations which work accurately only around a working point or where other system characteristics are taken into account when defining the boundaries of the realization. This means that the module is dedicated for a certain use in just the automation system. This fact enables usually simple and effective realizations without extra generality.

In the realization of a software module the used programming language has to be decided. When comparing the software development costs with hardware costs, it can be concluded that the use of high level languages is preferable. The choice of language depends, however, also on many other factors, such as the used processors, available software support etc.

System development procedure

The development of an automation system contains a sequence of different steps. Although the system specification is important there is a lot of other work to be done before the system is really working. A top down approach can however, be used in the design procedure. For example, the following steps could be regarded to describe this procedure:
- definition of functional requirements
- functional design
- functional specification
- distribution of tasks
- internal communication definition
- module design
- module specification
- module implementation
- module testing
- system integration
- system testing
- system documentation.

A DIGITAL INSTRUMENTATION SYSTEM

System features

The development of a new digital instrumentation system started with the definition of the intended scope of the system. The system should mainly be used for different applications in the process industry. It should then have the same capacities as conventional systems. Moreover the use of newer technics enables the integration of new functions in the same system. Hence, it was in an early stage of the development project decided that the system should contain both conventional control and binary control of pumps and valves.

The intended system can from the information processing point of view be divided into the following functional parts

- process interface
- process control
- man/machine interface
- data transmission.

The process interface connects the system to the process and this is done using different transmitters and actuators which in the new system should have features as in conventional systems. Hence, the system should contain at least the following process connections:
- standard analog input and output
- standard binary input and output
- on/off control of motors and valves
- valve control.

Moreover, there should be possibilities to adapt also other newer process connections such as pulse inputs.

The information from the process interface is processed by the process control part which can be an arbitrary mixture of at least the following functions
- analog control
- logical control
- sequence control.

The analog control contains all conventional analog signal manipulations such as different types of PID controllers and other mathematical signal manipulations. The use of digital calculations and memories enables for example, the calculations of very complicated mathematical functions in the system. Hence, it was decided to include new functions which can be used in the process industry.

The man/machine interface provides the communication equipment for the operator and was chosen to be mainly performed by visual displays and functional keyboards. It should however also be possible to use conventional control room equipment such as recorders, indicators, manual control stations, etc.

Information transmission between the control room and the process is taken care of by the data transmission system, which also manages the information exchange between other subsystems.

The development of a new digital instrumentation system with these features was done in cooperation with a process instrumentation manufacturer. The system has been described in more detail by Wahlström and coworkers 1982.

System structure

The realized system includes all functions mentioned above. Hence it includes a large number of both hardware and software modules. The functions of the system are distributed both geographically and hierarcially which can be seen from the system lay-out in Fig. 1. The functions are performed in so called substations which are of the following types
- control room stations
- traffic stations
- process stations.

The control room stations are usually placed in the control room. The traffic stations, which take care of the bus traffic, are placed in the room for equipment. The process stations which are used for continuous and sequence control, are usually placed near to the subprocess they are controlling.

For the communication between the substations a coxial cable is used. One branch of the bus may contain up to 14 substations and one bus controller can control two buses. The bus controllers can be connected together in the traffic station which enables data transfer between different bus controllers.

For continuous and sequence control the process stations are used and they take also care of the connections to the process. Control on a higher level, for example, co-ordinating control of different control loops can be performed in the same stations but usually it is done in a separate station on the serial bus or a process computer which is connected to the system.

Operation is normally done in the control room. Each process station has, however, an asynchronous transmission channel to which local operation equipment can be connected.

Hardware realization

All substations in the system are realized around a central processing unit (CPU). This is built using the Zilog Z-80 microprocessor and the CPU board includes 12 Kbytes EPROM and 4 k bytes RAM memories and other necessary electronics. For communication a serial communication unit (SCU) is used which contains two 250 kbaud and one 500 kbaud synchronous channels and an asynchronous standard channel. The synchronous channels are using the HDLC protocol with a biphase modulation for the 250 kbaud coaxial cable connections and no modulation in the 500 kbaud twin wire connection.

The use of the CPU units in the process station is given in Fig. 2, where the CPU is provided with two memory expansion unit. The RAM units are protected against power failures by accumulators on the boards.

In many applications redundant hardware systems are used. Hence, the CPU and memories are doubled in the process station in Fig. 2. There is an automatic switch over in case of failures in the processing units. The serial bus can also be doubled with an automatic connection in case of failures.

For connections to the process different types of interface units have been developed. These units communicate with the CPU using a parallell bus. In these units hardware fault recognition and localization are used. The maximum number of different process interface units in one process station is 128 units. The load of the information processing in the process station may, however, give

other restrictions on the interface and in a typical process industry application the interface contains
- 32 analog inputs
- 16 analog outputs
- 32 binary inputs (4 units of 8 inputs)
- 32 binary outputs (4 units of 8 outputs)
- 32 motor controls
- 16 motor valve controls.

In the other stations other types of interfaces are used. For example, in the control room stations interfaces to keyboards, videounits, etc. are used.

Software realization

The system software contains all software modules which perform standard tasks of the system. These modules contain, for example
- operating system
- communication software
- control room software
- control software.

The operating system has a simple synchronous structure, where the tasks are executed during fixed execution intervals. Using this structure there are no needs for the evaluation of the priorities of concurrent tasks. The synchronous structure also simplifies the synchronization of separate CPU:s and message collisions can easily be avoided. The overhead of this operating system is naturally very small. The operating system contains three different execution bands which have updating rates of 40 ms, 400 ms and 2 s. The bands are used to obtain the sampling rates for the control loops and the control room displays. The operating system can handle a total of 255 tasks and about 20 % of this amount is used for system software. The rest is available for later extensions of the system.

The communication software was built around the HDLC protocol and was tailored for typical needs in process automation. Hence, there are the following types of messages
- block messages (35 bytes),
- random messages (7 bytes),
- commands, receipt (1 byte).

Processinformation is mainly brought to the control room or the process computer by block messages which contain a block identification and 32 bytes of process information. From every process station 8 block messages can be transmitted and the regular updating interval is 2 s.

The random messages are used for transmission of randomly generated data such as control room commands etc. These short messages include all address information necessary to route the messages to the proper destination and to identify the memory location within the addressed substation. The messages also contain information about the origin of the messages in order to enable a check of whether the originator is allowed to operate on the addressed memory location.

For system commands from the bus controller (traffic station) one byte messages are used. This type of message is also used to acknowledge the received data.

The transmission errors are checked by the HDLC protocol using two cyclical redundancy check (CRC) bytes. The reliability of this check is improved by the use of fixed lengths of the messages. All short messages are acknowledged by a receipt message which indicates whether the message contains errors. The transmission of an errorness block message is **not** acknowledged but an errorness message is marked invalid. The regular updating of the message ensures that a new message normally appears within 2 s. Because the lowest control **levels** are at the process station a transmission error does not affect the operation of the process.

The control room software generates the displays and accepts operation information from the keyboards. The control room contains the following displays
- general display
- group display
- trend display
- alarm display
- sequence program display
- sequence step display.

The general display gives an overview of the process status and consists of eight display fields. One field contains the information of one group display and if all groups are, for example, control loops, one general display gives the overview information of 64 control loops.

In the group display there are eight fields and to each field one of the following types of information can be connected
- one controller
- two analog signals
- group or individual control of four motors.

In the controller display local, remote and computer setpoints are available. Moreover the controlled signal and the control output are displayed. These signals are given both as bars and as digital values. In the display area are also a control area for display and control of control parameters and alarm limits.

Analog signals are also displayed both as bars and numerically and in the control area alarm limits are displayed and controlled.

The state of four motors can be displayed in one field. The motors can be operated separately and information about individual motors is displayd. The information consists of motor status, interlocking and fault indications, etc. The motors may also be controlled by group control and then they are started and stopped in some programmed order by one group order. In this case the group status is also displayed.

The trend displays have capacity to present signal information in a graphical form. In one display page there may be four different signals and three different time scales are available (30 min, 8 h and 32 h).

Alarm displays are used to display up to 30 alarms on one page. Alarms are also indicated by colours (yellow and red) in the other displays.

For the sequence control the program display is an overview display showing the status of the sequence programs. For more exact information about the status the step display can be used. This display contains the status of all conditions of the step.

In the development of the control software an easy transfer for the application planner from the use of conventional analog systems to the new digital system was wanted. In order to achieve this goal control software was built of modules called functional blocks which are similar to the blocks of control schemes. The system program is an interpreter which is using a list of these functional blocks to specify subsequent module calls. The functional block is given in Fig. 3. The value of the input signals x and y together with the value of the pseudoaccumulator "acc" are used to calculate a new value for the output signal z. This value is also left in the pseudoaccumulator for possible use in subsequent blocks. The application planner makes the interconnection of the blocks in a similar way as the wiring diagrams of analog control systems. In Fig. 4 and Fig. 5 simple examples of the use of the functional blocks are given.

The control software communicates with other software parts using the signal areas. This approach where the signal areas are buffers enables and independent execution of the different software parts.

Application programming

The system software can be applied to the application process by the application programming which has to be done for all parts of the system software. Due to the structure of the system software this programming is mainly definition of some parameters for each part and for this reason some fill in the blanks forms have been developed.

Forms have been developed for the definition of the following items,
- signals
- messages
- control configuration
- sequence control configuration
- displays.

The definitions are given by mnemonic commands and in the signal definitions absolute, relative or symbolic signal names can be used.

The compiled and linked application programs have a structure which is very similar to the source code. Hence, the function can easily be checked in the application and temporary changes can be made on-line, for example, during the start-up phase.

A SEEING AND LEARNING SORTER

System features

Together with a Finnish company with different types of products in the bulding industry an automatic sorting system was developed. The sorting is based on visual information and the system should be applicable for different sorting purposes. It was also seen that there was a need for high capacity.

From these requirements if could be seen that the system which has been described in detail by Mäenpää and coworkers 1982 should have the following features
- modular
- adaptable
- hardware preprocessing
- parallel processing.

The need to use the system for different applications which were not specified indicated that the system should have a modular structure with a possibility to replace modules.

The system should easily be applied to new situations without too much tuning needs. This is important because the visual sorting algorithms may be very difficult to understand for the operators.

Visual images contains very much information. Software processing of this type of information may give capacity problems. Hence, a hardware preprocessing approach was chosen. During this phase the information could be reduced and the need for information transmission could be reduced.

From the capacity requirements it could also be seen that parallel software processing should give some advantages.

Realization

The sorting task can be divided into the following subtasks
- measurement
- classification
- sorting.

In the sorter the same functions are specified. This enables a natural division into subsystems which is seen in the structure of the system (cf. Fig. 6).

The measurement is done by a line-scan solid state camera, which scans a conveyor belt, where objects to be sorted occur randomly. The objects are detected from the background by the sweep analyzer. This analyzer also performs extraction of wanted object features. The analyzer is realized by hardware and the extracted features can be changed by replacing plug-in cards in the sweep analyzer.

The classification of the objects is done by the classification units. Each classification unit has its own object which it recognizes during the sweep. The classification is done by combining the features from different sweeps and comparing the features of the whole object to a prototype vector.

The classification units are realized by one-chip microprocessors and necessary electronics and external memory. The number of classification units in the system is dependent of the number of objects occuring in parallel on the conveyor belt, but the maximun number is 255 units.

The sorting of classified objects is supervised by the master processor which controls the reject mechanism. The same master processor also supervises the function of the classification units and the sweep analyzer and it takes care of the communication with the operation panels. The master processor is realized using a single board microprocessor.

Learning ability

In order to make the adaption to new circumstances the system calculates itself the sorting criteria. The operator can switch the system into the learning mode when he wants. In this mode typical samples of the classification classes can be fed to the system which calculates the prototype vectors from the features of the samples. These prototype vectors are used in the classification mode.

If the user is not quite satisfied with the sorting result he can change the prototype vectors in wanted directions by small steps using his operation panels.

By these arrangements the system needs very little application programming. The main adaption to new applications is done by the use of different features extraction modules. These modules are realized by hardware in the sweep analyzer. On the software side application programming is only the determination of some parameters for the software.

CONCLUSIONS

The described development projects have both resulted in commercially available systems. The general purpose instrumentation system has been applied for about 70 different processes since 1979. The main users are within the pulp and paper industry but other application fields are the petrochemical, chemical, food and power industry. In the largest application there are more than 100 CPUs. For example, the first application of the system was in the paper industry and contained 101 CPUs.

The first application of the sorter was for rock sorting, where it was used for sorting of lime stones. One licence agreement for rock sorting applications was achieved between an international company and the customer. In connection with this agreement six new systems were ordered. Two of these have been working since August 1981 with very satisfactory results.

A successful distribution of tasks in a system requires engineering skills to see
- what is relevant
- where are the possibilities
- where are the restrictions.

To make these judgements there are no general rules and the development process can be seen as tree of design decisions. At each branching point a design decision is made and this decision requires a judgement skill because the decisions are basically irreversible and the final information needed for the decision is available only at the end of the project.

REFERENCES

Mäenpää, I., Malinen, P., Ollus, M., Saukkonen, E., Wahlström, B and Uotila, E (1982). A computer systems for on-line sorting based on visual images. 6th International Conference on Automated Inspection and Product Control. 27-29 April 1982, Birmingham, U.K.

Wahlström, B., Juusela, A., Ollus, M., Närväinen, P., Lehmus I., and Lönnqvist, P. (1982). A distributed control system and its application to a board mill. Automatica 18 (accepted for publication).

Damatic

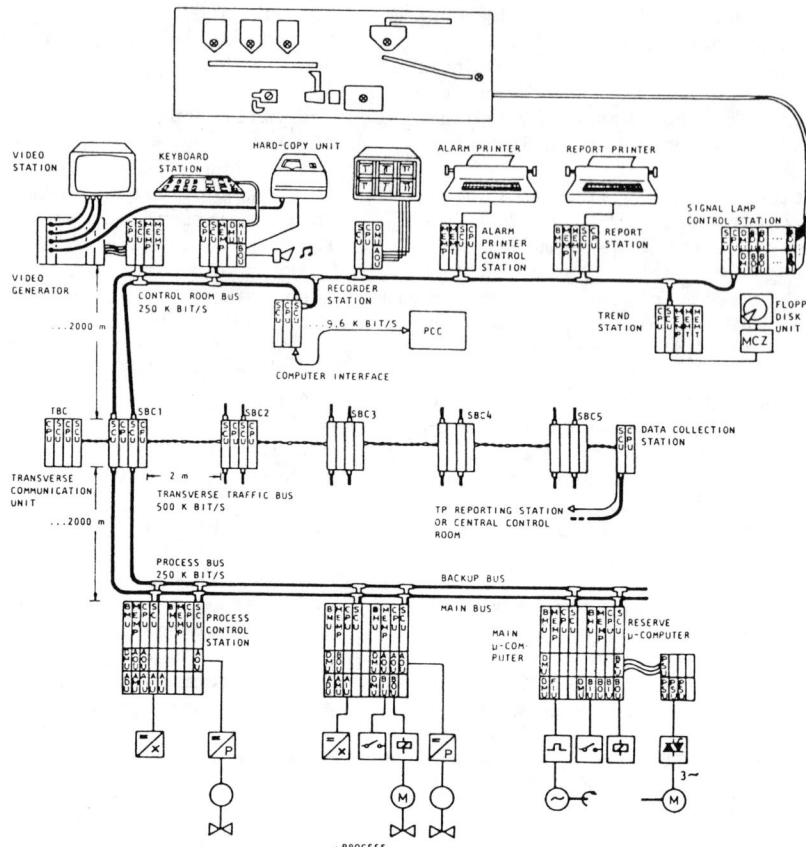

Fig. 1. General lay-out of the Damatic system

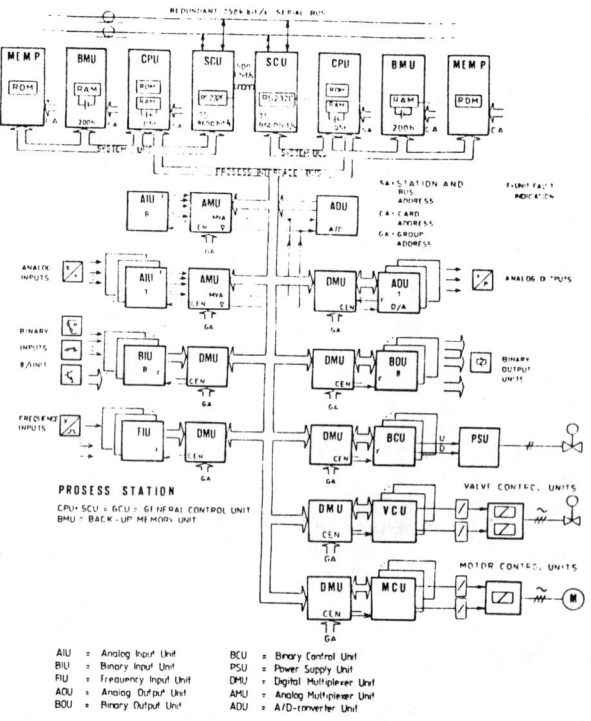

Fig. 2. Structure of the process station

Fig. 3. The functional block

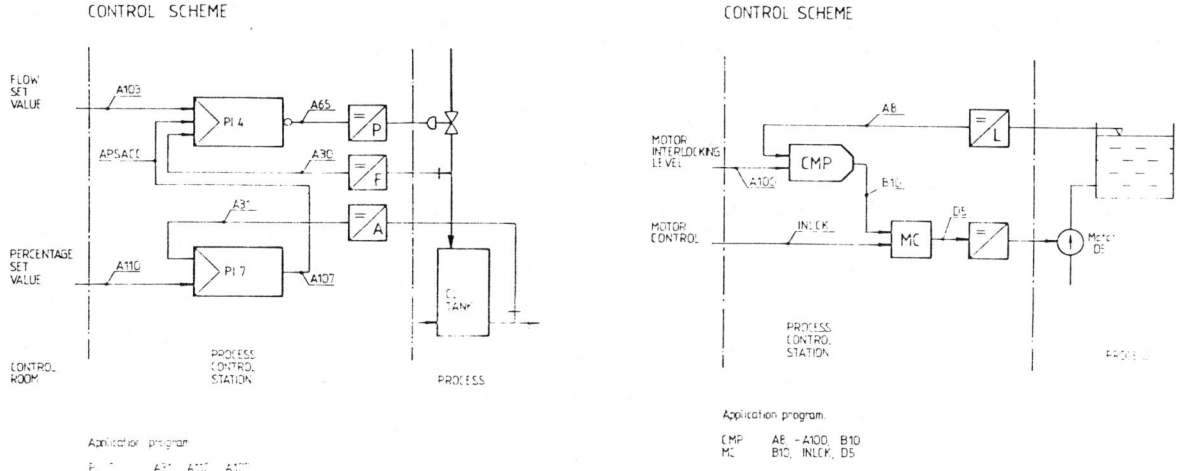

Fig. 4. The control of the clorine percentage by a cascade connection of controllers

Fig. 5. Pump interlocking from the tank level

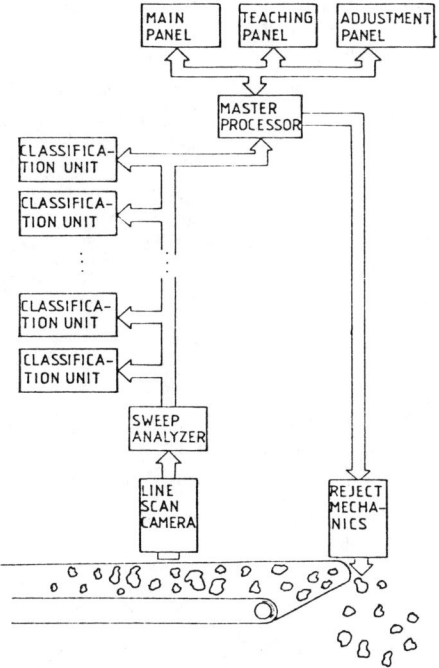

Fig. 6. Structure of the sorter

DISCUSSION

Shreiber: Could you give an example of how one of the managers can give a task to your systems? What language is used for that?

Ollus: The system itself does not have any facility for the manager because it is an instrumentation system. But the system can be connected to the process computer which, in its turn, can be connected to the factory computers for management information.

Tavast: Each data communication module in DAMATIC may be incident to three different buses. What media access methods are used in each of them and what priority assignment and addressing possibilities are there throughout the system?

Ollus: There are no priorities on the data communication, this is a synchronous cyclic system. You can address throughout the system but every substation is given a slot of time. During this time the communication with this substation is done. This is a very simple structure of the communication system.

Werner: Can you give any information about the reliability of your system, for example, the MTBF of the process station.

Ollus: I don't have any figures and I think that they are not counted. But there are one hundred installations of the DAMATIC system and from these, I think, you could get some information. The information that I have got is that it is more reliable than an analog instrumentation system.

Inamoto: I am interested in your classification unit. Is this classification unit a standard one-board processor that could be connected to any other computer after some modification, if necessary? Secondly, does the system dynamically allocate the tasks to the classification units connected to the main bus?

Ollus: I think I'll start with your second question. The classification units are dynamically allocated by the multiprocessor. Multiple classification units are in a waiting queue if they do not have any jobs. The number of these units needed in the system depends on the number of objects occurring in parallel on the conveyor belt. You must have more classification units in the system than there are parallel objects on the belt because one classification unit is handling one single object; the units don't have any specific places on the conveyor. Thus, when a new object appears in front of the camera, the master processor sees that no classification unit has taken this object, and the job is given to a classification unit in the waiting queue. After that the classification unit knows that this is its object. When the classification is done and the information is given to the master processor, this classification unit will return to the waiting queue again.

As to your first question, classification units are realized using a single board microcomputer around INTEL 8748 and there is a lot of special hardware on the same board. The master processor is INTEL 8080. This system is commercially available for rock sorting systems from a Canadian company. But if you are talking about some general standards, then there are no standards. It is a system made for this type of application.

Inamoto: What is the approximate response time or calculation time for one classification job?

Ollus: The response time interval between sweeps is one millisecond. During this millisecond the features are extracted for the objects which are in front of the camera. Nobody has measured the real classification time after that. During this millisecond the features are obtained using a recursive algorithm. After the object has gone, the classification is done, using the very simple nearest neighbouring algorithm, and it takes no more than one or two milliseconds.

SOFTWARE TEST FACILITIES WITH DISTRIBUTED ARCHITECTURE

K. Takezawa

Fuchu Works, Toshiba Corp. 1, Toshiba-cho, Fuchu-shi, Tokyo 183, Japan

Abstract: Applications of industrial computers are covering a wide variety of requirements such as labor saving, energy saving, protective monitoring and so on. As they come into use in various industrial fields, their hardware or software malfunction exerts a greater influence on the society, taking on extra urgency. The software quality assurance efforts we promote are twofold, one is production technology approach represented by software test techniques and production tool support throughout a life cycle, and the other is quality improvement through the quantitative grasp and control of software quality.

This paper describes the summary of the SWB (Software Work Bench) system and the test system which is one of the subsystems of the SWB system.

Keywords: Software test system; Software engineering; Software work bench system.

INTRODUCTION

Application fields of our industrial computers are far and wide. They include electric power systems such as nuclear power generation systems and power dispatching systems, public utility systems for waterworks and sewage treatment, building equipment management, etc., industrial systems like motor drive systems and energy saving systems, traffic systems including railroad systems and new transportation systems, and so on.

Approximately 2,000 software engineers of ours are engaged in industrial process control computer software production at the rate of 300 sets/year or 60,000 kilosteps/year for these fields in total. In order to produce such a large volume of software with high reliability and productivity we built the software factory and have developed software tools.

Fig. 1 shows an overview of the software factory. The two front buildings provide software design and manufacturing facilities and the building in the back provides software testing facilities. In this software factory, physical environments such as intelligent terminals, distributed processing computers, general purpose computers, test support computers, etc. and software development tools are installed in the manner that provide optimal circumstances for system analysts, system designers and programmers.

T-SWB which aims to provide integrated support throughout software life cycle as a software production tool, unites these computers organically, and is effective to the improvement of quality and productivity of our software production.

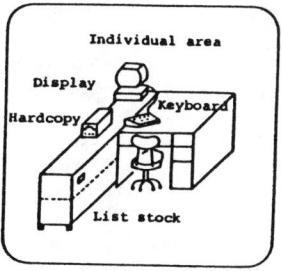

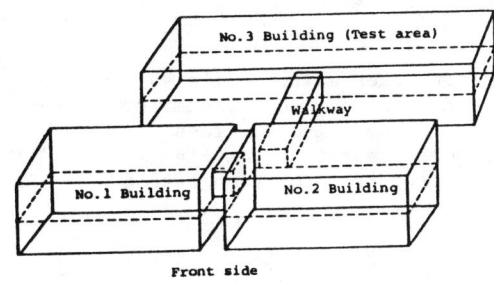

Fig. 1 An Overhead View of the Software Factory

T-SWB SYSTEM

T-SWB aims to improve quality and productivity by supporting software production with a consistent philosophy throughout software life cycle. Starting with the development of support tools for the implementation stage, we have almost developed tools for the test and maintenance stages and we are now developing systems to support the software design stage.

The overall T-SWB configuration is shown in Figure 2, and the software life cycle and representative support tools are in Figure 3.

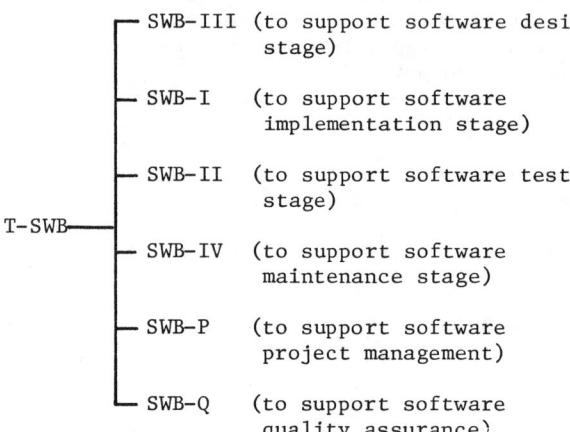

Fig. 2 T-SWB Configuration

SWB-I: This is the first subsystem developed and integrated for T-SWB and consists of support tools for use in the software implementation stage. The majority of these tools are executed by cross processing on general purpose computers. In our software factory, we use many types of mini-computers and micro computers for optimal plant control. The SWB-I tools provide a uniform interface, so software engineers can use them with a unique procedure.

SWB-I includes editor, file system, cross compilers, cross assemblers, cross linkage editors, cross debuggers, test tools, system generators, etc. for each target computer.

As for language processing for microcomputers whose demand is rapidly escalating these days, microcomputer language processors on which compilation and/or linkage editting etc. are executed are connected to the host computers.

SWB-II: This subsystem supports the test stage and its details are described in the next chapter.

SWB-III: This subsystem consists of tools and methodologies for software design stage support. The software design stage is divided into four phases, requirement definition/ system design, software design, program structure design and program detail design, and SWB-III supports each of them by defining formal design languages and design charts.

SWB-IV: This subsystem supports the maintenance stage, that is, maintenance information management and program library management. It consists of a program information retrieval system and database management system. Modification history, version control, error history and other information for each program are able to be retrieved easily.

SWB-P: This subsystem supports project management in software production. We are a private company; it is necessary to realize moderate profit by producing computer software. Production without a profit is out of the question. To control a lot of project properly, SWB-P supports project progress control, cost estimation and cost control.

SWB-Q: This subsystem supports software quality control, analyzing actual quality data collected using various tools. We use coverage as the metric of quality, path coverage, module coverage are calculated by this subsystem.

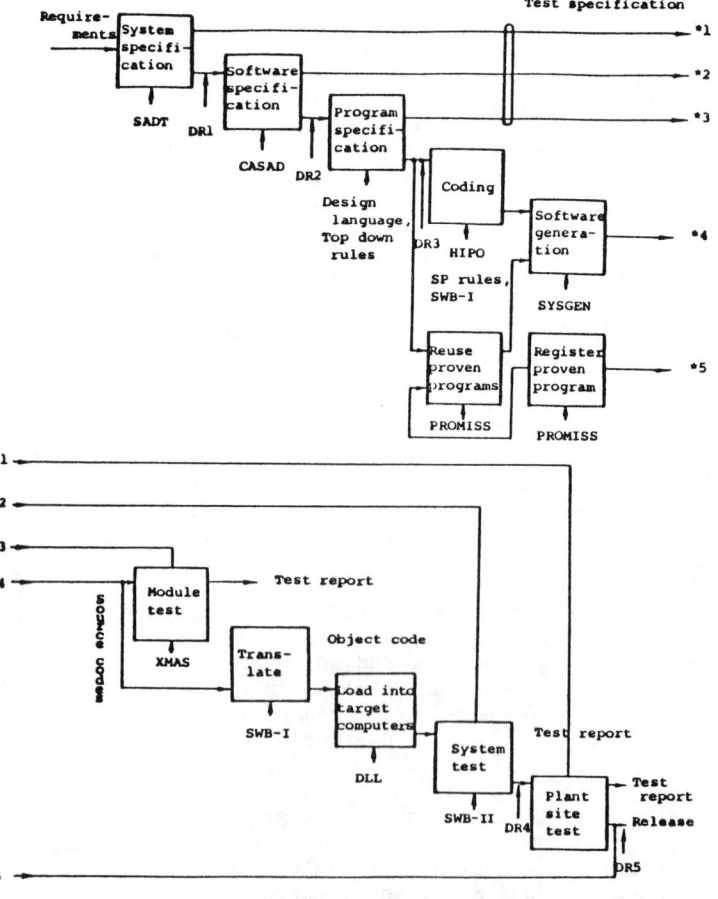

Fig. 3 Software Production Life Cycle, Milestones and Tools/Methodologies

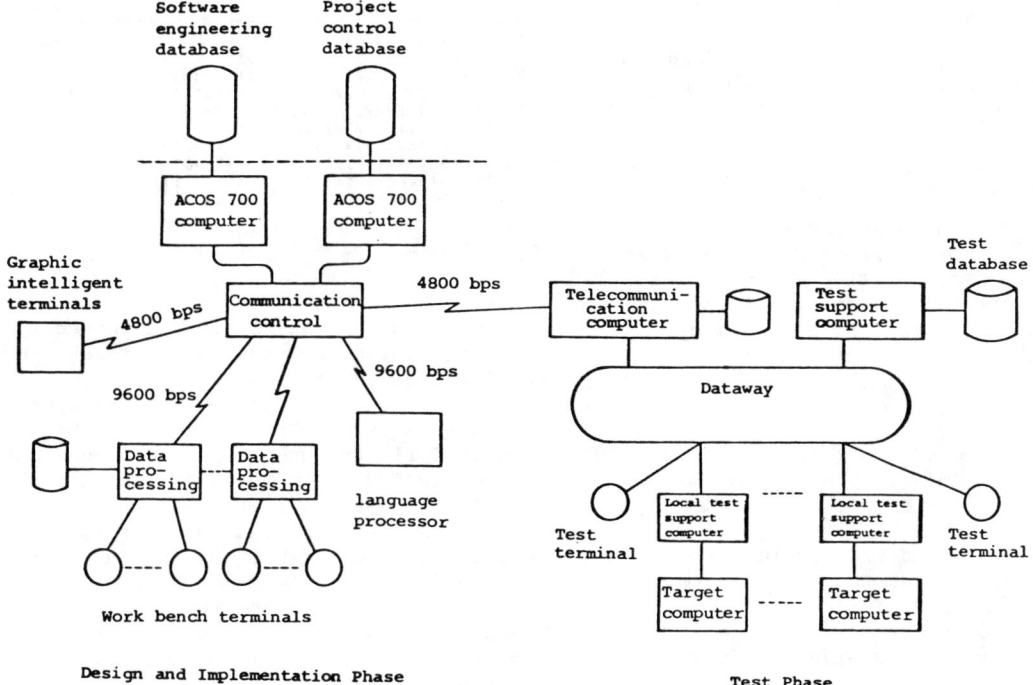

Fig. 4 T-SWB System Configuration

T-SWB SYSTEM CONFIGURATION

Fig. 4 shows the hardware configuration of the T-SWB described in the preceding chapter. Broadly speaking, the design/implementation and test phases are divided functionally and general purpose computers equipped with software engineering data bases provide collective information management between the two.

The left half of the T-SWB configuration shown in Fig. 4 supports design/production phases, while the right half supports the test phase. Three ACOS 700 have been installed to perform load sharing. The work bench terminals are connected with ACOS 700s through the data processing computer and communication controller. The connection procedure of terminals determines ACOS 700 selection.

Many intelligent terminals are connected with ACOS 700s through the communication controller. They perform data input and local editing and transmit information to the selected ACOS 700 for batch processing.

Computers which support the test phase are connected with the ACOS 700s through the telecommunication controller. Object modules and load modules generated by ACOS 700 are loaded into the target computer through the telecommunication computer and dataway, and tested. Test system details are introduced in a later chapter.

TEST FUNCTION WITH T-SWB

Software test for industrial computer is divided into four steps, unit test mainly check of the logic of individual programs, integration test check of interface between tasks, total functional test to conclude factory test activities and on-site tuning check of the performance of an industrial computer system connected with an actual plant.

T-SWB provides support tools for each of these four test steps.

Unit Test

The implementation stage is performed on an ACOS 700 large scale general-purpose computer as a virtual machine. XMAS 70 is a test tool for simulation execution using the object program generated on the virtual machine. Individual program logic is tested in unit test, so the environments installed permit many system designers and/or programmers to make unit tests at a time.

XMAS 70 test functions are as follows,

(1) Type check of arguments between subroutines and argument quantity check

(2) Check of accessibility to other areas than allocated

(3) Type check of assignment statement right side and left side variables

(4) Check of system resources utilization

XMAS 70 is also equipped with debugging functions and operates both in the batch and interactive mode.

Integration Test

Interfaces between tasks are mainly tested in this step. Object programs which have finished the unit test step are downline loaded into the target computer through the telecommunication computer and dataway. These are subjected to integration test using ASSIST, a test tool on the target computer.

The operation procedure of ASSIST is the same as that of XMAS 70. ASSIST has debugging functions in the same way as XMAS 70.

In XMAS 70, files and process inputs/outputs are all simulated by means of software. But it is possible for ASSIST to use actual files and P I/O (process I/O device) in the integration test in addition to software simulation.

Total Test

This test is the final step of factory test activities.
The SWB-II test system is provided as a support tool. It describes a model of subject plant behavior using a simulation language so that it is tested in a closed loop using a target machine.

On-Site Tuning

After factory tests are over, industrial computer software testing enters this on-site tuning step which mainly aims to test software performance at an actual plant. PROMOS (PRocess oriented MOnitoring System) is provided as the tool for actually measuring, analyzing and evaluating performance data such as CPU load, response time and other data.

An example of analysis reports produced by PROMOS is given in Fig. 5, showing the CPU load factors and related running tasks.

Performance data collected by PROMOS and information obtained from performance analysis conducted as part of on-site tuning activities make base data of computer load which will be considered in the future expansion of the subject plant. They are also used to create a plant model for performance estimation in the design stage.

SWB-II TEST SYSTEM

As mentioned above this SWB-II test system is a tool to support the total test phase. The total test is the final phase of function

Fig. 5 Example of Performance Evaluation Report

testing, and its quality is determined by the extent to which the subject plant behavior can be simulated accurately.
As shown in Fig. 6 (a), the process control computer is usually connected with the subject plant chiefly through an process input/output devices (P I/O). In the traditional method, the plant behaviour is simulated using analog computers, simulation panels, etc. as shown in Fig. 6 (b), but this approach was handicapped in terms of accuracy and test repeatability. The SWB-II test system has a function amounting to part A of Fig. 6 (b) built into its test support computer so that the subject plant process is simulated in terms of software.
The SWB-II test system has been developed, chiefly aiming at the following.

(1) Digital simulation of dynamic behaviour of plant process

(2) Plant model description in simulation language

(3) Coexistence of software simulation and hardware simulation

In order to achieve these objectives the SWB-II test system is essentially configured for distributed processing as shown in Fig. 7.

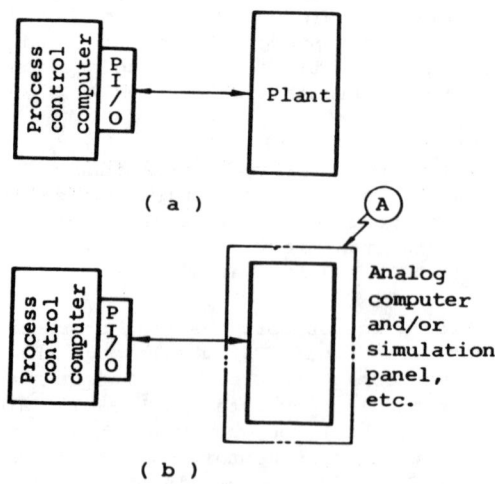

Fig. 6 Connection with Plant

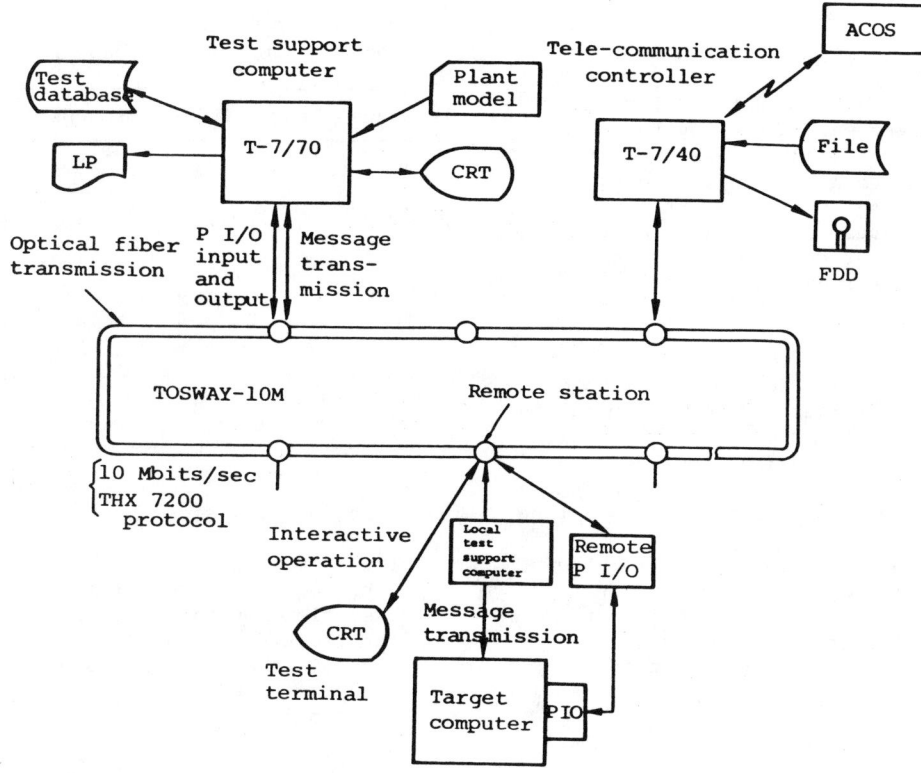

Fig. 7 SWB-II Test System Configuration

The test support computer (upper left in the drawing) on which plant behavior models run, and the process control computer (target computer) on which application programs run are connected through the dataway. The test terminals for total test are also connected with each remote station, located beside target computers. As mentioned previously, the general purpose computers are connected to the dataway through the telecommunication controller.

The plant model for plant process behavior simulation is called by the test terminal and runs on the test support computer to generate digital and/or analog signals. These signals are sent to the target computer through the dataway. Conversely control signals output from the target computer are received by the plant model and stored in the data base on the test support computer.

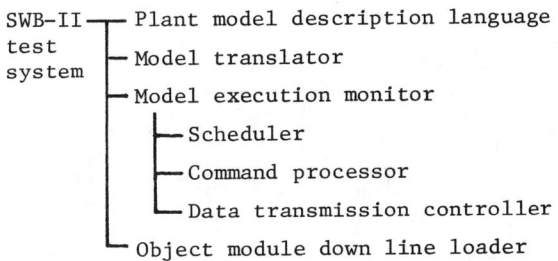

Fig. 8 SWB-II Test System Software Configuration

At the test terminal, the user can supervise the test process and start or stop execution of a plant model.

The local support computer is located between the remote station and the target computer. This arrangement realizes faster plant process behavior simulation by reducing test support computer load. Fig. 8 shows the SWB-II test system software configuration.

Plant Model Description Language

The plant model description language is a language exclusively used to describe a plant behavior. It is coded in almost the same way as programming languages in general.

The plant model description language consists of five divisions as described below.

(1) Identification division

This division is used for model identification. Author, identification name (order-ID), etc. are defined in this division.

(2) Environment division

This division is used to define environmental conditions for the model to run. The target computer type, execution cycle, number of times of execution, etc. are defined in this division. The execution cycle is specified in multiples of 50 ms.

(3) Data division

This division is used to declare data areas and/or values used by the model. Data area declared here are used to store results of macro functions. This division is also used to specify that data area contents should be stored in the file at certain intervals while the model is running.

(4) PIO division

This division is used to define information which is required for the model to communicate with the target computer. As shown in Fig. 6 (a), the model and target computer are connected through Process I/O whose information is define in this division.
Data reception by the model from the target computer is assumed as input and data transmitted by the model to the target computer is assumed output. They are usually expressed in the following formats.

  Input: data = FUN (input list);

  Output: FUN (output list) = data;

where FUN indicates a Process I/O type.

As shown in Fig. 7, information transfer between the test support computer and target computer takes the form of message transmission or remote Process I/O accessing, and FUN names differ between them. They are listed in Table 1.

For instance, analog output through the remote Process I/O is described as follows.

$$RAO\ (i_1,\ i_2,\ r_1,\ r_2,\ r_3,\ r_4) = S$$

Described this way; an analog value of $r_1 S^3 + r_2 S^2 + r_3 S + r_4$ is output at point i_2 of controller i_1.

Table 1 Support Process I/O

Message	Remote P I/O	Process I/O type
AO	RAO	Analog output
PO	RPO	Pulse output
DO	RDO	Digital output
RO	RRO	Relay output
AI	RAI	Analog input
DI	RDI	Digital input

(5) Procedure division

This division is used to describe model behavior using macro instructions. The Execution Monitor interprets and executes the procedure thus defined on pre-set cycle in the Environment Division.
The general macro instruction format is shown below.

  Stat = FUN (paralist 1) [(paralist 2)]

  Stat: Status quantity save area name

  FUN: Macro function name

  Paralists 1, 2: Parameter lists 1,2

This expression means that arithmetric operation should be performed in accordance with the specified macro function using parameters defined as the paralists 1 and 2, and that the result should be stored in the area 'Stat'.
More than 50 macro instructions are available, dealing with primary delay, time integration, waveform generation, conversion function, arithmetic function, etc.
Fig. 9 shows a turbine start control model drawing, and its block diagram is presented in Fig. 10.

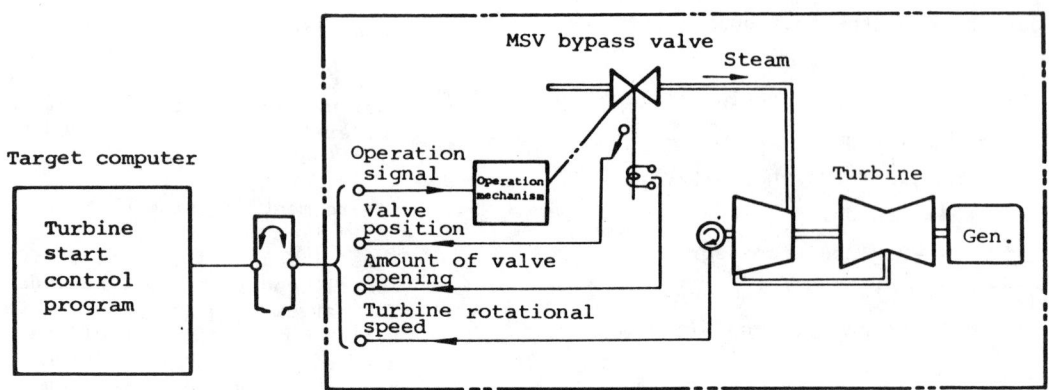

Fig. 9

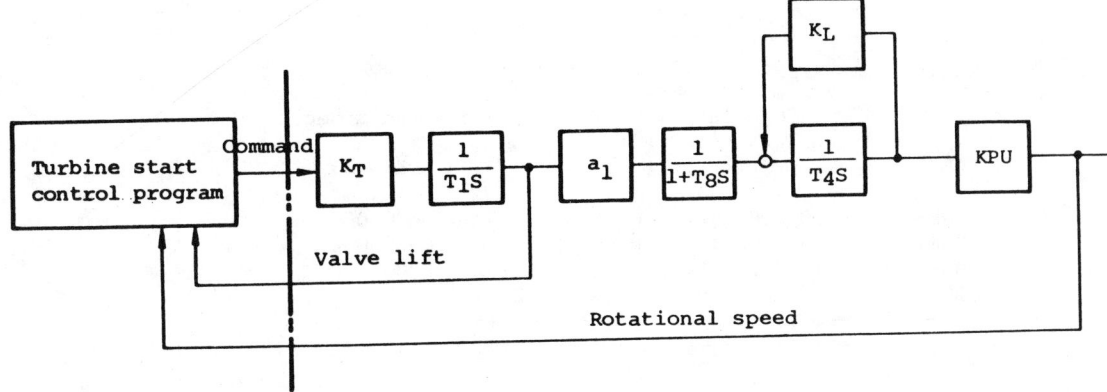

Fig. 10

Based on this control block diagram the model is described in the simulation language as shown in Fig. 11.

Command Processor

This processor enables the user at the test terminal connected with the remote station to carry out interactive processing with the simulation model which runs on the test support computer.

Representative commands are listed in Table 3.

Table 3 Command Menu

Command	Function
EDIT	Edits model source program described in simulation language.
STRT	Starts model execution.
STOP	Stops model execution.
HOLD	Holds (discontinues) model execution.
REST	Restarts model execution.
DSPY	Displays periodically data of model under execution.
DUMP	Dumps and patches data of model under execution.
STAT	Displays status of model under execution.
SCAN	Displays variables specified for saving.
NEWU	Passes control to another model.
BYE	Stops interactive processing.

The editor called by the EDIT command has the same command format and operation procedure as the editor used in the software implementation stage. This makes it possible to edit data easily.

Fig. 13 gives an instance of the use of the DSPY command to display areas named TSPEED, U, V, and digital input value of controller 0 and group 0.

```
IDENTIFICATION DIVISION;
          ORDER-ID = TBN;
          AUTHOR   = MIURA;
ENVIRONMENT DIVISION;
          RUNNING-CYCLE = 2 ;
          RUNNING-TIMES = 3000;
DATA DIVISION;
    - - - - -
PIO DIVISION;
    HAI (0,2), AO(0,1), DI(0,1), RO(0,1);
    - - - - -
PROCEDURE DIVISION:
  DLIFTRM = CSIGN (DLIFTR);
  DLIFTR = SW (DLIFTRM,DLIFTR)(ROSIGN = 1);
  DL = MULT (DLIFTR,KT);
  RLIFT = INTG (DL,T1K);
  LIFT = FIX (RLIFT);
  VOPMM = SW(0,1)(RLIFT    KOPMM);
  VOPN  = SW(0,1)(RLIFT    KOPN );
  RLIFTZB = SUB (RLIFT, KOPN);
  RRLIFT = SW (0.0, RLISTZB)(VOPN = 0);
  TPO = MULT (RRLIFT, A1);
  TPU = LAG (TPO, 0.0) (T8, K8);
  L = MULT (RRPMPU, KL);
  LM = CSIGN (L);
  TPUL = SUM (TPU, LM);
  RRPMPU = INTG (TPUL, T4k);
  RRPM = MULT (RRPMPU, KPU);
  RPM = FIX (RRPM);
END:
```

Fig. 11 Example of Model Description

```
SYSTEM?DSPY

VAR.? TSPEED , U , V , DI(0,0)  ; specify the symbols to be shown

DISPLAY CYCLE?   2              ; define the display cycle (unit:sec).

              TSPEED            U                 V              DI(0.0)
10.123    0.0000000E+00   0.0000000E+00     0.0000000E+00     00000000
10.124    0.0000000E+00   0.0000000E+00     0.0000000E+00     00000000
10.125    0.3000000E+03   0.1652012E+03     0.1321610E+02     00000000
10.126    0.3000000E+03   0.2727844E+03     0.5419333E+02     00000001
++++++    ++++++++++++    ++++++++++++      ++++++++++++      ++++++++
++++++    ++++++++++++    ++++++++++++      ++++++++++++      ++++++++
```

Fig. 12 Example of Use of DSPY Command

Simulation Execution Procedure

Fig. 13 shows the simulation execution procedure.

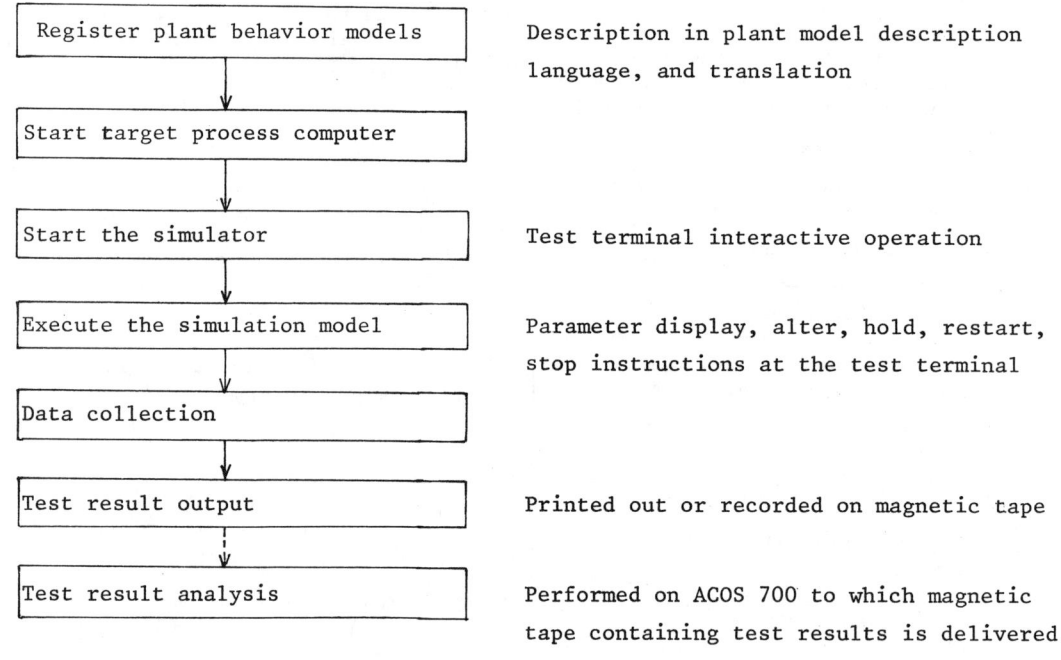

Fig. 13 Procedure of Simulation

As shown in Fig. 13, the test support computer is used up to model registration. But subsequent operations are directed at the test terminal entirely.

SWB-II DATA TRANSMISSION METHOD

Fig. 14 shows a simulation data transmission.

Software Test Facilities with Distributed Architecture

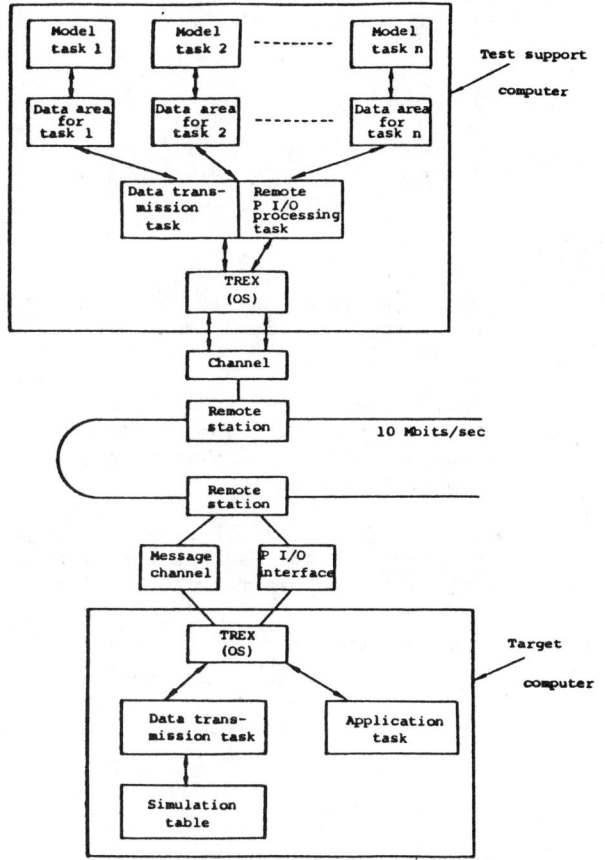

Fig. 14 Simulation Data Transmission Conceptual Drawing

Two methods are available for data transfer between the test support computer and target computer. One uses actual P I/O devices. This kind of data transfer is dealt with remote P I/O processing on the test support computer and is supported by TREX (OS). The other method does not use any actual P I/O but data generated during model execution are transmitted as messages to the target computer.

The data transfer task on the target computer receives a message, and references the simulation table for identification. If it is a message for simulated P I/O device, it is handled as if an actual P I/O were accessed, and data is transmitted to the application task through TREX. As shown in Fig. 13, the target computer, when started, waits for signals from the test support computer. By starting the model, firstly the simulation table information are transmitted to the target computer.
Simulated test starts after that procedure is over. As the model runs, plant behavior data is generated and stored in the data area reserved for it, and the transmission task is woken up at the same time. First of all, this task receives plant control information from the target computer and informs the model of its contents. Data stored in the data area of the model are next sent to the target computer by the transmission task.

This sequence is repeated as many times as specified in the model. If an error occurs during transmission, the subject model outputs error information on the test support computer console and processing is discontinued.

T-SWB HEREAFTER

T-SWB consists of six subsystems as shown in Fig. 2. Presently five of these subsystems (except for SWB-II) run on the ACOS 700, and the ACOS 700 shows sign of overloading gradually in response time and other operations.

As for the future we plan to develop T-SWB into a completely distributed processing system as shown in Fig. 15 so that each of its subsystems may be executed on an independent satellite computer with an information exchange network in the factory.

CONCLUSION

The SWB-II test system realizes more substantial factory test operations. For instance, about 60% of errors which were not detected until the on-site tuning phase are detected in the factory test phase. This contributes significantly to the improvement of quality assurance of our factory. Two years have passed since the development of SWB-II and its functions have been expanded in the meantime. Hereafter we will continue efforts for the enrichment of the macro instruction library, increase the number of types of compatible target machines, support to microcomputers, etc.

IMAGE OF DISTRIBUTED SWB PLAN

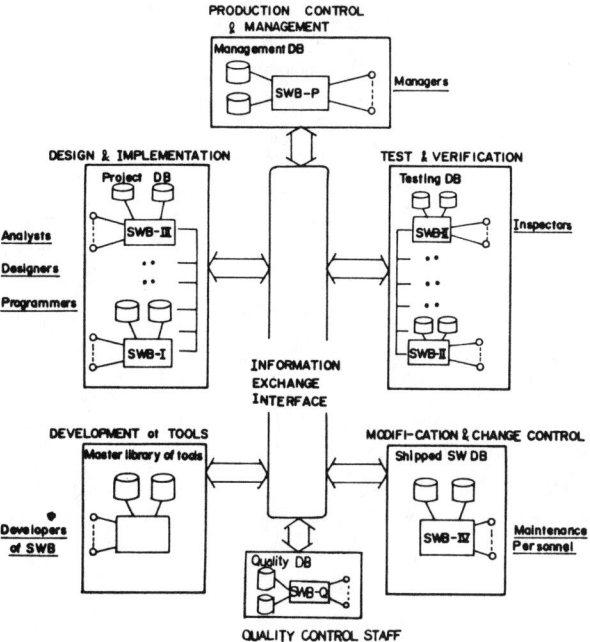

Certainly test support tools keep on improving in terms of both quality and quantity, yet they are still insufficient for improvement in terms of reliability evaluation as they are today. In the circumstances we are promoting the review of the effectiveness of the quantitative quality metrics such as coverage and examination of other quality control standards available. We fully recognize the usefulness of evaluating software quality quantitatively by establishing certain standards. In this connection, we will continue to promote the accumulation and analysis of test data while refining software production support tools.

ACKNOWLEDGEMENT

The contents of this paper are the results achieved by the SWB-II development project of Fuchu Works in Toshiba Corporation. The author wishes to thank Dr. Y. Matsumoto, Mr. O. Sasaki, Mr. S. Ebihara, Mr. H. Kurii, Mr. S. Nakajima, Mr. S. Yamamoto, Mr. H. Miura, Mr. T. Kitado, Mr. S. Tatebe and other project members for their approval to present at this workshop, and Mr. T. Sumi for his helpful suggestions to the work.

REFERENCES

(1) Y. Matsumoto et al.,"SPS: Software Production System for Mini-Computers and Micro-Computers", COMPSAC '78, pp.396-401

(2) Y. Matsumoto et al.,"A Distributed Processing System and Its Application to Industrial Control", NCC '78, pp.1273-1279

(3) Y. Matsumoto et al.,"SWB System: A Software Factory", Symposium on Software Engineering Environment, West Germany, 1980 June

(4) Werner Geiger et al.,"Program Testing Techniques for Nuclear Reactor Protection Systems", Computer, August 1979, pp.10-18

DISCUSSION

Harrison: I'm very impressed by the software productivity numbers that are contained in the 2nd paragraph of your introduction and I would like to ask you for some definitions. So, if I may, let me ask a couple of very small questions. Does "set" mean the number of computers shipped?

Takezawa: Yes

Harrison: Then this means each machine has, on the average 200 kilolines if shipped code?

Takezawa: Yes, it includes new and reused code.

Harrison: Are the 60,000 Ksteps/year equivalent to 60,000 K executable instructions?

Takezawa: Yes, essentially. They are the source lines at the assembler level.

Harrison: Do the 2000 software engineers, include quality engineers, test engineers, simulation engineers, as well as software development engineers?

Takezawa: Yes

Harrison: Thank you very much. These are expressive productivity numbers.

Joudu: In the large programs there are many inputs and each input has many levels, therefore there are very many situations possible in a large program to test. How do you select the situations for testing of large programs?

Takezawa: We divide test stages into four steps: unit test, integration test, total test and on-site tuning. We can increase maintainability and testability by dividing large programs into as small modules as possible. In this way we can test each module easily.

A COMMUNICATIONS SYSTEM FOR USE IN AN INDUSTRIAL DISTRIBUTED CONTROL SYSTEM

M. G. Rodd*, N. J. Peberdy**, H. F. Weehuizen*** and D. P. A. Bean****

*University of the Witwatersrand, 1 Jan Smuts Avenue, Johannesburg, South Africa
**Alpret (Pty) Limited, P.O. Box 391475, Bramley, Johannesburg, South Africa
***University of Cape Town, Private Bag, Rondebosch, South Africa
****Council for Minerals Technology, Private Bag X3015, Randburg, South Africa

Abstract. This paper discusses 'HYDRA', an experimental communications system specifically designed to support investigations into the use of Distributed Computer Control Techniques in the Metallurgical field. 'HYDRA' was aimed at supporting the control of processing plants which can be partitioned into a multiplicity of geographically dispersed sub-processes which are separately controlled. A thin-wire communication system links the controllers and enables their activities to be integrated and co-ordinated. Essentially 'HYDRA' is a bidirectional, packet-switched loop network comprising control computers which are connected into the ring via dedicated communication processors. Information is passed between nodes over single, twisted-pair cables which operate at 38K baud in half-duplex, synchronous mode. The system is described in the context of the particular requirements of an industrial data transmission system, with emphasis on topological issues and system security. The work has been accompanied by extensive simulation which has allowed the system's performance to be predicted and subsequently compared to the physical system. This modelling aspect of the work is reported on, as is the use of a fibre-optic bus as a replacement for the twisted-pair cable.

Keywords. Distributed systems, Loop networks, Industrial data transmission, Network simulation, Fibre-optic transmission.

INTRODUCTION

"Suddenly, there are lots of choices, and the first consideration from now on will be configuration". This is the view of that pragmatic editor of Control Engineering, Edward Kompass (Ref. 1), when summarizing the present trend in process-control engineering. This remark clearly reflects the maturity being attained by distributed computer control systems: the word 'distributed' is no longer hailed as the magic potion to solace all ills, but rather as indicating another class of solution.

The point is that, until a few years ago, all process-control systems shared the same form, with a centralized control room, and sensors and actuators sending and receiving analog information to and from this one point. There were, of course, some Direct Digital Control Systems, but they were generally very much in the minority. The pattern was upset by Honeywell's TDC-2000 system. Now, dedicated (typically digital) loop controllers could be used, communicating over a digital-transmission link back to the central computer(s) complex. Alternative configurations now became possible - from those which were conventional, but possibly had digital elements instead of analog, through to totally-distributed systems with local controllers mounted directly at the points under control.

In essence, the problem is one of definition, the word 'distributed' means different things to different people. To a purist, a distributed computer system is one in which all hardware and software are distributed throughout the computing structure, and the acid test of a distributed structure is that the failure of any part (be it hardware or software) should not affect system operation, except perhaps for a slight degradation in performance. On the other hand, any computer system with more than one processor is often termed distributed - Joseph (Ref. 2) lists some 25 terms which are used to categorize various systems which are in some sense distributed. Thus, a single minicomputer with one intelligent terminal may be regarded as a distributed one. Perhaps one should not be pedantic, but should adopt a more flexible and sensible approach.

ESSENTIAL ASPECTS OF A DISTRIBUTED CONTROL SYSTEM

A distributed control system can be conveniently stratified into three levels:

1. The underlying communications subsystem supporting inter-processor communication. It consists of 'cables' which interconnect the nodes, and a mechanism for reliably passing messages between controllers.

2. The operating system which has effective control over all of the systems' resources. It provides facilities at a high level for communications, synchronization, data base management, scheduling, input/output, file handling, etc.

3. The outermost level which is the plant control level. Here the process control engineer formulates a global control strategy and programs this into the distributed computer using the operating system facilities.

Level 1. has become a well established technology within the last few years. Most industrial distributed systems use the communications system to relay the values of process variables, setpoints and other information to a central display station, and then to transmit directives from the operator in the form of new setpoints back to the outstations. In some instances, the operator may be replaced or augmented by a computer which calculates setpoints based on global information. Level 2. is fairly well developed in the data processing environment and these techniques are being applied in the process control environment. However, some of the fundamental issues specific to real-time control have been, to some extent, ignored. The issue of time, in particular, has been glossed over. A notable exception is the MARS (Maintainable Real-Time System) being developed by Professor Kopetz at the University of Berlin, in which time messages are broadcast to all modules which then synchronize their local clocks (Ref. 3). Every message transmitted within the system contains a time-validity field which specifies at what point in time the validity of the message expires. The International Electrotechnical Commission is currently attempting (with the active support of the major process control vendors) to formulate communications standards which better meet the needs of process control applications than do existing protocols. (Ref. 4).

As yet, Level 3. - the distributed plant control strategy - remains at an early stage of development although much research is being carried out by process control engineers. (Refs. 5 and 6).

The work discussed in this paper was initiated some four years ago, at which time the communications aspects were still at an early stage (Ref. 7). The need for the work arose from Process Control investigations undertaken by the Council for Minerals Technology. These investigations, which have yielded many major breakthroughs, especially in the area of the automation of large submerged arc-furnaces, used in the manufacture of various ferroalloys (Ref. 8) showed that a distributed computer control system was a natural solution. However, in this industry the environmental conditions are at their harshest and reliable, secure data communications were difficult to achieve using the solutions available at that time. The goal of the project under review was to explore the communications area, concentrating initially on level 1. aspects, and to propose a suitable solution. The system adopted, based on a variety of parameters, was a packet-switched loop network, using bidirectional transmission. To assist in the evaluation of the solution, extensive simulation was undertaken and used to support the creation of an experimental 5-node system. A final aspect considered was the actual transport media - initially a simple twisted-pair configuration was adopted, but later investigations covered the use of a fibre-optic structure.

The rest of this paper describes the various aspects of the work, and highlights the main points which have emerged.

DESIGN GOALS

1. Absolute Security: The system should behave in a deterministic, completely predictable and safe fashion. From the communications point of view, a typical design goal is that not more than one error may escape detection in 1000 operating years - this figure is typical of state-of-the-art industrial control systems.

2. Ultra Reliable: The system should be extremely reliable and rugged. When it fails, it should fail to a safe condition. It should contain redundancy (for example on the communications bus) so that a high level of 'survivability' is attained.

3. Modularity: It should be modular, flexible and extensible. In particular the underlying data transmission system should place no unwanted restrictions on the control engineer. He should be able (ideally) to place controllers anywhere, be able to implement any control strategy, etc.

4. Performance: The performance of the system should be deterministic - that is, message transfer times and hence response times should be guaranteed at all traffic levels. It should be possible to calculate traffic levels for a given application so that the system can be configured to meet target response times. It should have a high level of availability.

5. Maintainable: Maintainability is a high priority in any distributed system. This includes self-checking, self-diagnostics, error logging and reporting.

6. Cost Effective

The communications subsystem is the backbone of any distributed control system and will determine how well the above requirements are met. Of prime importance is the structure of the system and the protocols it uses, i.e. the procedures by means of which information is exchanged.

LOOPS AND RINGS : A BRIEF SURVEY

From the outset of the project a loop or ring configuration was selected as offering many inherent advantages. Of key importance is the industrial requirement of retaining very simple transport media - this clearly excludes multiple-buses and complex configurations.

Four categories of loops, based on the mechanism for message transfer, can be distinguished:

1. In the Newhall loop (Ref. 9) a control token is passed from one node to the next in round-robin fashion. The token represents authority to transmit onto the loop. Transmission is unidirectional. The most well-known example of a Newhall loop is the Distributed Computing System which is a novel approach by the University of California to provide a campus with computing facilities. This consists, rather than one large mainframe as is typically the case, of five smaller computers which share resources and load connected as a Newhall loop. Communication control is decentralised.

2. In the Pierce loop (Ref. 10), communication time on the loop is broken up into a number of fixed length time slots which circulate continuously in one direction around the loop. Loop control is centralised. A transmitting station may 'grab' an empty slot simply by transmitting into it as it moves past.

3. The Distributed Loop Computer Network (Ref. 11) is of the delay-insertion type in which a node which wishes to transmit buffers an incoming message while it transmits its own message onto the loop. When it has completed transmitting its own message, it then transmits the buffered message. It has been shown that this technique is more efficient than the Newhall and Pierce loops.

The above three types of loops are unidirectional and hence are vulnerable to a link or node failure. Whilst various techniques have been devised to improve reliability, an alternative approach is described below.

4. The 'HYDRA' loop employs a bidirectional packet-switched approach which greatly improves the failure-effect characteristics at the expense of greater complexity. To the knowledge of the authors this technique has only been adopted in one other case, which is a process control application in the British food industry. (Ref. 12).

HYDRA - A BIDIRECTIONAL PACKET-SWITCHED NETWORK.

The goal of 'HYDRA' was to provide communications facilities for a number of geographically dispersed control computers. The approach adopted was one that is commonly used; each control computer being front-ended with a special-purpose communications processor (Fig. 1). The communications processors are then connected by means of data links over which information is transmitted. The main advantage of this front-ending technique is that the communications functions are separated from the control function. A communications sub-network can thus be set up, consisting of identical modules (the communications processors at each node). It is thus not necessary to program the different types of control computers with extensive communications software. It becomes easy to interface to the network since this interface is specifically designed to be simple. The control computers are largely unaware of the lower level communications network.

'HYDRA' consisted initially of five nodes connected into a loop. The maximum number of nodes will be limited by practical issues, such as target response time. The link between adjacent nodes consisted of a single, bidirectional, twisted-pair cable. The link operated in half-duplex synchronous mode at a speed of 38,4 kbps. Transmissions over each link in the loop take place independently of each other - there may thus be as many simultaneous transmissions in progress as there are links. These transmissions may be in either direction (clockwise or anti-clockwise). Each node contains two processors:

1. The Node Processor, which is interfaced to a portion of the plant it controls. The Node Processor may be any computer. The test system used microprocessors (INTEL SDK-80's) for four nodes and a minicomputer (Varian 620/1) for the fifth node.

2. The Communications Processor, which handles the communication tasks for the node.

The two processors are connected by means of a shared memory technique.

The Communications Processor is a custom-designed microprocessor based on the INTEL 8080A microprocessor (Fig. 2). It is wire-wrapped onto a single board and is configured to be compatible with INTEL's Single Board Computer series.

The features of 'HYDRA' are listed below:

1. All interaction between nodes (data transfer, synchronization, etc) is message-based. This uniformity greatly simplifies the system, and it has also been widely used in other systems. The contents and type of messages are transparent to 'HYDRA', except those which are initiated by 'HYDRA' for error control and

reporting, and diagnostics. All interaction can thus be monitored by reading messages.

2. Packet switching is employed by Communications Processors to relay packets from the source node via intermediate nodes to the destination.

Messages may be any length, up to 64 (or 128) bytes. If transmitted in this variable-length form through the network, additional overheads would be involved in catering for the arbitrary length of messages. For this reason, a message is placed within the framework of a 'packet' which is fixed in length at 64 (or 128) bytes. The packet contains additional information for the communications sub-network and concerns itself only with fixed length packets. Whilst this is somewhat inefficient, it simplifies matters significantly and allows for the 'automation' of such functions as transmission and reception of packets by using hardware DMA controllers so relieving the CPU of these tasks.

3. Process addressing is used as opposed to processor or node addressing. This implies that addressing is position independent. A loop structure allows this because a packet transmitted in either direction around the loop will eventually reach the node in which the destination process resides. It implies that reconfiguration of the network (perhaps due to expansion or during maintenance) does not impact on other nodes, requiring configuration tables to be updated. The intention is to minimise the amount of global information stored in a node and to structure the system accordingly. The maximum number of processes in the network is 256. Process addressing thus also serves to simplify the system significantly.

4. A bidirectional loop was selected because this suits a message-based, process-addressed system. It is essential that the loop be bidirectional so that two paths exist between any two nodes, providing one level of redundancy in the communications path. The incorporation of a redundant loop would imply that four-link or -node failures would be required to break the path between any two nodes. The survivability characteristics of a loop structure are thus excellent. The only other serious contender for a message-based, process-addressed communications system for loosely-coupled control applications is a global bus of the MARS type. (Ref. 7). Logically speaking the bus and loop are very similar. The primary advantage of the loop is its better failure-effect characteristics relative to the communications medium. The main disadvantage of the decentralised bus is that the bus is a shared resource which is crucial for system operation, and it is a potential bottleneck. Comparing a single bus system to a single loop system, it is evident that one bus failure will be catastrophic whereas two breaks in the loop are needed before communication is terminated. 'HYDRA' is designed so that one break in the loop will be transparent, messages will traverse the operational sector of the loop. This is done by means of the inherent nature of the protocol, and does not require reconfiguration whereby the break is located and masked out. The primary disadvantage of the 'HYDRA' loop is that there is not a direct connection between all nodes as is the case for a bus. The distribution of complexity (between hardware, data link control, path control and end-to-end control) is different for a bus and a loop but it is not certain which is simpler overall.

6. Protocol refers to the agreed procedure for interchanging information.

The mechanism by means of which 'HYDRA' transfers a message from one node to another is fully described in Fig. 3. Before examining this figure, it should be noted that each Communications Processor contains a list of processes which are currently resident in its local host Node Processor.

A Communications Processor does not hold tables of processes for other nodes - only its own. On receipt of a packet the Communications Processor compares the destination process name against its list; if no match is found the packet is forwarded onto the next node; alternatively, if a match is found, the packet is reformatted into the original message and passed 'up' to the Node Processor. Routing thus follows very simple rules.

As far as the Node Processor is concerned, the remainder of the system, namely the other control nodes together with the communication sub-network, appear simply as an area in memory. To communicate with a remote process the Node Processor places the message into shared memory. Similarly, it receives messages from other nodes via shared memory. The communications sub-system is transparent to it. The Node Processor is entirely unaware of network structure.

Interfacing to 'HYDRA' is thus a simple process, being an interface to shared memory, which may appear anywhere in the memory space of the Node Processor. The Communications Processor occupies a single board; the edge connector contains selected address, data and control lines which may be configured to meet the bus configuration of the particular Node Processor. The Communications Processor has been designed to slot directly into INTEL's standard SBC bus so that it is possible to interface any equipment using this bus standard directly into 'HYDRA'. Although not implemented it is envisaged that the Communications Processor would also have a serial TTY/RS-232C link as an interface to the Node Processor thus providing a simpler, standard interface.

'HYDRA's' protocol is stratified into a number of levels (Fig. 4). Handshaking is employed at the data link control level between adjacent nodes, and also between source and destination nodes at the end-to-end level.

7. The message and packet formats are shown in Fig. 5.

8. Data link control is of the byte oriented, stop-and-wait, request-for-retransmission type.

This was done for reasons of expediency and implemented only weakly because it was evident that more efficient bit-oriented protocols would be available in silicon within the foreseeable future. This prediction has materialized so that future implementations would incorporate hardware data link control. This approach is in line with the stated goal of the project, which was to investigate the issues.

SIMULATION OF THE COMMUNICATIONS SYSTEM

Objectives

Communications structures to be used to support a distributed control system are complex so work of this nature can only be done on an experimental system at considerable expense. Simulation lends itself to this application, offering great flexibility in the monitoring of system performance. It also provides the designer with facilities which would not be available on a physical system. Through the use of pseudo-random sequences, it allows the application of identical random stimuli to different systems, enabling useful performance comparisons to be made.

One problem with simulation is that one is working with a model instead of a real system. If the model differs from the actual system then the results obtained from the simulation could be invalid. Thus it is good practice to validate results on a physical system where possible. Thus the parallel development of 'HYDRA' (with its inherent flexibility) together with computer simulations, makes for the ideal situation.

This simulation can be used to study various aspects of the final design.

One of these is the network configuration. What are the exact differences between various methods of system interconnection? Another variable concerns the type of link used between various nodes. Does the increased flexibility of full duplex communication outweigh the added complexity that it introduces over a half duplex communication system?

Higher data rates obviously reduce the message transit delay, but does the increased susceptibility to noise offset this advantage? A similar consideration applies to packet size.

The method of acknowledgement of packets has an effect on the rate at which reliable data can be handled in noisy environments.

Another variable is routing strategy - the method used for directing a packet from source to destination. Choosing an arbitrary route is the simplest, but what speed penalty does this incur compared with a sophisticated system that monitors traffic flow and routes the packet through a lightly loaded route?

Simulation Methodology

The communications system of a distributed computer network has the properties of a queueing system. Entities called packets arrive at a service facility which is the communications processor of a node and wait in queues called buffers until they are serviced.

Queueing problems can be solved using Multiclass Queueing Networks and have been done very successfully (Ref. 13) but when the problem becomes complex with multiple queues, the problem is solved more easily using simulation.

Simulation can be done using a standard computer language such as Fortran. This would mean that the final program would be more portable than it would be if a simulation language were used. This was borne out when the work was transferred from England where it was initiated, to Cape Town where it continues.

The disadvantage of using a standard language is that considerable effort has to be expended in developing queueing and timing functions which are essential in simulation work. Therefore the use of one of the special purpose simulation languages was adopted - and 'SIMULA' was the choice preferred by the group with whom the work was started in England. It is a 'super set' of the Algol language with additional facilities being included in the form of procedures. The language is designed to model discrete event systems and describes a sequence of actions rather than a set of permanent relationships. For example, a service counter serving a line of customers can be viewed as a series of service completions. The permanent relationship view-point for the above example is that the customers are passive entities acted upon by the service clerk. This has the drawback that the interactions between passive entities cannot be studied. An example of this is the impatient customer who balks from the service line because of slow customers ahead of him. This would preclude investigation of the effects of noise on the data transport system in the present simulation. Since the main processor of the node takes no active part in the communication process, its behaviour is not considered. The communication processor can be considered as having two processes, one handling incoming packets and the other the outgoing packets. Thus the operation of the node would be represented by two types of activity, one having the attributes of the receiver while the other has the attributes of the transmitter. A receiver activity and a transmitter activity form one port. Because there are two ports, four activities would have to be created. These four activities share a data buffer which holds the packets pending possible retransmission. Each activity shares status flags with its opposite number in a neighbouring node as these form the communication links in the actual system. The passive activities in the program are the packets which are examined for information and operated on accordingly by the transmitter and receiver activities.

Other sections of the program generate the packets during the simulation and document the results on completion of the run.

At the moment packets are generated randomly

within a certain time interval and originate at random nodes, thus resulting in a nominally even spread of traffic throughout the network. This strategy can be easily modified to suit any required input pattern.

The operation of the program will now be described. Initially the node activities are created which in turn create the ports. These entities are activated, initialise themselves and cease operation when the packet buffers are tested and found empty. The system is now operational but inactive.

The section of the program that generates the communication packets is activated. This generates a packet, gives it a source and destination node and places it in the communication buffer of the source node, provided it is not full. The activity that generates the packet then checks to see if the node is active and activates it if necessary.

The node activity then assigns the packet to its first available port and activates the port. This sets its busy flag which is also common with the port of the remote communicating node. Thus the use of this link for transmitting other packets is prevented while the transmission is in progress.

After a delay equivalent to the transmission time of the packet, the buffer is cleared of this packet and the busy flag of the port is reset, freeing it for communication in either direction. If the remote node is not active, then the transmitting port activates it.

The remote node delays action for a period equivalent to the node processing time. It then examines the data in the packet to determine whether it is destined for this node. If it is not then the other port of the node is activated to continue the transmission to the next node.

When the packet reaches its destination, the statistical data is stored and the packet is deleted from the run.

While the above describes the simulation of the Hydra Distributed Computer System, the program has been made as general as possible so that a range of networks can be simulated and compared. The number of nodes, the number of ports per node and the method of interconnection are all fed in as data at the beginning of each run. At this stage in the development of the program certain aspects such as the routing strategy within a node are written specifically for the Hydra system but these will become generalised as development continues.

Some Simulation Results

Whilst work is still continuing and full results will be published later, it is useful to describe some of the more relevant conclusions.

Five hundred packets are transmitted per simulation run in order to obtain a reasonably normal distribution of packet origins and transmission distances. The number of packets successfully transmitted is output as an indication of the success of the run. At very high throughputs the system becomes deadlocked and packets fail to arrive at their destination.

Packet transit delay is the average delay of the second half of all packets successfully transmitted. Only the second half is considered so as to minimise the effects of start up conditions.

Packet throughput is the average throughput during the second half of the simulation. This is also done to minimise the effects of start up conditions but has the drawback that the throughput prior to deadlock conditions cannot be monitored. This will be modified for future runs. Normalised buffer utilization is the average buffer utilization of all communication's buffers in the system.

The output of several runs is tabulated in Table 1. Plots of some of these runs are given in Ref. 15.

In brief, the system behaves linearly as the applied load is increased up to the point where the buffers are fully utilized. At this point, system throughput drops to zero and no further communication is possible.

This agrees well with other published data (Ref. 14), but as the applied load increases, the system shows some recovery. This appears to be anomalous and could arise from the random nature of the origin of the packets generated. At this point where the applied load is 6,25 pkts/sec the buffer utilization factor of 0,55 is a bit misleading as the actual utilization increases steadily to a deadlocked situation.

USE OF A FIBRE-OPTIC TRANSMISSION MECHANISM

As was described previously, one of the major goals of this project was to produce an extremely reliable, but rugged, communications system. In order to improve on the twisted-pair mechanism previously used, it was decided to investigate the use of fibre-optic techniques. The system investigated is described in Ref. 16. In essence it comprises a standard 500 meter run of commercial fibre-optic cable with appropriate drivers and receivers. These are controlled by an 'intelligent' controller, consisting of high-level protocol controllers, an 8085 microprocessor and associated memory. (Ref. 16). It is a direct replacement for the CMP as described earlier.

The system produced was subjected to a variety of environmental tests - performance figures proved to be difficult to obtain because of the high quality of transmission - in real terms no transmission errors could be detected! The only source of errors, besides physical damage, arose from electrical noise at the two ends of the link!

EVALUATION

The bidirectional loop configuration has characteristics which are highly suitable for real-time control. It offers reliable communications facilities which will continue to operate in the presence of faults such as a break in the loop. It has excellent failure characteristics, where network performance will tend to degrade in small steps with an increasing number of faults as opposed to sudden catastrophic breakdown. This is mainly because communication does not depend on a shared resource such as a common bus system or star network.

The loop is a simple structure which allows the implementation of some very powerful concepts. The fact that messages may pass through all nodes implies that routing is not required. Process-addressing means that nodes are not position dependent, i.e. it does not matter where in the loop a node is situated. Maintainability of the system is greatly facilitated. Additional nodes may be inserted anywhere into the loop. Cost modularity, the cost of adding another node to the loop, is good, being simply the cost of node hardware plus the cost of cables to the two nearest nodes. The initial capital investment in a 'HYDRA' loop is low with the cost being proportional to the number of nodes. It provides a high throughput because many packets may be in transit at one time. A single message can be broadcast to more than one node (although it is questionable whether broadcasting is a desirable policy since the number of required acknowledgements is unknown). A message can be addressed to a process without regard to the processor in which it is resident. Because the loop structure is inherently non-hierarchical it is capable of supporting a fully distributed computer system. On the other hand, a two-level hierarchical distributed computer system can be supported, where one node contains the master processor. The system could further be expanded by allowing a number of loops to intersect or by having a large number of local loops connected by one or more global loops. In a similar way an hierarchy of loops could be constructed.

In real-time process control systems, perhaps the most important parameter of system performance is response-time. It is estimated that the average point-to-point message delay for a five-node loop between any two nodes would be approximately 100 milli-seconds at a transmission rate of 38,4 kbps. Thus a node can expect a response from a remote node (say, for a request for data) within 200 milli-seconds, or well within a quarter of a second. Simulation studies are presently being carried out to quantify system performance and hence to show how this can be improved.

It is worth noting that by far the largest component of response-time is the time taken to transmit a packet from one node to the next, the overall processing time needed to format messages into packets, and vice versa, and for servicing transit packets, is at least an order of magnitude less than actual transmission time. Hence response-time could be reduced by an order of magnitude if the data transmission rate were to be increased by an order of magnitude. Whilst present response-time is adequate for many metallurgical applications, the applicability of the network could be considerably broadened by use of higher data rate techniques. The goal of the work described here was essentially of an explorative nature - the aim was not to construct a fully operational network. The principles of operation have been implemented and tested. However, further work would be needed before 'HYDRA' could be used in a process control application.

ACKNOWLEDGEMENT

The financial support of the Council for Minerals Technology is gratefully acknowledged, as is the assistance of the University of the Witwatersrand in the production of this paper. The work was performed at the Universities of Cape Town, Witwatersrand and Canterbury, Kent. Their support is gratefully acknowledged.

REFERENCES

1. Kompass, E : "The Configuration Considerations" and "The Configuration of Process-Control:1979" Control Engineering, March 1979. pp. 41-54.
2. Joseph, E C : "Distributed Processing Architecture - Past, Present and Future Trends" Infotech State-of-the-Art Report, Distributed Systems, 1976. pp. 321-333.
3. Kopetz, H et al : "An Outline of Project MARS - Maintainable Real-Time System". Report MARS-R.79/1 of the Institut für Technische Informatik der Technischen, Universität Berlin, 24 July 1979.
4. International Electrotechnical Commission WG6 of SC 65A : "Process Data Highway (PROWAY) for Distributed Process Control Systems".
5. Hassan, M J et al : "Stability, Stabilization and Performance of Multilevel Controllers under Structural Perturbations". IEE Proceedings Vol. 127, Pt.D, No.5, Sept. 1980.
6. Williams, T J : "Hierarchical Control for Large Scale Systems - A Survey". 7th Triennial World Congress, IFAC, Helsinki, June 1978.
7. Peberdy, N J : "Distributed Computer Control of Industrial Processes". MSc dissertation University of Cape Town, July 1979.
8. Sommer, G et al : "The Cancer Project : Summary of the Computer Aided Operation of a 48 MVA Ferro Chromium Furnace". Report 2032 dated 28 November 1979.
9. Newhall, E E & Farber, W D : "An Experimental Distributed Switching System to Handle Bursty Traffic". Pine Mountain Conference Proceedings.
10. Pierce, J R : "Network for Block Switching of Data". Bell Systems Technical Journal, Vol. 51, No. 6, July-Aug. 1972, pp 1136-1145.
11. Mafner, E R et al : "A Digital Loop Communication System". IEE Trans. Com. June 1974.
12. Technology - Government Lab Works on Tomorrow's Factory. New Scientist, Nov. 1978. pp. 359.
13. F Viviers, R Reinecke, A Krzenski : "Analysis of Cyclic Construction Operations in Multiclass Queueing Networks". SACAC Symposium on Simulation of Dynamic Systems in Applied Science, Management & Engineering. Aug. 1979.
14. Haenle, J O, Giesler, A : "Simulation of Data Transport Systems of Packet Switched

Networks". MIT Technical Report, Packet Communications, No. TR 114.
15. Weehuizen, H F : "Simulation of data Transportation Systems for Distributed Systems" PhD Dissertation (in preparation), University of Cape Town, 1982.
16. Bean, D P A : "A Fibre-Optic Industrial Data Link". MSc Dissertation, University of The Witwatersrand, 1981.

TABLE 1   Simulated System Performance

Applied Load (pkts/sec.)	Successful Packets	Transit Delay (secs.)	Throughput (pkts/sec.)	Buffer Utilization (normalized)
0,60	500	0,437	0,60	0,05
2,00	500	0,472	2,00	0,1
5,00	500	0,543	5,00	0,26
5,71	500	0,583	5,71	0,32
5,88	64	0,611	–	1,00
6,06	98	0,554	–	0,95
6,25	378	0,595	3,29	0,55
7,14	63	0,873	–	0,95

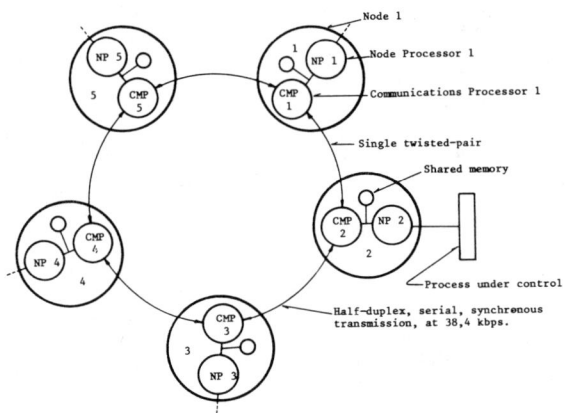

FIG. 1 : HYDRA TOPOLOGY

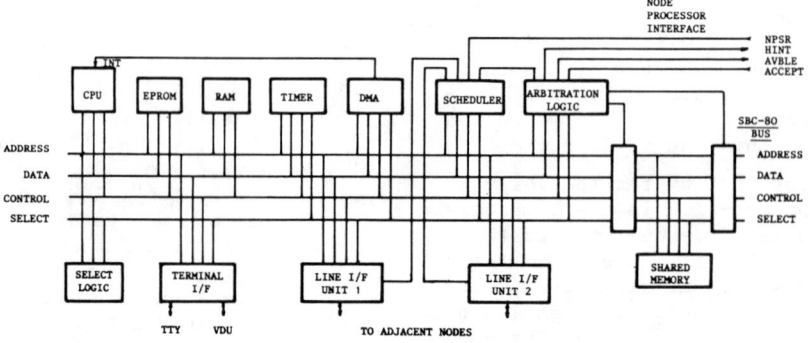

FIG. 2 : HYDRA HARDWARE STRUCTURES

Industrial Distributed Control System

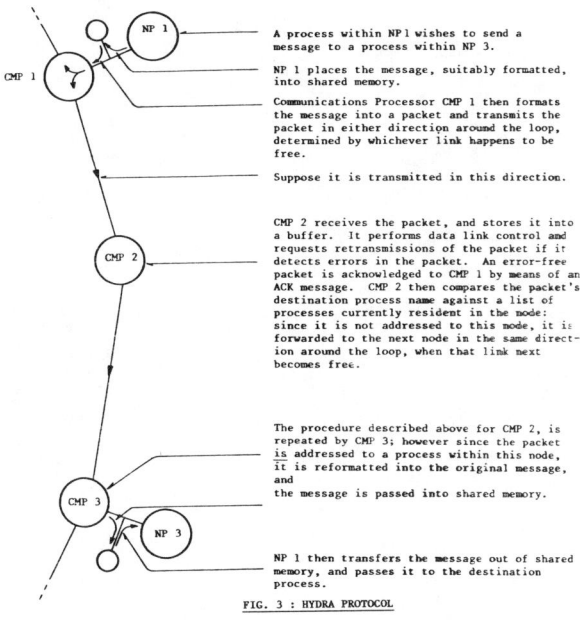

FIG. 3 : HYDRA PROTOCOL

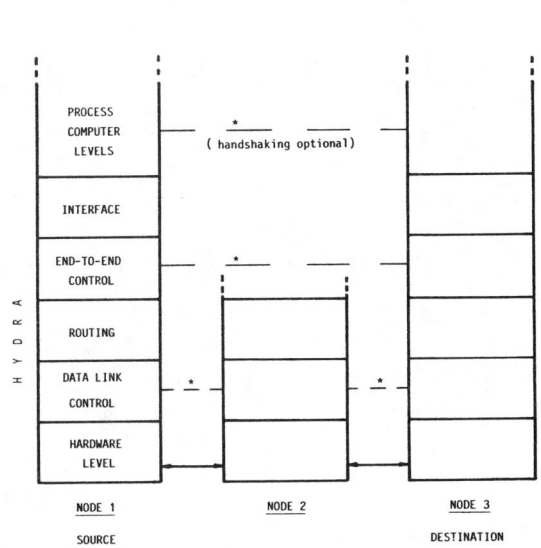

FIG. 4 : HYDRA PROTOCOL LEVELS

* Handshaking by means of ACKNOWLEDGE messages

MESSAGE FORMAT	BYTE NO.	FUNCTION	PACKET FORMAT
MESSAGE HEADER	1 & 2	Destination Process Name	PACKET HEADER
	3 & 4	Source Process Name	
	5	Message Name	
	6	Packet Type (blank in message)	
	7	Message Type	
	8 & 9	Message Length	
INFORMATION FIELD (variable length)	10 --	Information Field	INFORMATION FIELD (fixed length)
ERROR CHECK	L = 126	Message Error Check	
		BLANK	
	127 & 128	Packet Error Check	PACKET ERROR CHECK

FIG. 5 : HYDRA MESSAGE AND PACKET FORMATS

DISCUSSION

Tani: Are the messages sent in both directions?

Rodd: No just one direction. If we get a reply saying there is an error then we know the next node, so we then retransmit. We retransmit three times, in fact. If we still can't get through then we send in the other direction.

Tani: How do you choose the direction the first time?

Rodd: Randomly. We've looked at various possible ways of choosing that direction and we decided, and I think our simulation results showed, that the best is just random.

Work: I have a question about the flow control in your ring. You do not use the token and you do not use the message slots either. Is there any free space in the shared memory - the only item that controls the flow?

My second question is: if you send the messge from the first node to the third one and it goes through the second one, what will the delay in bit units be?

Rodd: Your first question was about the flow control. We started off on the basis that the amount of information being pushed around the system was not all that high and we hoped to redesign the buffer size such that we could never overload the network. Clearly one could overload the network on the structure. I think all we had to show was that, having got to the point where we can simulate the network and we know nothing about its behaviour, we could then estimate optimum buffer size assuming that the network never becomes saturated and starts losing information. Clearly we just can't afford to lose information.

Your second question was about the delay time in bit units. There is a delay time but I can't tell you what the units were. It is of the order of about 20 instruction cycles of the processor. We give incoming data the highest priority, so whatever the system is doing it will attend to incoming data. The problem is that if it also wants to transmit a previous message, which one does it handle? Again we assign priority to getting the incoming message that was going to go out straight away. We gave that high priority. It becomes obviously stochastic at this moment, that is why it becomes non-deterministic, unfortunately, because it does depend on the load of the system. But, in any case, we can get a message in and out within about 20 instructions.

DEVELOPMENT AND ANALYSIS OF PROTOCOLS FOR DISTRIBUTED COMPUTER NETWORKS

E. A. Yakubaitis, Ya. A. Kikuts and S. V. Rotanov

Institute of Electronics and Computer Science, Academy of Sciences of the Latvian SSR, Riga, USSR

Abstract. The paper is concerned with the problems of development, formal description and verification of protocols for the Experimental Computer Network (ECN) of the Latvian SSR Academy of Sciences. It describes the basic functions of protocol layers used, the methods of protocol formalization on the basis of state diagrams and algorithmic constructs and the methods of protocol validation based on representation of the protocol layer as a set of asynchronous processes.

Keywords. Computer network; protocol; formal description, protocol verification.

INTRODUCTION

The development of distributed data processing systems has brought about a new discipline of computer science - the analysis and synthesis of communications protocols. The main tasks of this field are the development, formal description and verification of protocols belonging to all levels of hierarchy. Extensive investigations being conducted currently in this area are geared to provide, first, the compatibility of hardware and software developed by different teams of designers and, second, to attain a reliable and efficient interaction of components of the distributed system. The basis for coordination of efforts on the protocols is the model of Open Systems Interconnection developed by ISO and the work being carried out in ISO and CCITT aimed at the creation of international standards and recommendations. However, certain features of the existing protocol standards do not allow to use directly these documents for protocol implementation. In particular, these features are: incompleteness of the description of protocol procedures, the lack of formal representation (which results in ambiguous interpretation), the presence of inconsistencies and even errors which require correction. In consequence, the international standards are to be treated only as source documents in developing protocols for national and regional distributed data processing systems.

The present-day protocols have a considerable complexity and a large volume, as well as a number of particularities, such as:
- the existence of three types of interactions between the protocol entity and its correspondent within the layer, and between the same entity and a higher layer entity and a lower layer entity within the system;
- the existence of concurrent operations of receiving and sending of protocol elements and service primitives;
- a sophisticated multiparameter structure of protocol elements;
- asynchronous type of operation between protocol entities;
- operation under conditions of non-reliability of the transmission medium.

All of this together with the absence up to now of the commonly accepted methods of formalization and the fairly convenient and universal methods of verification allows to define the tasks of protocol formalization and verification as non-trivial and requiring further investigation.

The given paper describes the basic results obtained in the course of development, creation of the formal description and verification of protocols for the Experimental Computer Network (ECN) of the Latvian SSR Academy of Sciences. The paper does not address such aspects as architecture, functional composition and topology of the network since they are covered fairly in detail elsewhere, for example (Yakubaitis, 1980).

PROTOCOL DEVELOPMENT

The ECN is designed as an open system which allows the connection of any advanced computers or systems whose architecture and protocols are consistent with the set of standards on the interaction of network elements. The basis for this set of standards are the ISO and CCITT documents. Currently the ECN uses protocol implementations of the following layers: the physical layer, the data link layer, the network layer, the transport layer and the session layer. The higher layers accomodate the already existing standard application program packages with built-in protocols (KROS, SRV, KAMA, OKA systems) and access the network services through the network access method which is functionally provided by the protocol entities of the above mentioned layers.

Physical Layer. The protocol of this layer determines the mechanical, electrical and procedural characteristics for establishing, maintaining and breaking the physical connection. At the moment two types of protocols are in use: one based on the CCITT Recommendation X.21 bis and a multi-channel interface with high-speed parallel transmission. The first one is intended for interfacing to the already existing now V.24-series modems and operates on the telephone circuits, while the other is used to interconnect computers within a single complex since it provides for high transmission rates (up to several hundreds of Kbit/s).

Link Layer. The protocol of this layer specifies the formats and rules of frame transmission over the physical channel. The basic purpose of the protocol is the enhancement of reliability of data communication between the network systems (the user systems and switching systems). The link protocol has been developed on the basis of the X.25 Recommendation (2nd layer - LAP/B). Since the description of LAP/B is essentially asymmetric (basically determines the rules of behavior of a switching system entity), the protocol has been symmetrized in such a way that it is equally applicable both to the user systems and switching systems. This has allowed to extend the field of application of the protocol and to use it inside the packet switching network rather than for the subscriber-network node connection only. The link protocol uses the duplex asynchronous balanced mode of operation. For data communication the I-frames with the data field size of 256 bytes are used (this size is intended for transmission of the standard data packets of 128 bytes and datagrams of 216 bytes according to X.25/3). Transmission control is performed with the frames RR, RNR, REJ, SABM, UA, DISC and DM. The distribution of commands and responses corresponds to LAP/B. The specific features of the protocol include the following: following the expiry of the timer which monitors the transmission of I-frames, the transmission of S-frames takes place with the P-bit set to 1. Following the reception of the response (i.e. when the number of the expected I-frame becomes known), the retransmission of I-frames takes place. Furthermore, the protocol incorporates a special procedure of correspondent busy monitoring which involves periodic requests of the busy correspondent by the S-frames. There are also insignificant differences in the DM-frame usage procedures. Despite these differences, the protocol is compatible with LAP/B due to the fact that there exist provisions for frame reception and processing to the full extent of LAP/B procedures. The link protocol is designed for data transmission over the physical lines with the error rate not exceeding 10^{-4} and it allows to reduce the probability of undetected error down to 10^{-8}.

Network Layer. The basic purpose of the network layer is organization and maintaining of the network channel (virtual circuit) in the packet-switching network. Two types of network channels are distinguished: permanent and switched. The paths of permanent network channels and their identification are determined by the network administration at subscription time. Switched network channels are established on the network subscriber's request for the time of his interaction with a remote correspondent only.

Evolution of the network protocol proceeded in two stages. At the first stage, where, on the one hand, the flow of requests for network channels was not too intensive and, on the other hand, the X.25/3 procedures have not been implemented to the full extent, permanent network channels were used only. In this

case the functions performed at the network layer involved the following:
- transmission of data packets with sequence number monitoring;
- packet flow control;
- monitoring of a full packet sequence;
- multiplexing and demultiplexing of network channels;
- network channel reset;
- network interface restart.

To perform these functions at the first stage the following packet types were used: DATA, RECEIVE READY, RESET, RESET CONFIRMATION, RESTART and RESTART CONFIRMATION. We should note that the RECEIVE NOT READY packet is not used here, moreover, when the standard window 2 is used, the flow control procedures operate fairly properly without using this packet on the basis of numbers in the DATA and RECEIVE READY packets.

At the second stage the packets and procedures for operation over the switched network channels and the INTERRUPT packets have been added. As a result of symmetrizing the procedures in the user and switching systems it has become possible to unify the network software and to use the protocol (with minor modifications) inside the communications network (Yakubaitis, 1979). The protocol widely uses diagnostic capabilities specified in X.25/3. Datagram service and end-to-end acknowledgement have been left for further study.

Transport Layer. The transport layer of the network provides its users, i.e. session entities, with the services for data transportation through the network, independent of the structure, protocols and specifics of the network layer. These services feature an end-to-end character, as well as such properties as a high reliability of message delivery while preserving their structure and sequence, the efficient control of flows and network resources, the protection against faults arising in the network. Currently there exists the trend towards standardization of the transport protocol. However, the existence of considerable procedural differences among the advanced developed or used transport protocols delays the preparation of international recommendations on transport procedures. In this context the probability arises that the protocol being developed for a specific network will not be compatible with the international transport protocol. Therefore in the process of development it is helpful to limit the list of functions performed at the transport layer so that the protocol be simple and therefore will not require major changes when the international standard becomes available.

The protocol developed for the ECN (Kikuts, 1981a) is applicable to the systems connected to the packet-switching network according to CCITT Recommendation X.25 and has the following functional structure:
- a flow control module;
- a data (message) transmission module;
- a data flow synchronization module;
- a transmit channel purge module.

The protocol is based on the document (ISO, 1980a), but does not belong to any class specified, since it uses permanent transport channels only. In using the protocol, the session layer is provided with the following list of services:
- several independent permanent transport channels (according to the protocol - up to 64K);
- data transmission while preserving their integrity and sequence;
- message flow control;
- message acceptance confirmation by the correspondent;
- transport channel purge (removal of all messages or their fragments from the channel).

The transport service is described in more detail in (Zibert, 1979, 1981).

Execution of transport functions involves the exchange of the following blocks:

PURGE, PURGE ACK::=⟨block type⟩, ⟨credit⟩, ⟨permanent TC number⟩, ⟨parameter⟩

DATA::=⟨block type⟩, ⟨reserve⟩, ⟨permanent TC number⟩, ⟨last block label⟩, ⟨block number⟩, ⟨data field⟩

DATA ACK, SYNCH::=⟨block type⟩, ⟨credit⟩, ⟨permanent TC number⟩, ⟨number of unit to be received⟩

Besides their basic functions, purge units are used for activation or deactivation of the transport channel, to which end the parameter field is used. At the transport layer the fragmentation of long messages into data blocks is performed. The transmission of data blocks is monitored by the timer. Flow control is performed with the help of credits. Each credit shows how many data blocks of the maximum length can be accepted, starting with the block expected to be received. The block acknowledgement scheme is such that the DATA ACK block is sent immediately following the reception of the error-free sequential data block which has been received within the limit of the credit. If an erroneous block is received, the transport channel is deactivated.

For data transmission recovery following the reset at the network layer, when the loss of blocks is possible, synchronization of the transport channel is performed, which involves the exchange of SYNCH blocks containing the number of the data block expected to be received.

The protocol in question is the basis for construction of a fuller version of the transport layer which will include the operation over the switched transport channels with multiplexing into the network channel, the transmission of expedited blocks and protocol identification.

Session Layer. Currently there is no unified international standard on the session layer protocol, although there is a number of proposals (ISO, 1980b, 1980c). These works have been used in developing the session protocol (Kikuts, 1981b), whose implementation has allowed to provide the following services to the users:

- establishment of a session with a remote user in the following modes:
  - one-way data transfer;
  - two-way alternate data transfer;
  - two-way simultaneous data transfer.
- normal termination of a session on user's request;
- urgent (abnormal) termination of a session on user's request or when it is impossible to maintain the session;
- simultaneous support of several independent sessions of one user with multiple correspondents;
- simultaneous support of several 'own' users;
- differentiation of user's data into:
  - simple data units (SDU);
  - quarantine data units (QDU);
  - interactive data units (IDU);
- provision of a safe delivery of data to the correspondent in the sequence determined by the user alongside with preserving their integrity;
- user notification of error situations unrecoverable at the session layer.

The SDU is an array of data which is sent by the user to its correspondent in a single interaction. If it is necessary to send an array of data in more than a single interaction to enable the correspondent to start processing this array when it is received as a whole, the quarantine data unit is used:

$$QDU ::= \langle SQDU \rangle, \langle SDU \rangle, [\langle SDU \rangle \ldots \langle SDU \rangle], \langle EQDU \rangle,$$

SQDU - start of the QDU,
EQDU - end of the QDU.

The possibility of using the QDU is specified at the session establishment time.

In the session with a two-way alternate data transfer an IDU is used which is a logical unit of interaction, on the end of which the reversal of direction of the user data transfer takes place. The IDU has the following structure:

$$IDU ::= \begin{cases} \langle SDU \rangle, [\langle SDU \rangle \ldots \langle SDU \rangle], \langle EIDU \rangle \\ \langle QDU \rangle, [\langle QDU \rangle \ldots \langle QDU \rangle], \langle EQDU \rangle \end{cases}$$

EIDU - end of the IDU.

The session protocol in question is oriented toward the service involving both the permanent and switched transport channels used for transmission of session protocol units with provision of their integrity and sequence, as well as end-to-end delivery confirmation.

FORMAL DESCRIPTION

The basic purpose of a formal protocol description is to provide a clear and unambiguous understanding of protocol procedures; the latter factor has a direct impact on compatibility of the network hardware and software developed by different teams of designers.

Currently it is possible to identify two major trends of formalization - those which involve the use of automata and the algorithmic models (Bochmann, 1979, 1980; Sunshine, 1979). In using either model, their modifications or combinations, it is necessary to satisfy the requirement of adequacy to the protocol. Basically this requirement reduces to two aspects: the reflection by the model of the hierarchy and parallelism.

From the point of view of network architecture, the protocol layer is a medium for interaction of the higher layer entities, i.e. each layer provides a certain type of services (dealing with data transmission, channel purge, etc.). The concept of the service must be treated as a composite part of formal description, which is indispensable both for providing the completeness of description and for determining the criteria of correctness of protocol procedures. Formal description of the service is the four-tuple:

$$F = \langle P, S, R, A \rangle,$$

where P is a multitude of service primitives treated as a set of parameters needed to provide the service; S is a multitude of states at the interface through which the service is provided; R are the rules of transmitting primitives through the interface as $R:S \times P \to S$; A is a multitude of assertions which determine the correspondence of parameters (including the types of primitives), which is attained in executing the service.

The second important aspect of formal description is a correct reflection of parallelism of the receive and send processes. A specific feature of protocols is an asynchronous character of interaction in the context of unreliable transmission medium, and therefore an incorrect (inadequate) use of the mechanisms of parallel process interaction can result in the errors in procedure execution.

Consider, for example, a possible incorrectness brought about by an incorrect use of shared variables as applicable to the timer type variable. According to X.25, the information channel is used to provide a full-duplex frame exchange. It means that in developing formal description of the protocol, the processes of receiving and sending a channel entity must be considered as parallel. Let the variable timer=(start, stop, end) be interpreted as a timer, while its meaning - as a start, stop and expiry of the timer, respectively. This variable is used both by the send process and the receive process. If $s1$, $s2$ are the states of the send process for a normal transmission of I-frames prior to expiry of the timer and following expiry (in recovering with the help of P-bit in the REQ frame), respectively, $r1$ is a state of the normal (error-free) reception of confirmations nrrcv, recov=OK is interpreted as recovery termination, lastnr is the last accepted nrrcv, varsend is the counter of I-frames being sent, then the axiomatic description of state transitions will have the following form:

A1: $timer \neq end(s1 \to s1) timer = start$,
A2: $timer = end(s1 \to s2) timer = start$,
A3: $timer \neq end\ \&\ recov = OK(s2 \to s1) timer = stop$,
A4: $timer = end\ \&\ recov \neq OK(s2 \to s2) timer = start$,
A5: $nrrcv = varsend(r1 \to r1) timer = stop$,
A6: $nrrcv < varsend(r1 \to r1) timer = start$.

The above processes are asynchronous and are executed over a finite interval of time. Let, for example, $(s1 \to s2)$ is being executed. In accordance with A2, the result of execution is the start of the timer. If, in accordance with A5, $(r1 \to r1)$ is being executed at that time, the situation is possible where in the s2 state the timer will not be started as a result of reception of the frame with nrrcv=varsend. Since, according to A3, the transition $(s2 \to s1)$ can be executed if recov=OK, while the latter can be true only following reception of the frame with the F-bit set to 1, the execution of $(r1 \to r1)$ according to A6 will not result in the execution of $(s2 \to s1)$. Here a deadlock situation arises, in which following a single transmission of the REQ frame the timer is stopped (the REQ frame will not be retransmitted), while the receive process will be waiting for a frame with the F-bit set to 1, which will not arrive due to its corruption (or corruption of the REQ frame) and will never be retransmitted. Such cases of incorrectness are encountered as a result of non-determinism brought about by the sharing of variables (Zave, 1976). Elimination of non-determinism can be attained by introducing the mechanism of process interaction which involves the exchange of variable values in the form of messages. In this case each process has a set of its own variables over which the assignment operations can be executed by this process only, and a set of non-proper variables which are set by a parallel process and communicated in the form of messages. With reference to the scheme of timer operation such an approach means that the timer start and stop operations will be performed by the receive process only.

<u>Automata description.</u> The description of the session layer protocol (Kikuts, 1981b) can be considered as a typical example of application of the automata approach. This is due to the fact that the session protocol (unlike the lower layer protocols) does not have its 'own' functions and essentially all the procedures are reduced to transformation of service primitives from the presentation layer into the protocol units and primitives of the transport layer, as well as the reverse operation. Formal protocol description consists of four parts:
a) rules of interaction at the presentation layer interface;
b) rules of interaction at the transport layer interface;
c) rules of interaction of the session entities without regard for the interface states;
d) session entity behavior with allowance for the protocol and interface interactions.

In describing the rules of interaction, the following notation is used:

$$(a(c)) \downarrow \uparrow A(b(d)),$$

where $(a(c))$ is the substate (c) of the state (a) prior to transmission of the element or primitive A, $(b(d))$ - following the transmission. Arrows ($\downarrow\uparrow$) denote the transmission of a primitive from a higher (lower) layer or the transmission (reception) of a protocol unit.

In describing the behavior of a session entity with allowance for the protocol and interface interactions, the correspondence is established between the actions at the interfaces and the actions of a session entity while receiving and sending protocol units in the following form:

$$(pi) \left\{ \begin{matrix} (hi) \\ (li) \end{matrix} \bigg| \begin{matrix} (hi) \rightarrow (hk) \\ (li) \rightarrow (lk) \end{matrix} \right\}$$

$$\left\{ \begin{matrix} (hn) \\ (ln) \end{matrix} \bigg| \begin{matrix} (hn) \rightarrow (hm) \\ (ln) \rightarrow (lm) \end{matrix} \right\} (pj).$$

In this notation (pi) is a source protocol state, (hi), (li) are the states of interfaces with the higher and lower layers. For transition into the (pj) state it is necessary that the source interface states be (hi) and (li). Execution of the transition $(pi) \rightarrow (pj)$ results in the change of the interface states: $(hi) \rightarrow (hk)$ and $(li) \rightarrow (lk)$. To allow the transition $(pi) \rightarrow (pj)$ to complete, it is sufficient that the interface states become (hn) and (ln), following which the transitions $(hn) \rightarrow (hm)$ and $(ln) \rightarrow (lm)$ are performed. Until the conditions (hn), (ln) hold, the protocol state remains (pi), while following satisfaction of these conditions it becomes (pj).

Consider, for example, the behavior of a session entity at the opening of the session. The interaction at a higher layer interface has the form:

$(hI) \downarrow$ CONNECT $<$param$>$ $(h2)$,

$(h2) \uparrow$ OK $<$param$>$ $(h3)$.

In this notation the (hI) state corresponds to the source state, while the $(h3)$ state corresponds to the termination of session establishment and the transition to data exchange.

Interaction at the transport layer interface:

$(lI(tI)) \downarrow$ SEND$<$data$>(lI(t2))$,

$(lI(t2)) \uparrow$ OK$(lI(t3))$,

$(lI(rI)) \downarrow$ RCV$<$buff$>(lI(r2))$,

$(lI(r2)) \uparrow$ OK $(lI(r3))$.

In this notation the (lI) state corresponds to the source state for data transmission, i.e. when the transport channel has been already established. Substates (tI) and (rI) mark the transmission and reception of the messages contained within the primitive fields. The termination of transmission and reception are marked by the substates $(t3)$ and $(r3)$, respectively.

Interaction of the session entities:

$(pI) \downarrow$ REQUEST$(p2)$, $(p2) \uparrow$ ACK$(p3)$.

In this notation the state (pI) corresponds to the source state, i.e. when the session has not been established as yet, the state $(p3)$ - to the termination of session establishment.

The session entity behavior with allowance for the protocol and interface interactions:

$$(pI) \left\{ \begin{matrix} (h2) \\ (lI) \end{matrix} \bigg| \begin{matrix} (lI(tI)) \rightarrow (lI(t2)) \\ (lI(rI)) \rightarrow (lI(r2)) \end{matrix} \right\}$$

$$\left\{ \begin{matrix} (h2) \\ (lI(t3)) \end{matrix} \bigg| \right\} (p2),$$

$$(p2) \left\{ \begin{matrix} (h2) \\ (lI(r3)) \end{matrix} \bigg| \right\}$$

$$\left\{ \begin{matrix} (h2) \\ (l8) \end{matrix} \bigg| (h2) \rightarrow (h3) \right\} (p3).$$

This notation corresponds to the presence of CONNECT(state(h2)), the transport channel being established (state(lI)) and means that for the REQUEST to be transmitted, it is necessary to make the transition $(lI(tI)) \rightarrow (lI(t2))$, to get ready for reception, i.e. to go over to $(lI(r2))$, and to receive confirmation at the transport layer (state(lI(t3))). The termination of establishment proceeds after the reception has been terminated (state(lI(r3))), following which the user is notified about the transition to data transmission (transition $(h2) \rightarrow (h3)$).

Modification of the automata description. An example of the description based on the expansion of automata models to the abstract machines is the description of the link protocol (Kikuts, 1982a). Graphically this description looks like the state diagrams in which each transition is marked by the conditions required for its execution and the actions to be performed:

$(stateI) \xrightarrow{\frac{condition}{action}} (state2).$

Conditions and actions are expressions over the set of variables (the timer, counters, sequence numbers, etc.), as well as special symbols of frame reception and transmission. A detailed description of the link protocol on the basis of state diagrams is given in (Kikuts, 1982b).

Algorithmic description. Algorithmic description has been developed for the network layer protocol. A feature of this protocol lies in a large number of actions over the variables which reflect the values of packet fields (identifiers, cause, diagnostics, etc.). Algorithmic description is based on the following principles:

- each group of protocol procedures is contained within an automata module which has no inputs or outputs to other modules;
- the connection of parallel processes is based on the exchange of messages (the proper variables are used);
- for the transmission and reception of service primitives and protocol units the same operations are used.

The description of each system consists of a set of the following modules:

<module> :: =

<packet-type-recognition-module>)
<restart-module>)
<connection-establishment-module>)
<disconnect-module>)
<reset-module>)
<data-transmission-module>)
<data-reception-module>)
<interrupt-transmission-module>)
<interrupt-reception-module>).

Each module contains:

<module-description> :: =

<data-type-description>
<entry-condition-description>
<action-condition-unit> .

Data types are determined by the module proper variables and the parameters being exchanged:

<data-types>::=<protocol-units>
            <service-primitives>
            <variables> .

Each module contains the description of conditions for entering this module:

<entry-condition-description>:: =

entry: < logical-expression > .

For example, for the restart module the entry condition will be: the reception of restart packets, the arrival from the link layer of the reset primitive or the primitive specifying the establishment of the information channel. In order to be able to make transition to the next module following the execution of the previous one, special variables are used. For example, for error signalling the following are used:

variables: transmit-restart=(yes,no);
           transmit-disconnect =
                         (yes,no);
           transmit-reset=(yes,no).

The modules which describe the restart, disconnect and reset have these variables in their entry logical expressions.

The actions performed over the variables are described in the condition and action units:

<condition-action-unit>::=unit-begin
                    <unit-body>
                    unit end,

<unit-body>::=if<logical-expression>
              then<action-list>
              else<action-list> |
              < repeated-constructs>.

A special type of the boolean variable used in the logical expression is the receive (element-type). The truth corresponds to the reception of the primitive specified in the element type. For packets the truth of a variable is established following the completion of procedures in the packet type recognition module. The action list includes also a special action send (element-type) which involves formation of a primitive or packet to be sent and sending it to its destination. The syntax of the description language allows the use of nested constructs owing to the use of the beginning and end delimiters and the procedures furnished with identifiers.

VERIFICATION

Protocol verification involves the solution of the following problems:
- selection of the formal criteria of correctness;
- development of the model of a protocol layer;
- development of the methods for anasyzing the model for correctness.

Let us consider each of these problems.

Correctness criteria. The most feasible correctness criterion is the degree to which the service provided by the protocol layer meets the requirements imposed by a higher layer. Such an approach is analogous to that proposed in (Bochmann, 1979). Let us consider, by way of example, the transmission of data from the entity A to the entity B. Formal description of the service is specified by the following service primitives:

SEND<idata> - data transmission request contained in the <idata> field,

RCV<jbuff> — request for data reception in the <jbuff> field.

The SEND and RCV primitives form the queues at the A and B entity interfaces, respectively, so that each member of the queue is uniquely identified by the i or j index (it is supposed that the values of indices $i,j = 0,1,2...$ are not limited). The service involves the transmission of data from one entity-user to another without loss, duplication and violation of the sequence of transmission. This means that when the states is3 and jr3 (the final states for the i- and j-elements, respectively) are attained, there must be a one-to-one mapping of the <idata> fields onto the <jbuff> fields, such that the following is true:

$I: (\forall i, \forall j, \forall m)[(\exists i=j) \text{sstate}=is3 \ \& \ \text{rstate}=jr3 \Rightarrow (\exists i=j)<\text{idata}>=<\text{jbuff}> \ \& \ (\neg \exists m>j)<\text{mdata}>=<\text{jbuff}>]$.

This expression is an invariant of correct procedures of the protocol layer. The correctness proof involves the proof of validity of the inference on condition that the assumption is valid (partial correctness), followed by the proof of validity of the assumption (full correctness).

<u>Model of the protocol layer</u>. The protocol layer can be considered as a set of interacting asynchronous processes. The process is described by its 'external' characteristics from the point of view of its interaction with other processes. Such an approach allows to simplify the analysis of complex protocols and to combine advantages of the axiomatic method and those of the method of analyzing the reachability graphs. The basic elements of the model of a protocol layer are the asynchronous processes and their couplings. An asynchronous process (AP) is the four-tuple $P = <V,D,A,B>$, where: V is a finite non-empty set of variables, D is a finite non-empty set of ports, A and B are the input and output predicates over V, respectively.

The dynamics of AP is interpreted in the following way. When a signal arrives to the input port $I \in D$ (the enable port) the analysis of the input predicate A takes place with allowance for the messages received by the receive ports $R \subset D$, and if the predicate is valid, the process P is initiated. It is thought that the P process is deterministic and does not contain inner loops and therefore following a finite number of operations over the variables it will be completed. The result is the appearance of signals at the send ports $T \subset D$, the transmission of the enable signal through the output port O and the setting of relations between the values of variables according to B. To specify the couplings between APs we will consider two APs, $P_1 = <V_1,D_1,A_1,B_1>$, $P_2 = <V_2,D_2,A_2,B_2>$. Let for these processes $T_1 \times R_2 = \emptyset$, $T_2 \times R_1 = \emptyset$ and assume that the output O_1 is identified with the input I_2 such that $B_1 \supset A_2$. Let us construct, if possible, the process $P_3 = <V_3,D_3,A_3,B_3>$ such that:
1) $V_3 = V_1 \cup V_2$; 2) $I_3 = I_1$, $O_3 = O_2$;
3) $D_3 = D_1 \cup D_2$; 4) $A_1 \supset A_3 \ \& \ B_3 \supset B_2$.

<u>Definition</u>. Sequential composition of the AP P_1, P_2 is called the AP P_3 obtained at identifying O_1 with I_2 such that $B_1 \supset A_2$, which satisfies the above requirements 1) - 4). The same method is used to specify the composition of the processes P_1 and P_2, for which either $T_1 \times R_2 \neq \emptyset$, or $T_2 \times R_1 \neq \emptyset$, and $V_1 \cap V_2 \neq \emptyset$. Assume the T-ports of one process are identifiable with the R-ports of the other process such that $B_2 \supset A_1$ and $B_1 \supset A_2$ and we can construct the process P_3 such that

1) $V_3 = V_1 \cup V_2$; 2) $I_3 = I_2 \cup I_1$, $O_3 = O_2 \cup O_1$;
3) $D_3 = (D_1 \cup D_2) \setminus T \cup R$, $T, R \subset D_1 \cup D_2$;
4) $(A_1 \& A_2 \supset A_3) \ \& \ (B_3 \supset B_1 \ \& \ B_2)$.

<u>Definition</u>. Parallel composition of the AP P_1 and P_2 is called the AP P_3 obtained at identifying the T,R-ports between the processes such that $(B_2 \supset A_1) \land (B_1 \supset A_2)$, which satisfies the conditions 1) - 4).

By making use of the asynchronous processes, the sequential and parallel composition, we can simulate a broad class of interacting entities. In analyzing the correctness of protocols, in addition to the properties of composition, the properties of the AP network are of interest, which allow to determine the reachability of some vector from V, interpreted as a finite state of the AP network in which the execution of service functions takes place.

<u>Analysis of the model</u>. Analysis of the model proceeds in two stages. At the first stage it is assumed that the finite state of the AP network

is attained. If the composition of processes occurring in the protocol layer is admissible, partial correctness is determined by the validity of the assumption:

$$S_f \ \& \ B_f \Rightarrow I,$$

where: S_f is an assertion which is true when the finite state is achieved, $B_f = \wedge B_i$ is a conjunction of the input predicates of all i-processes in which S_f is valid, and I is a service invariant.

Let us consider the properties of the AP network graph in which each node is interpreted as an AP, while the edges are interpreted as couplings between ASs which are determined by their sequential or parallel composition. For each AP the l-edges corresponding to the I,O-ports and the r-edges corresponding to the T,R-ports are identified. We will assume for simplicity that each process is either a receiving or sending process. The finite state of the AP network will be attained if and only if the network graph does not contain deadlocks and closed loops which do not include the finite state.

Definition. A set of the receive processes $M = \{P_1^r, P_2^r, \ldots P_N^r\}$ is called a deadlock set if for each pair $\langle P_2^r, P_j^r \rangle \in M$ the following is true:
a) initiation of the P_j^r process depends on the completion of the P_i^r process;
b) the initiation condition for P_i^r, P_j^r is the arrival of a message over the r-connection;
c) between P_i^r and P_j^r there is a finite sequence of r,l-connections, such that the r-connections on the basis of (b) are included into this sequence.

According to the definition, the necessary condition for the existence of a deadlock is the presence of a loop in the AP graph so that this loop includes all the vertices interpreted as the processes from M. The presence of the cycle, however, does not mean yet that the deadlock situation will necessarily arise, since the latter depends on a mutual displacement of the receive and send processes.

Proposition. If the AP network does contain a cycle which includes all the receive nodes P_j^r, this cycle will be free of deadlocks, if in the graph:

a) there will be found the send node P_k^t such that P_k^t does not occur in the cycle, and the $\langle P_k^t, P_j^r \rangle$ belongs to the graph;
b) for each receive node P_j^r there will be found the send node P_k^t such that the $\langle P_k^t, P_j^r \rangle$ edge is the l-edge and does occur in the cycle, and there will not be found the send node P_l^t such that the l-edge $\langle P_j^r, P_l^t \rangle$ also occurs in the cycle.

Determination of the presence of closed loops which do not include the finite state of the AP network reduces to the following.

Definition. Graph G built on the basis of the AP network graph such that each $\langle P_i, P_j \rangle$ edge of the AP network graph will belong to G when P_i, P_j belong to the set of finite processes is called the reachability graph.

Proposition. If the AP network permits the composition of processes and is deadlock-free, the set of finite states is reachable if and only if the reachability graph is acyclic.

Interpretation of these propositions can be illustrated by the following simple example: in the flow control procedures a deadlock arises when the receive process does not receive the confirmation which would allow to shift the send window. As a result the send process is temporarily halted and its initiation can be performed only if there exists an independent process (process-timer). Frame retransmission on the expiry of the timer may proceed infinitely, since the completion of the recovery process depends on the completion of the receive process. If, however, the retransmission counter is used, the completion of the recovery process will not depend on the receive process, i.e. the reachability graph will be acyclic.

Example of incorrectness. One 'interesting' incorrectness involves the occurrence of a cycle with the use of procedures for transmission of supervisory frames to provide recovery in the link protocol. The formation of a cycle involves the following. If the timer for I-frames expires and the recovery with the help of S-frames is started, the possibility arises for a frame to be received with the F-bit set to I, in which no I-frame is acknowledged. However, the recovery will be completed, the retransmission counter will be reset and the changeover to the transmission of I-frames will take place. If the timer expires once again, the process will be repeated (and so infinitely, since the retransmission counter will be reset within this cycle). This incorrectness is accounted for by the following.

Since the probability of frame corruption is proportional to its length, the probability of corruption of I-frames of a standard length will significantly exceed the probability of corruption of S-frames. When the degradation of the communication line is encountered, the probability of recovery with the help of S-frames can become much higher than the probability of transmission of the I-frame, which will result in the situation in which the recovery completes successfully, but no I-frame is transmitted over the line, the lattaer circumstance results in timer expiry and the recovery procedures are initiated.

Such an incorrectness can be eliminated in three ways. First, the retransmission counter can be introduced, which will be reset on the reception of confirmation if only for a single I-frame, rather than on the exit from the 'abnormal' states. Second, the entry into the 'abnormal' states can be viewed as indication as to the necessity of diminishing the length of the I-frame and the adaptation of the length is to be made on the basis of the number of passages through the cycle. Third, it is possible to delimit the number of retransmissions of S-frames so that the probability of recovery becomes comparable to that of the transmission of an I-frame. The latter method is obviously more preferable, since it does not require reconsideration of the protocol procedures and can be used in the systems already implemented (the maximum number of retransmissions is a systems parameter, rather than an in-built value). This method has been proposed and described in detail in (Rotanov, 1981).

CONCLUSIONS

The paper describes the basic results of the efforts concerned with the development, formal description and verification of protocols for the Experimental Computer Network. The development of protocols on the basis of international standards and recommendations is compounded by their major shortcoming - the lack of formal description. The development of the formal description which is convenient, adequate and suitable for the designers involves the solution of two basic problems: the selection of the mechanism of interaction of the protocol processes and the methods for mapping the hierarchy. The use of the message passing technique (in the explicit or implicit form) allows to eliminate incorrect situations which arise as a result of the asynchronous nature of process interaction. For hierarchy mapping it is feasible to apply the automata-based methods of formal description of the service with the help of service primitives.

Verification is to be viewed as an indispensable stage of protocol development. One approach to verification involves representation of the protocol layer as a network of asynchronous processes followed by the analysis of the network graph. The use of correctness criteria based on the formal description of the service allows to detect the abnormal situations which are not erroneous from the point of view of protocol procedures, but do result in non-execution of the service. One such incorrectness is a possible endless loop occurring with the use of supervisory frames in the link protocol recovery procedures.

The authors express their gratitude to their colleague Mr. Yu. S. Podvisotsky for a number of valuable remarks and suggestions made by him in the course of prolonged discussions.

REFERENCES

Bochmann, G. V. (1979). Architecture of distributed computer systems. Lect. Notes Comp. Sci., 77, 179p.
Bochmann, G. V. and Sunshine C. (1980). Formal methods in communication protocol design. IEEE Trans. Commun., 4, 624-631.
ISO (1980a). Contribution by ECMA on transport protocols. ISO/TC 97/SC 6 No 247, 41p.
ISO (1980b). Proposal for session layer protocol. ISO/TC 97/SC 16 No 256, 26p.
ISO (1980c). A session layer protocol. ISO/TC 97/SC 16 No 318, 35p.
Kikuts, Ya. A., Yu. S. Podvisotsky and S. V. Rotanov (1981a). Data transportation in X.25 network. In: Packet-switched computer networks, Proceedings of the 2nd All-Union Conf., Riga, 39-43.
Kikuts, Ya. A., Yu. S. Podvisotsky and S. V. Rotanov (1981b). Session protocol for the experimen-

tal computer network. In: Packet-switched computer networks, Proceedings of the 2nd All-Union Conf., Riga, 44-50.

Kikuts, Ya. A., Yu. S. Podvisotsky and S. V. Rotanov (1982a). Information channel control protocol. IEVT A - 2, Part 1, Riga, 47p.

Kikuts, Ya. A., Yu. S. Podvisotsky and S. V. Rotanov (1982b). Information channel control protocol. IEVT A - 2, Part 2, Riga, 39p.

Rotanov, S. V. (1981). Selection of systems parameters for the information channel control protocol. In: Packet-switched computer networks, Proceedings of the 2nd All-Union Conf., Riga, 67-72.

Sunshine, C. (1979). Formal techniques for protocol specification and verification. Computer, 10, 20-27.

Yakubaitis, E. A., A. F. Petrenko, Yu. S. Podvisotsky and S. V. Rotanov (1979). Protocols for the experimental packet-switched computer network. In: Data Communications 79, Proceedings of the 4th Internat. Conf., Prague, 154-158.

Yakubaitis, E. A. (1980). Computer Network Architecture. Statistika, Moscow, 278p.

Zave, P. (1976). On the formal definition of processes. Proc. 1976 Int. Conf. Parallel Proc., 35-42.

Ziebert, M., Yu. S. Podvisotsky and S. V. Rotanov (1979). Transport service on the X.25 basis. Kommunikation in Rechennetzen, Proc. Int. Symp., Potsdam, 163-176.

Zibert, M., Yu. S. Podvisotsky and S. V. Rotanov (1981). Transport functions in the X.25 network. Automatic Control and Computer Sciences, 3, 64-70.

DISCUSSION

Wood: You are doing work in the analysis of the bottom two layers which is where actually most of the standards definition work is occurring on the PROWAY and the 802 projects. I'm concerned that we will need something at the network, transport and higher layers.

Ratanov: I think that on the bottom layers there are some problems too, as was shown here - the problem of incorrectness, despite the fact that this is a standard CCITT and ISO protocol.

Wood: In the documents published for the X-25 they have started to use state transition diagrams. Have you any comments, or are you in fact finding some of the problems you have identified, or are the problems being corrected, as they do in the state diagrams?

Rotanov: We use state diagrams too, but I think that the approach based on state diagrams doesn't allow to verify the protocols, especially the data transmission procedures of protocols, because the state diagrams do not cover signals, numbers, concurrency etc.

Wood: In a PROWAY definition we have been trying to use state transition diagrams as a more effective method of defining a protocol.

RING COMPUTER NETWORKS FOR REAL TIME PROCESS CONTROL

A. Gościński, T. Walasek and K. Zieliński

Institute of Computer Science, Stanislaw Staszic University of Mining and Metallurgy, Cracow, Poland

Abstract. The objective of this paper is to examine the problem stated as follows: In what way the production process parameters and the distributed computer system elements influence the DCCS performance. The analysis of the production process and the discussion of the distributed computer system have been presented herein. The simulation model of the DCCS has been described and discussed. The simulation results have been given. Some design aspects of the DCCS have been presented.

Keywords. Real time control systems; distributed computer control systems; ring structured local computer networks.

INTRODUCTION

There are many motivations to develop distributed computer control systems (DCCS). With these in mind, a variety of network topologies have been proposed and implemented so far in the hope of achieving efficient control of production processes. One of the widely known topologies proposed to be used in control systems is a ring. Ring or loop networks have a number of desirable properties. There are two classes of loop systems: decentralized and centralized. As the response times of the decentralized system are faster then the ones of the centralized systems, we will restrict our consideration to the former class. Various properties of the ring structured computer networks have been studied and presented elsewhere (Jafari,1980; Yu,1979; Giessler,1978). It seems to us that it is worthwhile to study a problem that can be stated as follows: In what way the production process parameters and the distributed computer system elements influence the DCCS performance? The objective of this paper is to examine the problem mentioned above. The starting point of our study is the statement that for real time process control applications computer networks require an efficient communication subsystem, the most important features of which are high reliability and short message transmission times.

The paper contains:
- a short analysis of the production process from the control computer viewpoint,
- the discussion of the elements of the distributed computer system influenced by the controlled process,

- presentation of the model of the distributed computer system, the production process, and their connections,
- the simulation results.

The object of our consideration was, that the performance functions of the DCCS have been defined on the basis of the analysis common to the process and the distributed computer system. To solve the problem characterized above we propose to use simulation methods, keeping in mind that the principles of simulation study and optimal design of computer networks for process control are often different from the principles governing the study of conventional computer networks. There have also been presented the results of simulation studies. Some suggestions concerning the design of the ring structured DCCS have been made.

PROBLEM STATEMENT.
DISCUSSION OF THE RESEARCH AREA

There has been proposed here an approach common to the distributed computer system and to the production process, i.e., they will be taken into consideration within one model as elements of its structure and/or its parameters. The interactions between them will also be characterized by a common set of performance functions.

Description of the Production Process

Production process analysis from the control computer viewpoint shows that the following parameters should be taken into account and modelled:

1. The location of information acquisition points and control influence points referred to as data sources and sinks- we propose to analyse the functional aspects of the location rather than the geometrical one, that is we are interested in: (i) which of the network nodes are connected with data sources and sinks? (ii) what are the interactions between sources and sinks?
2. The frequencies of the information acquisition and control influence- we propose to restrict the studies to two cases: (i) the arrival rates are the same for all nodes, (ii) two nodes have much higher arrival rates than others;
3. The numbers of information generation points and control influence points - this element is very closely related to the element pointed in 1 and will be modelled in the functional sense;
4. The lengths of the gathered data messages and process control messages. The worst conditions for the DCCS have been taken into account assuming that the message lengths are all equal and relatively high.

Another two elements of the production process have been taken into account: (i) the frequency of the control algorithms activation and the computational time of these algorithms and (ii) the location and level of disturbances that can interfere with the transmitted information. The effect of (i), (ii) on the distributed computer system is being studied at present and the results will be discussed in the next paper.

Distributed Computer System Elements

The following elements of distributed computer system should be analysed while considering its application to process control:
a) the communication subsystem and communication protocols - these influence the reliability, message transmission times, and the cost of the whole control system;
b) parameters of several computers and transmission lines - the effect of these elements on the control systems is the same as in (a);
c) the decomposition of the data base - this element very closely connected with the reliability, transmission times

and the cost of control system, though very important, will not be discussed here because it concerns other aspects of the distributed computer system ;

d) the cost of the whole DCCS - this can be considered at the quality level only and will not be modelled explicitly.

More detailed description of the distributed computer system elements is presented in Section 3.

Performance Functions

The simulation model has been used to evaluate the performance of the whole DCCS. This is done in terms of three groups of performance functions:

(i) the first group is related to interactions between the distributed computer system and the production process:
 - the average message delay,
 - the maximum message delay,
 - the average message waiting time in the input buffer;

(ii) the second group defines the quality of the communication subsystem services:
 - the average throughput,
 - the average data transmission efficiency,
 - the average retransmission coefficient,
 - the number of rejected messages;

(iii) the third group enables to estimate the utilization of several of the network elements:
 - the average node processor utilization,
 - the average channel utilization,
 - the longest queue in the transient buffer,
 - the longest input queue,
 - the longest output queue.

The performance functions given above are significant because of their direct relations with the reliability, response time and the cost of the whole control system.

The objective of this paper is to show how the efficiency of DCCS is effected by the production process factors and by the distributed computer system parameters. It is also our intention to give some suggestions which may be helpful in designing the ring structured DCCS.

DISTRIBUTED COMPUTER CONTROL SYSTEM MODEL

Distributed Computer System Submodel

Let us consider the ring structured local computer network in which various nodes are connected together by a full duplex communication lines. Messages are handled in a store-and-forward way and are segmented into packets of fixed size. It should be noted that all transmissions are point-to-point.

We also assume that transmission facilities enable transmission which is serial, synchronous, and transparent. Moreover, each node performs both typical communication functions (error control and recovery, flow control, serial to parallel and parallel to serial code conversion) and the functions connected with the message processing (destination address checking, data field processing, buffer management). This implies the choice of the model. The scheme of the node is depicted in Fig.1. Due to such a model it is possible to analyse the parameters of the node elements, such as follows: size of the buffers (input, output, and transient), speed of switching operation, and the parameters of the protocols, especially the line protocol.

It has been assumed that the channels are full duplex and synchronized. The channels are defined by two parameters: throughput $[\text{bit/s}]$ and propagation time $[s]$.

The model of the distributed computer system actually contains some elements of the three lowest layers of the ISO Open System Architecture (seven layer protocol model) (ISO,1981). Some physical aspects are manifested by the model of the node. The second data link layer pf the ISO protocol model, i.e. line protocol, is realized in our model by DDCMP protocol (Digital,1974). We propose to use this protocol because of its practical application in computer systems. The DDCMP is designed to operate on synchronized full-duplex or half-duplex channels, switched or direct links, point-to-point or multipoint networks, and serial or parallel transmission facilities. DDCMP protocol allows transparency and accomodates both synchronous and start-stop modes. Moreover, it can be implemented in many operating systems. Thus, the DDCMP fulfilles, the requirements of the distributed computer system model discussed herein. The assumed line protocol is characterized by the following parameters, which have been taken into consideration in the constructed model: time-out, window size, data and control pocket lengths.

The third network layer is represented in our model by two groups of algorithms:
(i) the access control mechanism; we consider three mechanisms:
   α - the ATDM access control mechanism which assigns higher priority to the messages arriving from the ring.
   β - the DLCN access control mechanism which assigns higher priority to the messages arriving from the production process.
   γ - the mechanism which assigns varying priority to messages waiting in transient or input buffers; it depends on the ratio of the queue length at the moment the node processor becomes free to the maximum buffer capacity. The buffer for which this ratio is greater is served first.

It should be noted that all buffers are served according to FIFO algorithm with no regard to access control method.
(ii) the direction of the information flow:
A - the transmission in one direction,
B - the transmission in both directions with the choice of the shortest path as far as the number of intermediate buffers and nodes is concerned.

The ATDM mechanism seems to be useful when we want to separate the distributed computer system and the production process, e.g. to avoid the network congestion. The DLCN access control method allows close and firm linking of the distributed computer system with the production process, when the communication subsystem is not overloaded. The third mechanism (γ) does not prefer either the input or transient buffer.

One direction information flow (A) is characterized by the simplicity and low costs of the solution (the transmission facilities are much simpler). This type of transmission may be treated as half-duplex one. The transmission in both directions (B) requires more complicated communication subsystem facilities but enables shorter transmission delays.

Detailed description of the assumptions made in the simulation model are contained in (Gościński,1981).

Process Submodel

All information sources are modelled by the stochastic streams connected with several nodes. The following information is related to each of the messages (data packets):
- the address of the destination node (generated in a stochastic way according to the given distribution),
- the moment the message arrives to the node, practically to its input buffer.

In our model it is also possible to define the length of the message. As we have assumed that the length of the message is constant, this possibility is irrelevant to the question discussed here.

Several functional structures of the production process have been taken into consideration. They are presented in Fig.2 and characterized in short:

1. The input stream is connected with one node only, e.g. the process is connected via this node with one central data base which is the information source for other elements, or one controller controls many elements of the lower intelligence level with the remaining nodes;

2. One node is a destination node and all the others address their messages to this one; the input streams are connected with these nodes, e.g. the information is gathered in one centralized data base. The information acquired from many points is processed in one node;

3. Superposition of 1 and 2 given above;

4. Each of the nodes has the same connection with the production process from the point of view of its functions. Each node has its own input stream and is a destination node; the arrival rate is the same for all nodes, e.g. the process consists of closely connected subprocess, the control system realizes many functions such as: the direct digital control, optimization, the manufacture process control, and the management;

5. The message arrival and destination structure is as in (4), but the two remotest nodes X, Y address the messages mutually to each other much more frequently than the others; the arrival rate is the same in all nodes;

6. The load scheme is as in (5) but the most frequent interactions occur between two neighbouring nodes X,Y;

7. The load is similar to the one in (5), so the messages sent by the node X to the node Y are more frequent than those sent to the remaining nodes. The only difference between (5) and

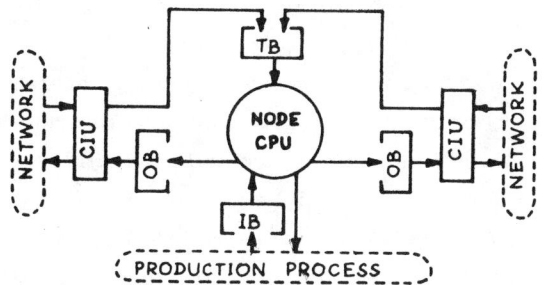

Fig.1 Node model; CIU-Communication Interface Unit, IB,OB,TB - Input, Output, Transient Buffers respectively.

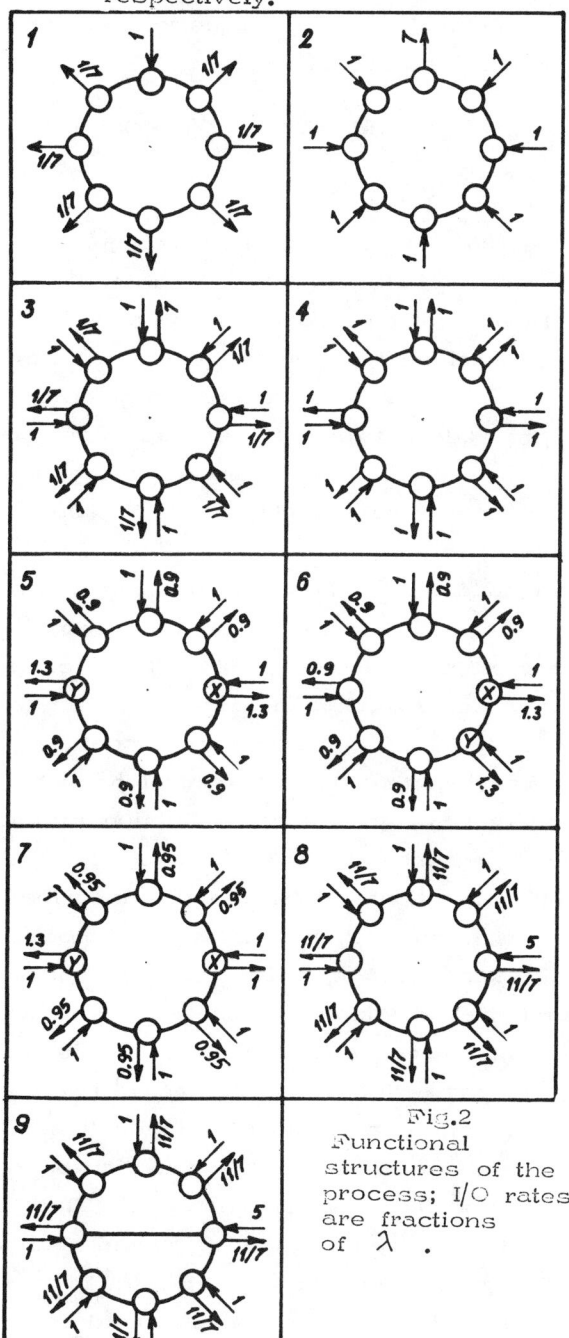

Fig.2 Functional structures of the process; I/O rates are fractions of λ.

(7) is that the node Y is loaded in the same way as the rest;
8. Destination addresses are generated in each node according to uniform distribution but the arrival rate in one node is much higher;
9. The load scheme is as in (8) but the indicated nodes are connected to each other by the additional channel. This channel is used only by the source node X. This is the only modification of the ring structure that has been discussed here.

SIMULATION STUDIES

Simulation Area

A set of simulation results has been presented below in order to show the influence of the production process and the distributed computer system parameters on the DCCS. We are considering eight node network with 1 Mbps communication channels. The carrier propagation time in transmission lines is 100 000 kmps.
All nodes, as it has been assumed, are identical. The buffer size (input, transient, and output) enables to store 64 packets. A packet size, chosen to send the longest message from/to production process is 128 bytes; the protocol control packet length is 8 bytes. The production process is represented by the messages streams. Their functional distribution is described by their connections with several nodes, by their message arrival rates in nodes, and their destination addresses. Message arrivals are assumed to be the Poisson processes with the parameter λ [messages per sec]. Message destinations are selected according to the distributions specified in "Process Submodel". The parameter has been changed in a wide range. This makes possible to study the performance of the DCCS in various load states. Each simulation run was stopped after the reception of 1000 messages in their destination nodes had been completed. The simulation studies have been made for 9 process functional structures presented in Section "Process Submodel".

Two elements of the distributed computer system effecting the DCCS performance have been taken into consideration:
- the directions of the information flow,
- the access control mechanisms,

i.e., both protocols for the network layer of the ISO protocol structure.

Simulation Results Discussion

The first part of this section is devoted to the performance comparison of the DCCS which are connected with the production processes of various structures. The discussion presented below concerns the DCCS performance when the information is transmitted in one direction.

The analysis of the process functional structures denoted as 1 and 2 (Fig.2) shows, that for the identical arrival rates both are equivalent for each performance function assumed. This correlation was not evident because the structure 2 is characterized by the conflict between the input and transient streams in each node.

The comparison between the structure 4 and the structure 3 which results from the superposition of the structures 1 and 2 shows that performance functions are dependent on the destination addresses of several message streams. Fig.3 shows the average packet delay and network throughput, both versus arrival rate per node λ .

There has been observed a similar kind of dependence for other performance function. This illustrates the fact that the process which has a homogeneous destination structure is much more effectively served by the system than the one which has a heterogeneous structure. The process structures 1-4 are characterized by the growing symmetry of the distribution of the message sources and

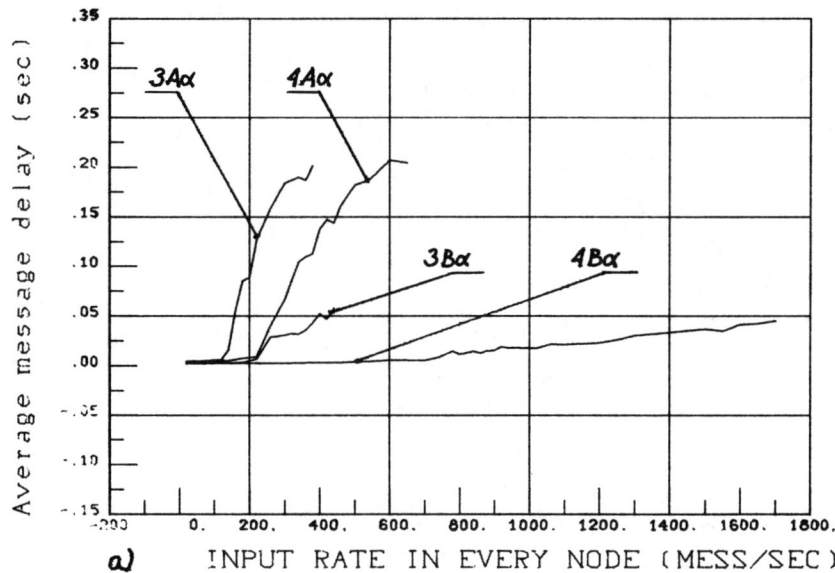

a)

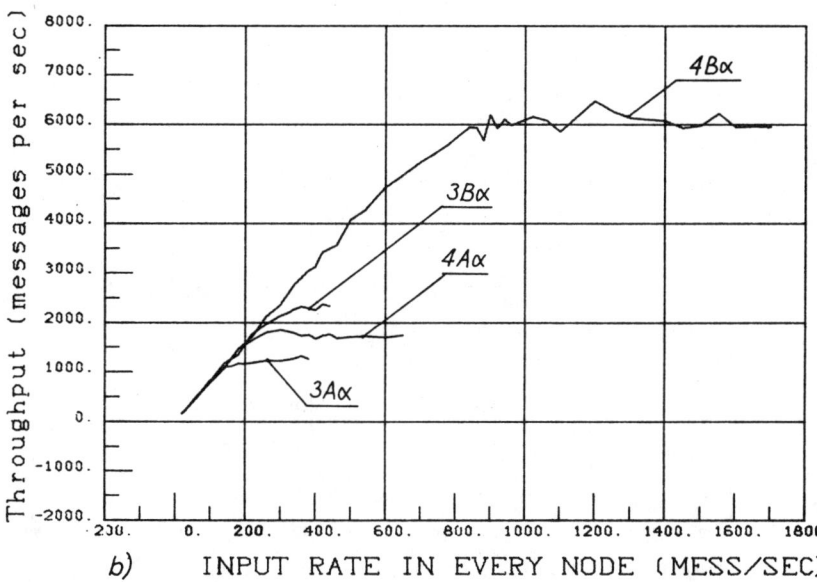

b)

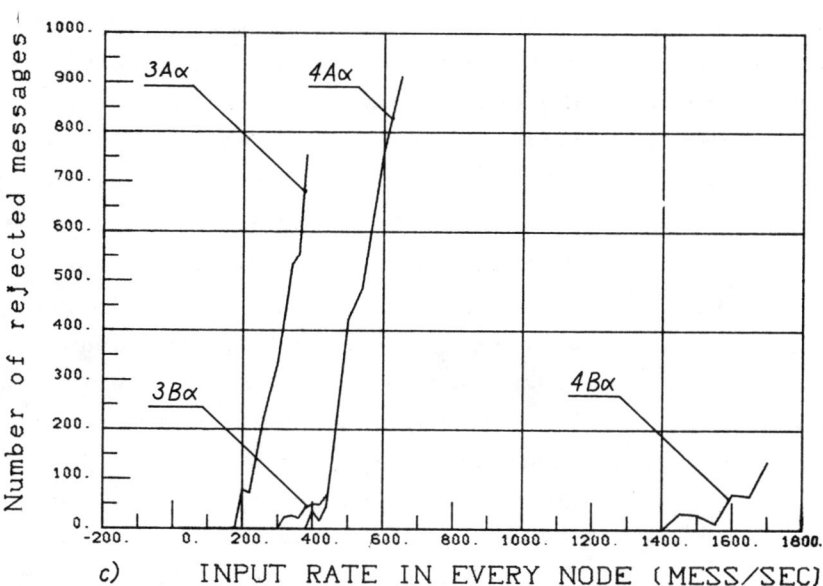

c)

Fig.3

System performance; N (N=1,9) - the index of the process functional structure; A,B - the index of the routing algorithm, α, β, γ - the index of the access control mechanism.

sinks. The analysis of these process structures shows the improvement of the DCCS performance.

Our next problem concerns the process structures 5, 6, 7 (Fig.2), characterized by the unsymmetrical message addressing in various nodes and by the same arrival rates as it has been assumed in case 4. It has been observed that the unsymmetrical addressing does not cause any significant difference between cases 5,6,7. It can be explained by the fact that in these cases the average packet lifetime is the same. On the other hand, the performance functions of the structures 5,6,7 are worse than the functions in the completely symmetrical case denoted as 4.

The influence of the unsymmetrical load (cases 8 and 9) on the system performance has been presented in Fig.4. The difference between the curves in Fig.3c and the curves in Fig.4b indicates that the communication network in case 8 has

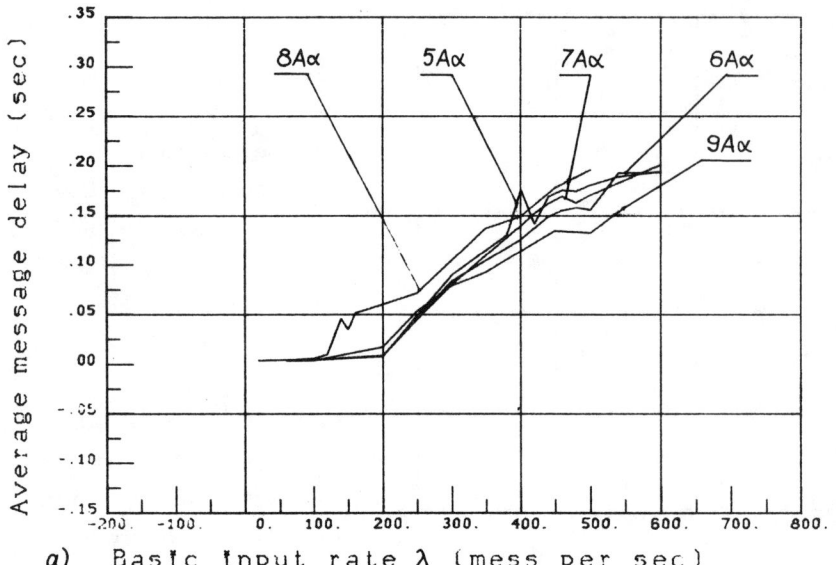

Fig.4

System performance

a) Basic input rate λ (mess per sec)

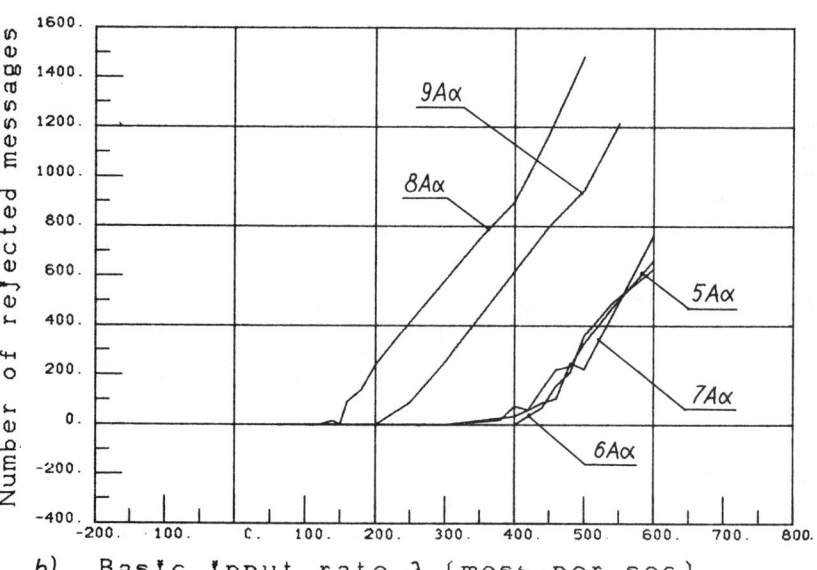

b) Basic input rate λ (mess per sec)

become congested under the load smaller than the one in case 4. When the input rate does not cause congestion the performance of the system in case 8 is much worse than in case 4.

We propose to use an additional channel in order to improve the system performance under unsymmetrical loads. The modified network topology has been denoted as 9 in Fig.2. This modification has caused the increase of the critical input rate, that is the input rate which produces communication system congestion. Moreover, the improvement of all performance functions is visible in case 9.

There have also been investigated the process structures 1-4 together with the shortest path routing algorithm (B). For this protocol the improvement of all performance functions has been observed (Fig.3 and Fig.5). In case of this protocol the performance functions are more sensitive to the functional structure of the production process than in case of protocol A. We have also studied the influence of access control mechanism on the system performance in case of symmetrical load (structure 4) and two alternative routing algorithms (A and B).

It should be noted that access control mechanisms do not have a great effect upon the performance functions characterizing both the interactions between the distributed computer system and the production process (Fig.5a), and the quality of the data communication subsystem (Fig. 5b). However, it has been shown that in the overloaded distributed computer system the access mechanism α prevents the system from congestion but there occur message rejections (Fig.5c) The mechanism β does not possess this feature however it causes packet retransmissions and the system instability which is manifested by the performance degradation. This instability is connected only with the heavy loaded system. The mechanism γ causes retransmissions and rejections. The access control mechanisms α, β, γ cause various degrees of utilization of the communication subsystem elements.

The application of the mechanism α requires the input buffers of large size and small transient ones. The access control mechanism β presents opposite requirements. The mechanism γ does not prefer any buffer (Fig.5d,e).

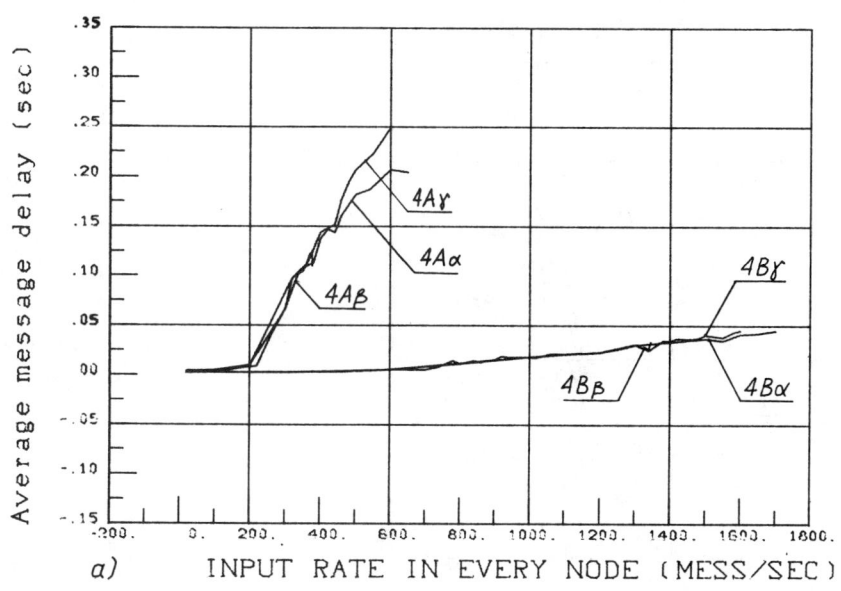

Fig.5a

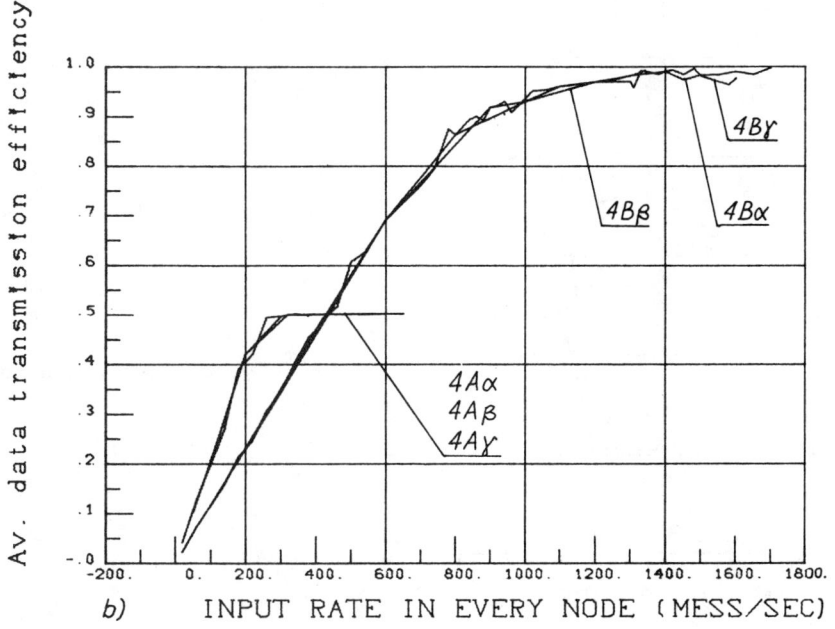

Fig.5b

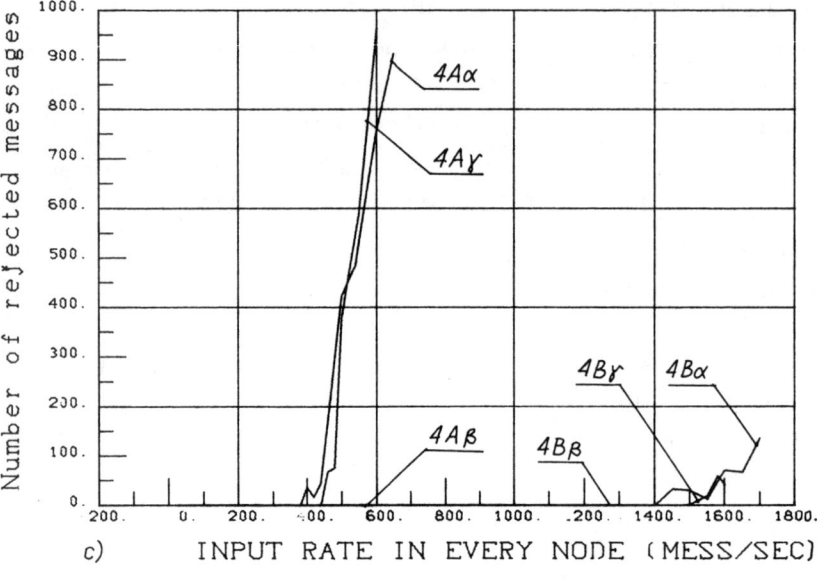

Fig.5c

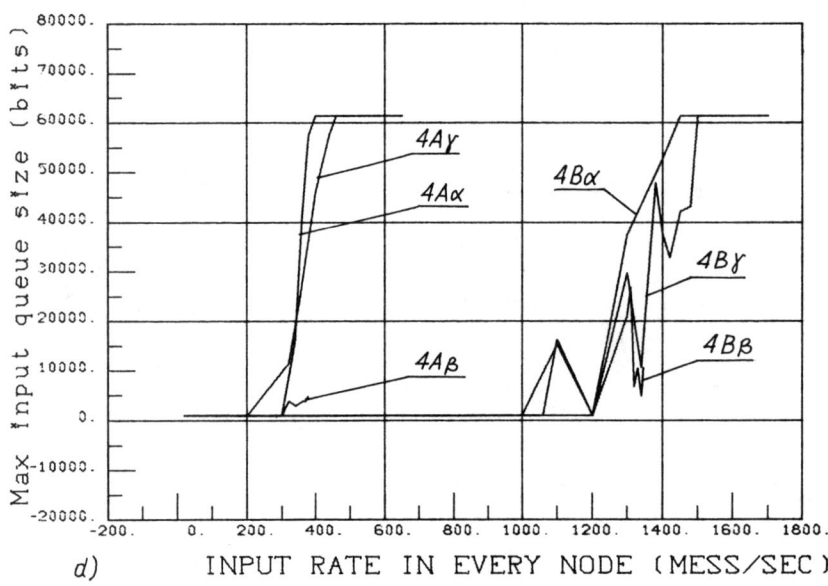

Fig.5d

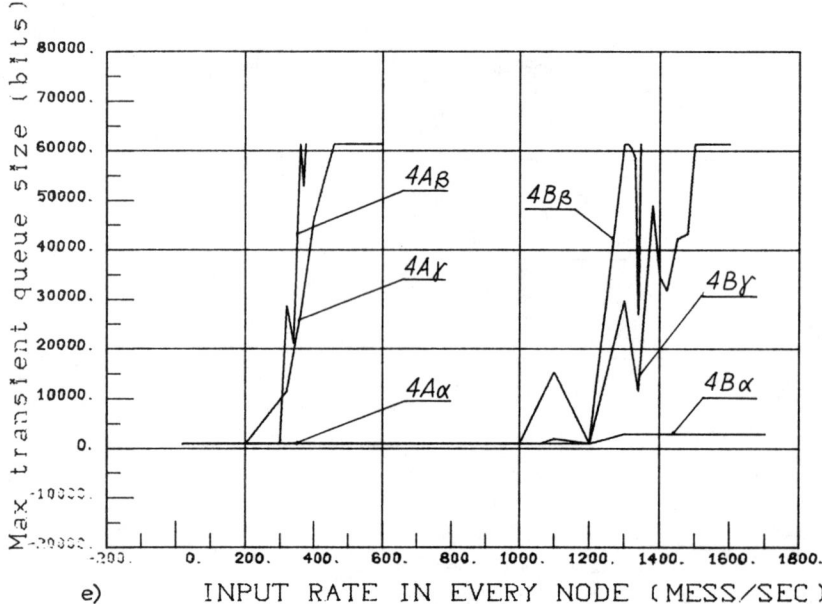

Fig.5e

System performance

CONCLUSIONS

The influence of the production process functional structure and the parameters of the distributed computer system on the DCCS performance has been discussed based on the common approach to the both factors given above. This problem has been solved by simulation. The simulation results obtained show: (i) the strong dependence of the DCCS performance on the distribution of the message sources and sinks, (ii) the system performance improvement obtained due to the introduction of the additional channel to the distributed computer system unsymmetrically loaded, (iii) the strong influence of the direction of the information flow on the DCCS performance functions, and (iv) the effect which is not to great, of the access control mechanism on the performance functions which characterize interactions between the distributed computer system and the production process and the quality of the data communication subsystem; the access control mechanisms cause various degrees of utilization of the communication subsystem elements.

REFERENCES

Digital Equipment Corporation (1974). Specification for DDCMP - Digital Data Communications Message Protocol; Reference Manual. 40 pp.

Giessler,A., J.Hänle, A.König, and E.Pade (1978). Free Buffer Allocation - an Investigation by Simulation. Comput. Networks, 2, 191-209.

Gościński,A., T.Walasek, and K.Zieliński (1981). Modelling and Simulation of Local Computer Networks. 5th Winter Scientific School on Computer Networks. Wrocław, Nov. 1981 (submitted to: Scientific Papers of Wrocław Technical University, in Polish).

ISO/TC9/SC16, 1981. Data Processing - Open Systems Interconnection - Basic Reference Model. Comput.Networks, 5, 81-118.

Jafari,H., T.G.Lewis, and J.D.Sparings (1980). Simulation of a Class of Ring-Structured Networks. IEEE Trans.Comput., 5, 385-392.

Yu,L.W., and J.C.Majithia (1979). An Adaptive Loop -type Data Network. Computer Networks, 3, 95-104.

DISCUSSION

Rodd: I'd like to thank you for a most interesting paper. This was one of the problems that we are all wrestling with.

Your graphs reassure me that the work we were doing in simulating our system in fact had some relevance. I wonder if you could try to give us some explanation of some of the odd characteristics of the graphs you showed. I'm not a mathematician and I have great difficulty in understanding why. As you load the system, clearly your delay time, message through put time gets worse and worse. Your graph shows the characteristics that we found that it suddenly picks up again and things get better. Then you are increasing the load and it obviously gets worse and worse again. Have you any explanation for this? Because I haven't.

Gościński: I can't give full explanation of this phenomenon. We have had some experience with the problems stated by Professor Rodd and I would like to show once more, Fig. 4b of my report. It is very important that we have carried out a simulation with a very wide input rate in every node and we can see three regions in these graphs. The first one is not interesting for us because the system is not loaded. The second region corresponds to the loaded nodes with zero rejection as compared to the third region, where there are rejections. From the computer control viewpoint it is impossible to recover any information. In this situation it is necessary to use another solution. We can see this proposition on Fig. 5c. It is oriented in the study of the direction of information flow. For example, it is possible to have very good results for one direction only. But in this area it is impossible to use that mechanism. We use two directions of the information flow and get good results with all mechanisms.

Wood: I think in a way it is unfortunate that you missed the earlier paper by Tom Harrison on the 802 system, because one of the approaches is using rings with a priority algorithm where you have at least three levels of stations. Some with high priority get a guaranteed percentage of the total bandwidth, and the others have low priorities and probabilities, whether they get in with an equal share of the bandwidths or not. Do you have any particular conclusions, or have you studied the effect of having some stations with different priorities?

Gościński: No, we don't have any distinct solution in this area. Now we have oriented our studies on the access control mechanism only with the ring structure. But in about four or five months we'll begin a study oriented toward the problem stated in your question. I haven't got any results now.

Inamoto: I am interested in the performance evaluation of the distributed system. You analyze a structure with identical nodes but a production control system has usually hierarchical structure where the nodes have unequal data flows. Have you any idea of the particular application field of your system? And my second question is whether this approach will also be effective with the distributed computer control system and with the hierarchical structure.

Gościński: A potential user of our approach is a chemical company - we are trying to find the best topology for their production control system.

I think that you are right talking about the hierarchical structure as the most useful for production processes. But I think that it is possible to use other structures as well.

BOTTLENECKS IN THE DESIGN AND IMPLEMENTATION OF DCCS AND THE WAYS TO FIGHT THEM

A PANEL DISCUSSION
T. T. Harrison

I.B.M. Corporation, Boca Raton, Florida, USA

ABSTRACT: This paper introduces a set of four papers which provide the basis for a round table discussion. These papers are short in length and incomplete in content. This is by design as the organizers of this Workshop feel that the essence of the panel discussion should be the extemporaneous statements of the workshop participants which, in somewhat edited form, are also published here.

INTRODUCTION

A panel discussion has been a feature of each DCCS Workshop. The format we use provides each panel member an opportunity to make a brief (5 to 10 minute) statement of his position on the subject selected for the panel discussion. Following these statements, the other Workshop participants are invited to ask questions of the panelists or to share their opinions on the subject. The resulting discussion is magnetically recorded and transcribed. Although edited and in some cases, summarised, the discussion is then published in the Proceedngs of the Workshop. In common with most verbal conversations, the resulting transcript may lack organisation; topics may arise in isolation, disappear, and then later reappear in the conversation. In addition, errors in transcription or interpretation sometimes occur as we all struggle with the English language. The reader is asked to bear with us as the transcript is read. It has been our experience from the past Workshops that the panel discussion often emphasizes the common experience of the practitioners of computer control throughout the world. That realization alone is valuable information for all of us as we pursue our independent tasks.

THE PROBLEM STATEMENT

Basically, we are looking for "bottlenecks", those things in the project development process that impede our progress toward the goal. In addition, however, we would like to discuss if, and how, we can "debottleneck" our projects.

The "if" in this paraphrased statement of the problem may be provocative since we usually assume that "debottlenecking" is always desirable. Perhaps we should remember, however, that the published bottleneck shape of the common Coca-Cola bottle turned out to be a solution to problems associated with shockwaves in high speed aircraft! With an open mind, perhaps some of our "bottleneckes" will become solutions, rather than problems.

The members of the panel, meeting to plan the discussion, agreed unanimously on one point: the problem described is very broad; there are many potential "bottlenecks" in any project, involving both hardware and software and ranging from theory to practical implementation. Rather than attempting to narrow the scope, however, the panel chairman has encouraged the panelists to pick their favourite bottleneck, wherever it may be within the broad scope. The chairman has also suggested, however, that the relation of the problems, and their potential solutions, be contrasted or compared to their equivalents in non-distributed computer control applications.

So, let us hear from the panelists and other participants -- but don't forget the "Coke" bottle!

TRADE-OFFS AMONG COST, PERFORMANCE AND RELIABILITY — A CASE STUDY

S. Narita

Department of Electrical Engineering, Waseda University, Tokyo 160, Japan

INTRODUCTION

The design and implementation of a distributed computer control system can be both easier and harder than that of conventional centralized systems. It can be easier since we can make full use of an infinite number of combinations of potentially useful hardware and software resources as well as communication facilities so that we have the possibility of tailoring optimum systems to meet a wide variety of user requirements. It can be harder since we have too many combinations and too many freedoms, leading to too many complexities.

The design approach of DCCS has two axes, vertical and horizontal. The vertical axis consists of a set of vertical levels or layers, i.e., subsystem/network level, nodal level, internal computer level and card level or integrated HW/SW level, from top to bottom. Horizontally, the design follows analysis, partitioning, allocation and synthesis phases. There are a plurality of factors or criteria to be considered in the respective design activity phases (cost, nodal constraints and computing capabilities, performance, RASIS, and so on.) In other words, the design of DCCS inevitably is a multiobjective optimizing problem and the system designer will confront the problem of trade-offs among many conflicting requirements.

The purpose of this short note is to demonstrate the difficulties of the trade-off problem through a case study on the hierarchical distributed computer control of power systems.

EVALUATION MODELS

In order to facilitate not only qualitative but also quantitative evaluation of DCCS, we need to have simple yet realistic models. The model building was a very time-consuming and tedious work, because we had to collect a vast amount of information from a total of nine utility companies in Japan about their existing and planned electric stations, computer control facilities and communication channels. For example, we needed to have quantitative information on such items as the number of control stations in respective hierarchical levels, the number of feeders and transformer banks at substations to calculate the information flow rates on "up-going" and "down-comming" communication channels such as cyclic digital telemeters. We also needed reliability data such as component failure and repair rates not only of power apparatus but also of communication channels and computing facilities including peripheral devices.

Figure 1 shows the three level hierarchical control system model constructed on the basis of the quantiatative data mentioned above. The model can be regarded to represent an average computerized power control system in Japan. For the purpose of comparison, a two-level model was also constructed that covers the same service area, i.e., the numbers of power stations of respective types (nuclear, thermal and hydro) were taken to be the same for the two models.

SYSTEM ARCHITECTURE AND DESIGN CRITERIA

Use was made of the power control system models to study how and to what extent the system configuration or system architecture was affected by the choice of design criterion.

Number of Hierarchy Levels

The choice of the optimum number of hierarchy levels in DCCS is an utmost difficult undertaking bacause so many factors are involved. For example, the Oita Works of the Nippon Iron and Steel Corporation essentially adopted a five-level achitectaure (partially six layers) while the Ogishima Plant of Nippon Kokan Company employed a four-level configuration. (Since the advent of high-performance microcomputer-based controllers, it has become common practice to reduce the number of control layers for reliability and simplicity.)

If we are concerned only with system reliability, disregarding other (possibly eqaully important) criteria such as cost, performance, and expandablity, the two-level power control system model is superior to the three-level configuration, where the amount of information loss caused by the failures and/or malfunctions of the components of computer control system including the data communication system is taken as the measure of overall system reliability.

Figure 2 shows the information flow rates and the availabilties of the control stations and communication channels. Table 1 lists the amount of informaton generated at or passing through the respective electric stations and the amount of information loss for the two- and three-level configurations. As seen from Table 1, the information loss of the two-level congifuration is about 50 percent less than that of the three-level configuration in this particular case.

When it comes to other criteria such as the ease of control and expandability, the three-level configuration may have many favorable features.

Number of Substations
--- Cost and Reliability

The next problem is : What are the optimum numbers of subdivisions at respective layers of hierarchy for the best economy or for the highest reliability? Or, what is the optimum grouping size of electric stations (controlled plants)? The two- and three-level power control system models were employed again to study this problem where the numbers of control stations at respective layers are not fixed but taken as variable parameters. In order to evaluate the total system construction cost, we needed several cost models and cost data for the components constituting the models such as computing facilities and communication channels.

Figure 3 plots the total system cost and the overall system reliability in terms of information reachability (=100 % - percent information loss) against the number of control stations (the number of subdivisions) for the two-level model. It is seen from this figure, for example, that if one specifies the system reliability level to be 95 percent, the optimum number of control stations is 16 or 17 and the total construction cost is 2.58×10^7.

Similar quantitative evaluation was performed for the three-level configuration. Figure 4 shows the plots of the information reachability and the total number of control stations against the total construction cost for the two- and three-level models. It is seen from this figure that, for the same system reliability, i.e., the same information reachability, the two-level configuraton is preferred.

Number of Subdivisions
--- Performance

Performance is another important issue in the design of DCCS. Once again, the power system control models were considered. The problem here is : What is the optimum number of subdivisions for the best control performance? A wide variety of control functions are involved in power system control such as switching operations, data logging, system state monitoring, load-frequency control, economic dispatch and voltage control.

The measure of control performance employed here is the loss of energy in terms of megawatt-hours caused by power failures. Figure 5 plots the probability of occurrence of power failure and the restoration time against the magnitude of power failure in terms of loss of load in megawatts.

The loss of energy is closely related to the number of subdivisions or the area for which each control station is responsible. Too many subdivisions give rise to an increase in the loss of energy because power failures whose magnitudes are greater than some threshold value require cooperative control operations among several control stations for system restoration. Too small subdivisions, on the other hand, have the problem that if the control station in charge of restoration happens to be "down", it may take a long time to restore the power failure section by manual restoring operation.

Figure 6 shows the plots of loss of energy or system performance against the number of control stations or subdivisions, where the parameter A stands for the ratio of the manual restoration time to the computerized restoration time. It is seen from the figure that the optimum number of control stations is around five, which is far smaller than the number of subdivisions for the highest reliability.

DISCUSSION AND CONCLUSION

In designing DCCS, the system designer may be confronted with the trade-offs among several conflicting requirements such as limited budget, highest reliability and best performance. The simple examples described so far demonstrated that the system configuration in terms of the number of subdivisions was greatly affected by the design criterion adopted. The approach most commonly taken may be to assign the highest priority to system reliability at the sacrifice of degraded performance.

The cost-performance-reliablity trade-off problem may exist in almost all DCCS. In the field of digital instrumentation/control of continuous or discrete processes, there have long been arguments regarding the choice of the number of control loops for which a multi-loop controller is responsible. Starting from over one hundred DDC loops at the early age of process control computers, the advent of microprocessors has reduced the number of loops, and a wide variety of single-loop digital controllers are in the market today.

Once again, the design of DCCS is easier and harder, bacause we have so many choices and so many degrees of freedom in the determination of best system architecture.

REFERENCE

1. S. Narita, et al.: Reliability Evaluation of Hierarchical Power Control Systems and Design of Component Redundancies, IEEE Trans. on Power Apparatus and Systems, Vol. PAS-95, No. 3, May/June, 1976.

TABLE 1 Information Generation and Reachability of Two- and Three-Level Models.

Power system model	Information amount	Arrival information	Reachability	Information loss
Three-level	76,250	75,321.369	98.7821 %	1.2179 %
Two-level	62,050	61,546.351	99.1883	0.8117

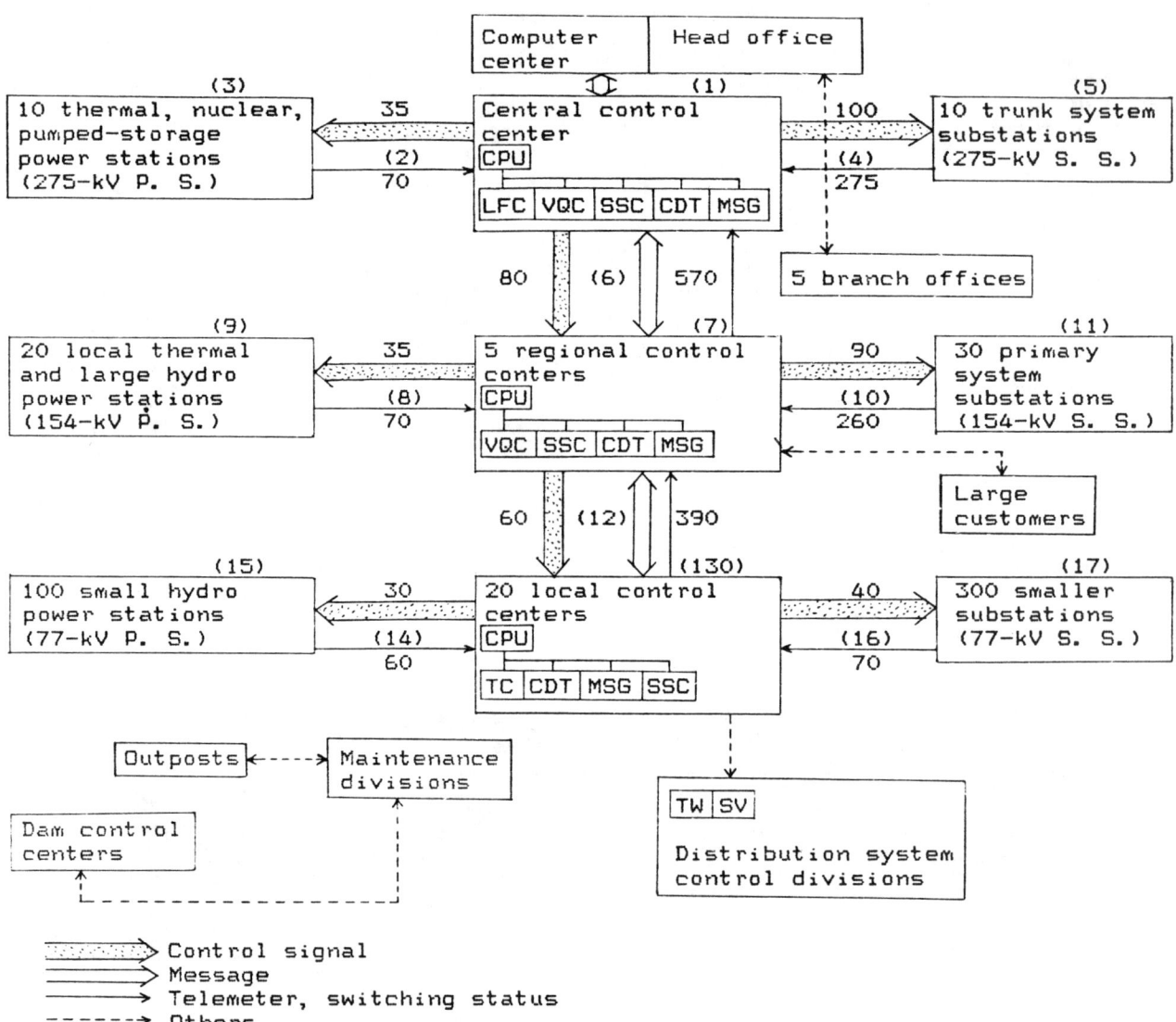

Fig. 1. Three-level computerized power control system model.

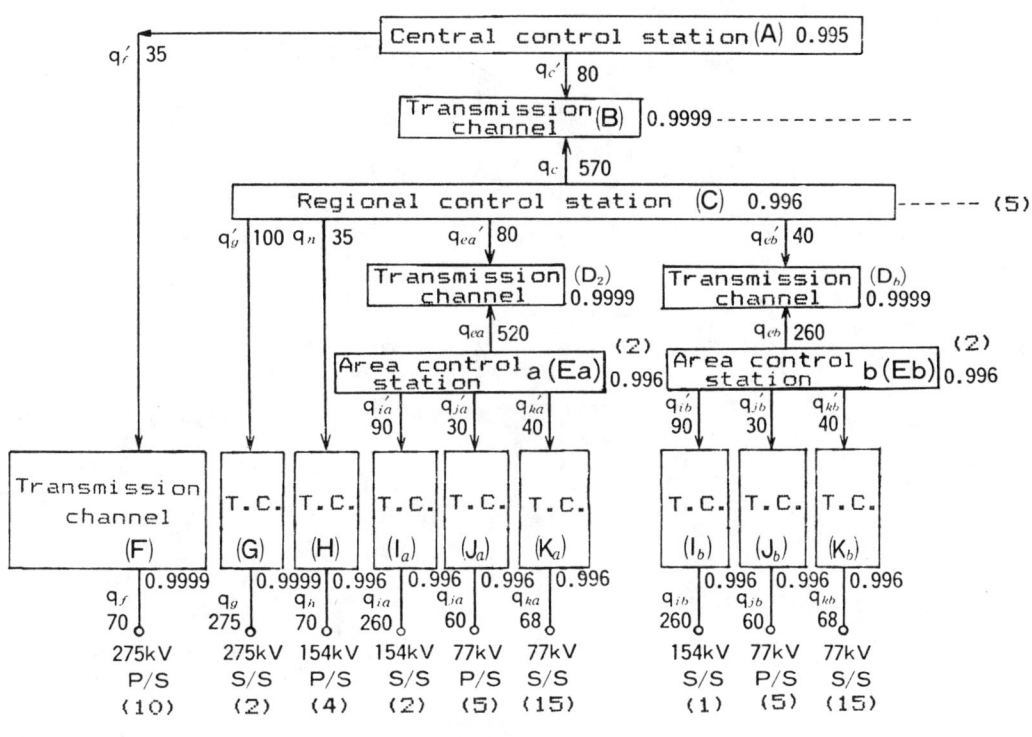

Fig. 2. Availabilities and information flows in 3-level model.

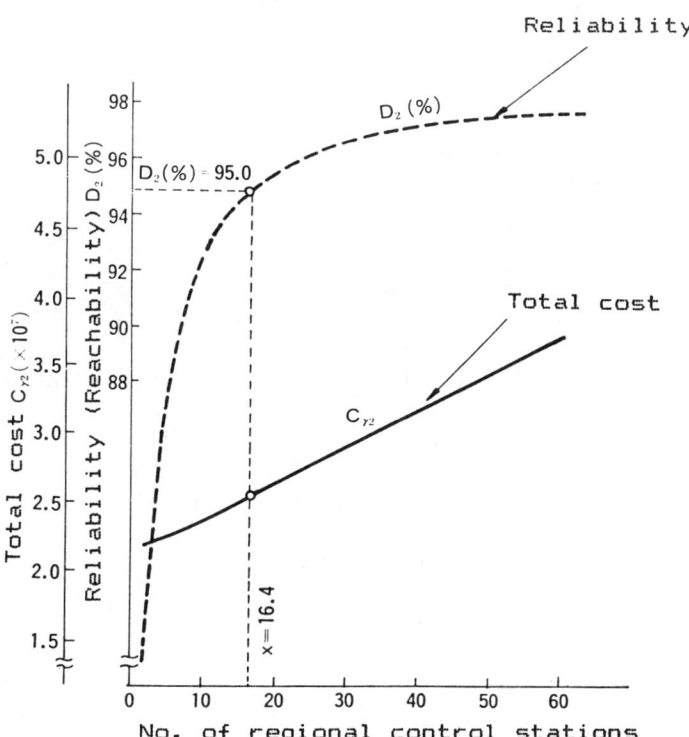

Fig. 3. Reliability and cost vs. no. of regional control stations.

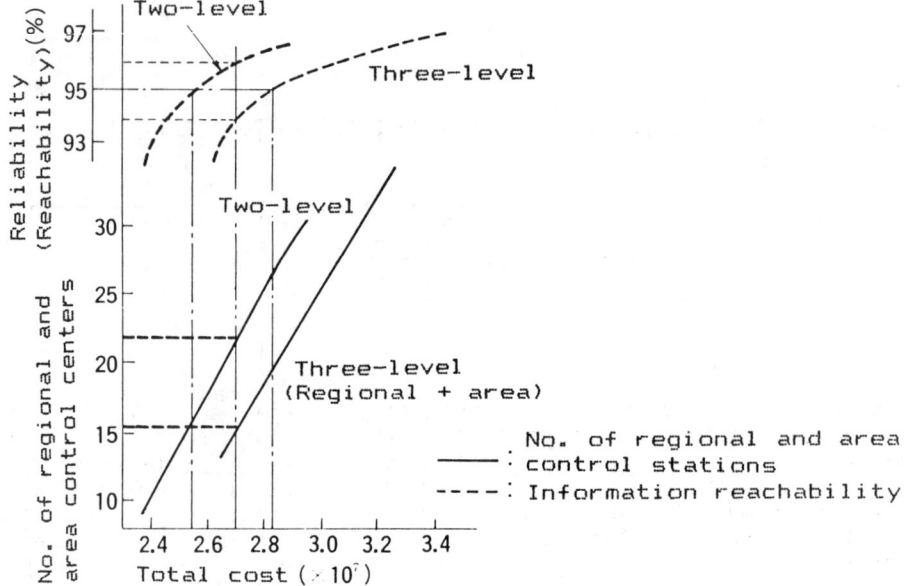

Fig. 4. Comparison of two- and three-level models.

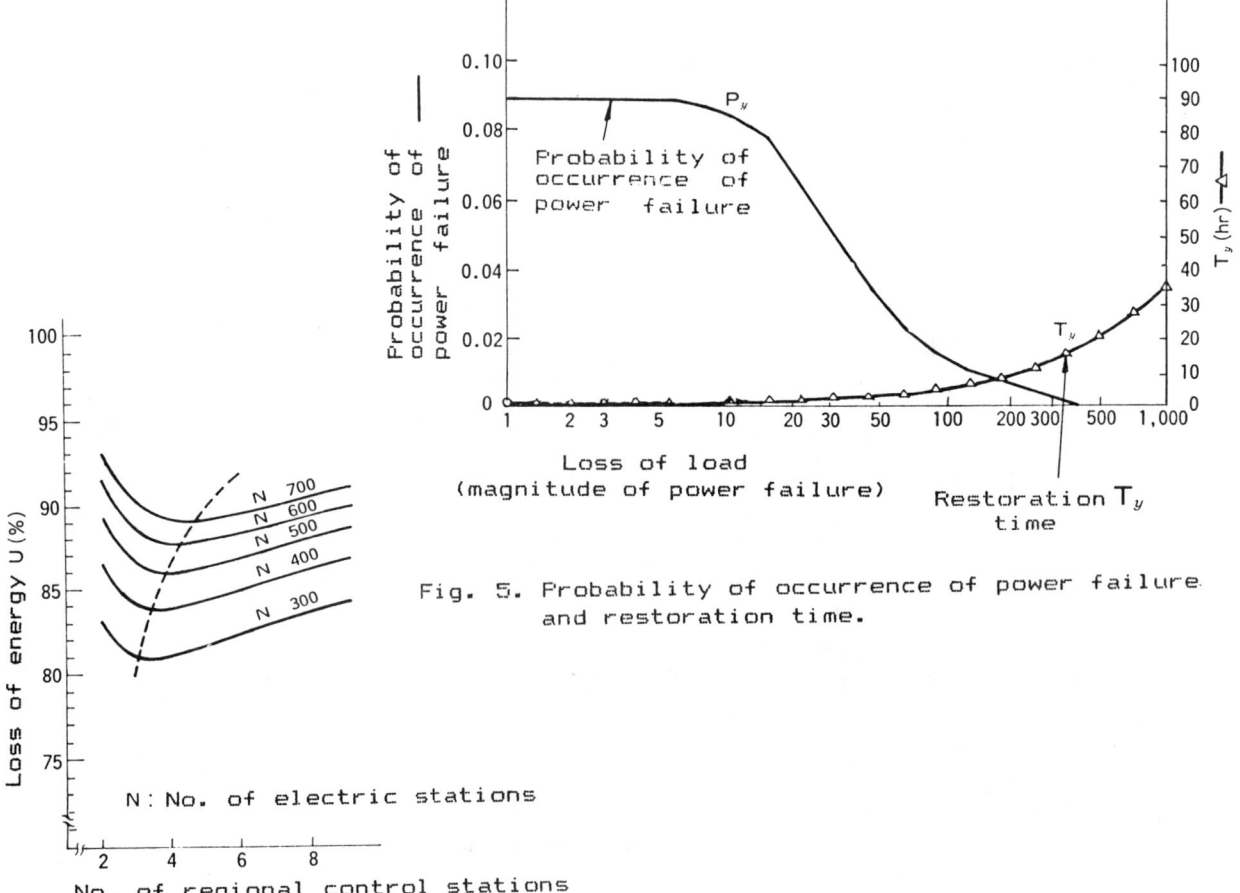

Fig. 5. Probability of occurrence of power failure and restoration time.

Fig. 6. Loss of energy vs. no. of regional control stations.

DISTRIBUTED VERSUS CENTRAL CONTROL

M. Maxwell

Control Systems, Colgate-Palmolive Co., New Jersey, USA

My business is making detergents by computers if profitable. Right now I'm involved in a decision about whether to implement a specific control scheme with central or distributed control. It will probably be central and the reason is cost. Let me elaborate on that a little.

Detergents are small quantity low profit high competition items. Computers are necessary for competition because they save materials and labour, improve control over quality, increase the throughput of the existing equipment. But computer investment must be consistent with profit range. Our computer investment, therefore, must be very small. With this in mind we are looking at the inevitable transition to distributed control as a matter of time. At present economics favour the central computer in our case because we have relatively short wiring runs and they make central control less expensive than it would be in a larger area system. The second reason is that we have a very highly dollar-efficient system at present and that makes it very difficult to find a substitute of any kind, whether it is distributed or centralized. So, we are watching the course of centralized versus distributed for our particular situation. When the cost of wiring rises high enough, we all go distributed.

THE CHALLENGE OF STANDARDS FOR PLANT COMMUNICATION

G. G. Wood

The Foxboro Co., UK

DISCLAIMER

This paper gives some personal opinions of the author. They do not necessarily represent the position of his employer or any standards committee of which he is a member.

THE BENEFITS

A distributed computer control system (DCCS) offers two important features for a user organisation:-
a) Small, modular control units allow partitioning of a process into separate local areas, each with a dedicated local unit control system.
b) The multi-master communication systems which are used within local areas can also provide connection to other plant areas. The result is a complete plant-wide integrated control structure.

Some benefits from the small modular approach:
- Low initial cost to try out the control structure,
- Faster pay-back when plants are commissioned in stages,
- Paying for each stage when it is built, gives better cash flow.

Some benefits from multi-master communication across the complete plant:
- A common message structure throughout the plant simplifies fault monitoring and reduces maintenance costs.
- All stations can readily obtain data from other stations.
- The system supports great control flexibility.

PRESENT CHALLENGES

One bottleneck which presently delays achievements of these benefits is the lack of standards for interconnecting process equipment from different manufacturers.

Typically, existing plants have different local computer systems in different plant areas. These local systems also have different forms of internal communication which are generally non-compatible. This problem also occurs when a plant is built in stages. The first stage control system is installed by Manufacturer A. Sometime later when stage 2 is purchased, Manufacturer B offers the preferred equipment, and the two sub-systems must be interconnected to form a DCCS.

Important issues which arise:
- Cost and manpower required to ensure protocol compatibility,
- Frame format conversions,
- Answering the question: Who carries system responsibility for the safe operation of the combined system of "Manufacturer A + Manufacturer B?"

WHAT IS HAPPENING

Suitable standards are not yet available to support interconnection of DCCS equipment presently on the market.

However, a large amount of background work has been done in several Standards Committee working on communications for DCCS[1].
- The next generation of VLSI chips for HDLC[2] and X25[3] will support multi-master variants.
- The IEEE 802 project [4] committee hopes to adopt a final Standard in 1982.
- The PROWAY [5] Working Group has a target date of 1985 for submission of its Standard to national committees.

These developments can be expected to have a strong influence on manufacturers of DCCS systems, and the next generation designs should begin convergence towards an industrial communication standard.

FUTURE CHALLENGES

One objective for a standard is to impose sufficient stability on the user environment to allow the user to obtain economic benefits from simplified design and easier purchase decisions. The challenge which remains is to adapt to the continuing fast movement of technology without destroying the original standard. Some areas which will lead to continuing

evolution of DCCS communication structures:
- Definition of standards for "Application Message Formats"
- Development of Network and Trasport layer protocols and standards in the X25 and packet-switching marketplace.
- Availability of fibre optic links to replace the physical layers of our industrial DCCS.

While these developments are not excluded from the present work of PROWAY and IEEE 802, they could become future bottlenecks when our present challenges are overcome.

REFERENCES

1. "Standardization work for communication among Distributed Computer Control Systems" by G.G. Wood - IFAC 4th Workshop on DCCS, Tallinn, USSR.

2. High Level Data Link Control. A frame format and control code procedure used by X25 and defined in ISO standards 3309 on frame structure and 4335 on elements of procedure.

3. International Telephone and Telegraphy Consultative Committee. Recommendations X25 covers Physical, Data Link and Network functions.

4. Institute of Electrical and Electronic Engineers, Project 802. Draft documents were published for comment at the end of 1981.

5. PROWAY is a Standard being produced by the International Electrotechnical Commission, Technical Committee 65C, Working Group 6 and drafts for national comment were circulated by TC65A.

USER'S VS. SYSTEMS ANALYST'S POINT OF VIEW IN THE DEFINITION PHASE OF INDUSTRIAL DCCS

R.-R. Tavast

Institute of Cybernetics, Academy of Sciences of the Estonian SSR, Tallinn, USSR

INTRODUCTION

The authors experience shows that the conceptual and language barrier between industrial user and designer of DCCS is high. The consequence of this bottleneck is computer system misspecification leading to expensive redesign or short life periods of the system. In the extreme the hardware and software designers tend to consider their product as an end in itself to which the production plant and the user organization are attached. The user is expected to adapt his needs to the CCS offered.

In contrast to that, industry people consider any control system just as a part of the whole production organization to improve the production effectiveness. The computer system is expected to suit the conservative production technology with minimum implementation effort and cost and to adapt to changes of the production situation.

Insufficiency of configuration management tools often caused large scale centralized CCS to go wrong. DCCS projects are less sensitive to specification errors due to flexible modular design and implementation. The DCCS designer, however, tackles with additional degrees of freedom of functional distribution, interconnection and coordination of multiple information processing entities, redundancy, software distribution, control equipment allocation, etc. At the same time the user's position has little changed, therefore adequate and timely DCCS definition remains a bottleneck.

Answers to this problem are plant-oriented system analysis and simulation, design of DCCS as an embedded modifiable computer system and computer-aided specification and prototyping support systems to ease user participation in the design process.

USER'S DEFINITIONS

Prior to DCCS design the project initialization is motivated for any group of industrial users (top managmeent, engineers or plant operators) by very general requirements:
- plant central monitoring and manipulation for fast response and quality of decisions by few operators.
- support to optimum production management decisions in changing environment
- safety guarantee
- smooth and reliable process operation by automatic compensation of disturbances of different source and frequency range
- availability not less than that of the plant equipment and instrumentation
- automatic generation of relevant production documents
- human-orientation to ease learning, run, maintenance and modification of the control system
- flexibility with respect to changes of, for example, product and feedstock specifications, plant equipment or instrumentation
- smooth and short implementation period, short payback time of investment

In these requirements DCCS is considered solely as a functional entity of the entire production plant, indifferent to its architecture. Important constraints like interfaces, computer equipment allocation, intrinsic safety, etc. are specified later on as the result of user participated plant analysis.

SYSTEMS ANALYST'S POINT OF VIEW

In the data processing field the user-designer gap is bridged by the system analyst whose activity is (top-down) formalization of data flows and transformations in existing and future system [De Marco, 1979]. In industrial computer control the scope of systems analysis is investigation of flows of material and energy plus the attached human organization, the environment and the flow of production documents. Derivation of

DCCS specification from the user's requirements is never a simple translation but merely contains a tentative design of the future system. The system analyst, promoting the user's interests, has on his mind, at least implicity, the following assertments:

A Correspondence, or duality principle (Syrbe, 1979), of physical and information processes in control systems
B Embedding of the control computer system on-line into the whole production organization through multiple interfaces.

The consequences of A are oriented toward analysis: the ultimate need for modelling, identification and simulation of physical processes and design and simulation of the correspondent control algorithms, interconnection structure of the corresponding information processes.

Heavy impact of assertion B on the design decisions is certainly a specific of computer control from which some important design features can be deduced (Fig. 1):
- information processing in a local control station with any functions is performed by possibly three interlinked entities: dedicated control equipment, computer and personnel. Therefore, a maximum of 9 types of interfaces are to be specified between the processing entities (in a station and between stations), the plant and environment.
- there are three different types of interconnections between local stations: data highway, voice and communicatons through physical processes and plant interfaces. Specificaton of communications (e.g. data highway) in terms of throughput, response time, station activity, application formats, etc. depend on task distribution as well as on the dynamics of the physical processes.
- plant control and control of information processes use the same processing entities and should be specified together.
- inevitable changes in the plant environment or control entities necessitate modification of control processes, hence modification management is an inherent function of the DCCS.
- prototyping and testing of the DCCS assume simulation together of all system components - plant, embedded DCCS and the environment.

CONCLUSIONS

Labour productivity of systems analysts of DCCS is to be increased by development of computer-aided specification and simulation support systems [Takezawa, 1982].

The gap between users and designers can be cleared by prototyping based on tentative design of the DCCS as the result of system analysis.

Standards for all DCCS interfaces are needed.

REFERENCES

1. De Marco, T. (1979) Structured analysis and system specification. Prentice Hall, N.J. 1979

2. Syrbe, M. (1978) Basic principles of advanced process control system structures and a realization with optical fibres-coupled distributed microcomputers. Reprints of the 7th Triennial World Congress, IFAC, Helsinki, Pergamon Press, N.Y. 393-401.

3. Takezawa, K. (1982) Software test facilities with distributed architecture. IFAC 4th Workshop on Distributed Computer Control Systems, May 24-26, 1982. Pergamon Press, N.Y.

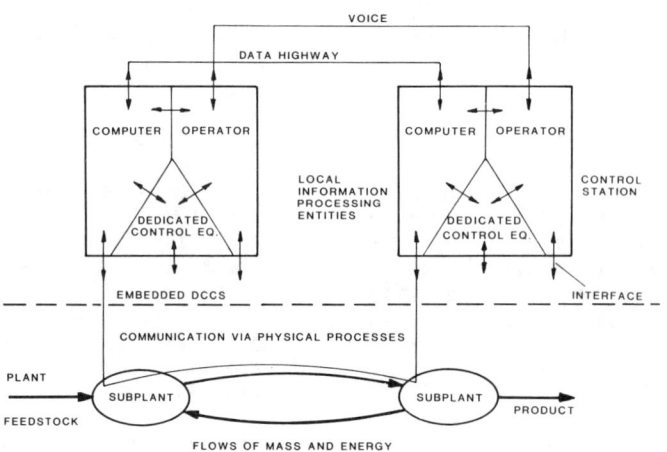

Figure 1. Interfaces and Communications in DCCS.

DISCUSSION

Harrison: We have at least four different problems on the table at this point. Professor Narita has raised the question of partitioning - how to partition the system. He brings up the interesting point that distributed control is both easier and more difficult than traditional or centralized control.

Raul Tavast has brought up very strongly the fact that there are different points of view that we run into and, perhaps, take personally depending on whether we are the user, the system analyst or the designer. He suggests that prototyping is a way to solve the problem.

Milton Maxwell really said that for him DCCS is not here yet, that the cost is not right for his application. The question that, perhaps, one raises at the point is: Will it ever be the right thing for certain applications?

Graham Wood has returned us to discussions we had off and on during the last two days on the question of interconnectability of equipment, be it with a single vendor or with multiple vendors, and whether or not communication standards are an aid in de-bottlenecking. I guess we might raise the question whether they will be here soon enough to do any of us any good.

I'd like to open the floor to questions.

By the way, Graham (Wood) you've been saying "standards" rather than "standard". Does the plural refer to a standard for each of the levels, or do you foresee that there might be multiple standards at a given level that are acceptable?

Wood: I think that you are probably right. There may be HDLC and IEEE 802 and PROWAY, or a combination of different standards at the bottom two levels. My comment was about the need for the network layer and the transport layer standards. I suspect that those standards will be confirmed in the packet switching and X25 market-place, and if we've got our OSI model correct, we should be able to just take them from the other technology and use them in our industrial plants. That is my hope. It probably won't be certain until we see what comes from the other technologies.

I see nothing peculiar about process control that strongly defines what happens at the network layer and, maybe, we'll find that. I think it is possible to have several standards if they are modular in the OSI sense. We should be able to cross-pollinate them, and we must pick and choose if we are to partition internal software similarly.

Work: A comment to Dr. Harrison's question. I think that one of the bottlenecks, while developing distributed computer control systems, is the fact that all the designers are human beings and they all have something like a religion. They were born with that religion and they die with it. For example, in designing a communication subsystem for a control system one must pick up one local area network, and it is exactly like religion: Having once picked up one type, let's say ring, they will continue to use this one till the end of their life. They always want to find some features in this topology that are somewhat superior over other solutions. The whole field is so new, and the people in this field are most creative, always wanting to add something new and not wanting to use what has been done already. The result is that now we have, maybe, too big a variety of very similar data communication systems or local area network types. There is, practically, no need to create so many of them; the variety already existing is too big, and that's why the standardization is badly needed. I think that in a few years we'll reach such a state that we'll take the data communication subsystem as a "given" or standard. Then we can really go nearer to the control object itself and think more on how to control the process.

Harrison: Is there an analogous situation if we look at computer architectures? Some people argue that some features of architecture are added only to make it different from other architectures. It is a debatable point as to whether that is good or bad. But we've learned to live with it. Does it detract, or does it add something to have some degree of variety? One might argue that the only computer in the world should be a PDP-11, or a 370, or -- take your choice!

Tavast: A good point is that the variety of options in the IEEE 802 or perhaps PROWAY standards leads to a high number of interconnection options that may optimally cover the requirements of most of the applicants with minimum restrictions on implementation technology. Yet optimum discrimination between the interconnection alternatives is obviously a difficult task.

Harrison: So you are really saying that there was an attempt to cover the spectrum and they did it intentionally, and that it is okay to have a set of standards. In fact, the options were a compromise in many respects. One of the requirements in the IEEE 802 standard will be that the equipment manufacturer must clearly label the equipment as to which of the various options they have implemented.

Vamos: Cooperation between DCCS modules requires connection in the communication sense, and only flexible protocol and interface standards can lead to the openness required for cooperation. My belief is that a compromise between PROWAY and IEEE 802 standards is needed with possible extraction of appropriate subsets for particular application. I stress that general standards are long-lasting and have tremendous impact on future technology, as we can see with the example of the telephone network.

Wood: I think several of you are, as I, concerned that we need standards very quickly. I wish I could find a way to have the international standards committees meet more frequently, but the big bottleneck is time and effort. To get an international group together for a week twice a year seems to be about the limit we can manage with the approval of our management. I think that is where IEEE 802 has a big advantage - it is all inside one country. The disadvantage for us in Europe is that it is more difficult for us to find out what is happening. I'm personally very appreciative of Tom's (Harrison) paper that has given us the latest insight so soon after it's been decided.

Harrison: Project IEEE 802 activities have been rather expensive. There have been in the order of 50 to 100 people who have met thus far, I believe, 12 times all over the United States. If you take one hundred people times 12 meetings, the cost of travel, and their net worth to the company, I suspect that it is a multimillion dollar standards development activity.

Lan Jin: I'd like to state some of my opinions, continuing Professor Narita's statements on partition. I think distributed computer control systems are typical distributed systems with a special feature - autonomy. The bottlenecks in distributed systems are communication links. How to de-bottleneck this problem? Distributed computer systems may be examined in different partitions. In horizontal partitions distributed computer systems are hierarchical systems - they are decomposed into several layers. In each layer there exists a multiplicity of processors. In horizontal partitions these processors, located in the same layer, operate on the basis of task distribution. The processors located in different vertical partitions operate on the basis of functional distribution. The principal of autonomy may be achieved by minimization of communication traffic between different layers and different processors in the same layer.

Harrison: I think that you bring up a good point that we have, in fact, been dealing with hierarchical systems now for many years and if you want to talk about "fads" in the control industry, the first fad that I was involved in was the digital computer. The second fad was clearly hierarchical control, and, perhaps, the third fad is distributed control. Although we attempted to minimize communications between levels for the reason Lan Jin points out, we did, in fact, do a lot of communications, and we made a lot of decisions about communications. Yet we are talking here that it is a bottleneck. Did we solve it in a hierarchical case sufficiently well? Is there a lesson here that we can learn and apply to distribution horizontally, instead of distribution vertically which we have apparently already done?

Narita: I think that autonomy is a very important issue in DCCS. A number of communications equipment manufacturers competing for speeding up the communication channel are proud of the very high communication speed of their system, say, 10 Mbit/sec. even 20 Mbit/sec. But the most important point is what kind of information they are transmitting between the processors. In some cases they may exchange meaningless information, or information of little significance. Therefore, it is very important to design the entire system with deep consideration given to the autonomy of the stations in the system.

Vamos: I fully agree with Mr. Work that DCCS solutions should not be a matter of religion. Religious orthodoxy always means lack of adaptivity. My belief, not religion, is that a single level architecture of cooperating control units may be superior to 2 or 3 level hierarchical control advocated by Professor Narita.

Narita: May I ask Academician Vamos to explain his statement about single level architecture. I think a one-level system is essentially a completely autonomous system, with few communications needed between subsystems. Right?

Vamos: Not exactly.

Harrison: Why don't you give us a definition?

Vamos: This is a free coalition of system components each of which is capable of surviving autonomously and which are cooperating by standards to reach common goals.

Harrison: Dr. Narita, are you saying that a single level is a pure connected autonomous subsystem configuration?

Narita: My understanding of a one-level system is that it is a very large centralized system, called a "compunication" system (a combination of "computer" and "communication").

Rodd: Can I pick on Mr. Maxwell? His question on economics is obviously a very battered one, but I think we should take this economics thing a little bit further. The point that is often made with distributed systems is the question of reliability, reconfigurability, maintainability. Didn't these come into his equation, or does he just look at the bottom line and say - after they had installed the plant - the same that they installed two years ago, that the plant works very efficiently, so we will just stick with that architecture?

Maxwell: Well, you are assuming that we are installing new facilities. But, more often than we install new facilities, we are adding a computer to an old facility. The reason for it is, in the first place, that we want to make money by the reasons that I stated in my opening remarks; Material conservation, labour saving and throughput for the given capital investment are certainly important money saving items. So, whether we put a computer in at all depends on the number that is at the bottom of the estimate; how quick we can get the money back! So it isn't a question of whether we will make it as reliable as we can. The question is whether you do it at all. Of course we want reliability and we take what I think are very important steps in maintaining our reliability. Namely, we buy good stuff, we try to get the electricians put it in right, and we try to train people not to step on circuit boards. Above all, we try to make this system as simple as we possibly can conceive, and, I think, that may be the key to the whole thing.

Wood: Mr. Maxwell, I'm really wondering if your plant model isn't similar to mine where I see a large job site with a number of separate detergent plants on it. Perhaps because of your constraints on very low capital investment due to a low-cost product you, in the first stage, are putting a computer in each detergent node. But do you not see in the coming future some need for integration between the computer in the detergent unit, the computer in the packing unit, and the computer in the additive subunit for overall coordination and integration? Is that where you'd get a plant-wide distributed system?

Maxwell: I think the integration may come through our master business computer net, using that as a communication medium. We are taking some steps to make our process machines communicate, but with business machines; we have been very slow in this. The computer control in our company came from the bottom; it came from the few guys who could see some potential in computers. We certainly didn't forsee what we have now; it just grew out of all proportions. But, it grew as we could afford it. We were not directed by top management to go out and get a computer, the biggest, fastest computer you can find connect it to every valve and every thermostat. We know of cases where that has happened, and it has caused a tremendous amount of waste. We also know of cases where the data processing people have tried to control all processes. They can't do it. So there are all sorts of foolishnesses going on in the world and they all result in expenditure and waste of money. That's why we started out small, with discipline. I believe we will get into distributed processing, only I don't know exactly how and when it will happen.

Suski: Again, picking on Mr. Maxwell. Actually, I guess, I have to support his point of view. It seems to me that he is really just facing the question we all face when designing a distributed computer based control systems; that is, at what point you start distributing. In fact, we can do these functions on a single computer, and he is at that level right now. He has found that point for his system and so, in fact, we are all facing the same questions.

If I could, I'd like to introduce some issues that weren't brought up until now. The first obvious issues are things which we've come across in the Nova system. The first was the lack of centralization in real-time programming. The PRAXIS language is something we've attempted to do, -- and that's worked up for us -- but we see Ada coming on line. We are really not too sure how that factor will influence the future plans. I'd like to hear some discussion about Ada.

One other point I have to bring up is that in a large system such as ours, we faced the problem of configuration management. What do we do when we start changing the system? We don't want to leave qualified people around, to start modifying software, and we don't see the solution to that right now. I know there is some alternative but we haven't found it yet.

Finally, I'd like to bring up an issue of overall programming productivity. It relates to a lot of different things, including real-time languages. What we need is an order of magnitude increase of programming labour productivity, and the answer is the impressive effectiveness of the software factory presented by Mr. Takezawa.

Motus: Actually my question is very closely related to that of Dr. Suski. To my mind, DCCS contains quite a wide spectrum of problems and I want to ask the opinion of the Panel on the software: Has anybody had any problems creating software? In some systems the cost of software is over 70 per cent of the overall cost of the system.

Harrison: I think we have a panel which is primarily hardware-oriented, but this is a matter of interest. Just raise your hands and let's see how many of you consider themselves to be software, systems, or hardware people (pause). It appears that we probably are more than two-thirds systems and software, and about equally divided between these two. About 20% are indicating that they are hardware people. So, maybe we can get into the software issue.

Vamos: Our experience with Ada shows that this implementation language is too complicated and a lot of users have lost their belief in Ada. Special software tools for DCCS are badly needed. An operating system that adopts distributed control concept is UNIX, with its parallel and concurrent processes. Software still tends to be harder than hardware, because of the lack of modification tools.

Maxwell: As a response to Dr. Suski's question, let me point out that we went through some machine languages and then we settled on BASIC. We have been programming in BASIC for eight years. I think that's one way to stay efficient. We have a software factory, this is about one quarter the size of this room [Editor: The room was about 200 sq.m]. We came up the hard way with a few people - we started out with 3 people about 8 years ago, getting so many jobs that we had to expand our section drastically. We now have 5 people. There has been some recognition that we need some more help. My only answer to the software problem is to get the right people. It really needs motivated people. If you dont' have them, then you have a real factory situation, and you will never beat it: You have just a bunch of guys sitting and doing software work as if they were making shirts. You cannot do it economically that way. If you have mechanical people doing software, you get mechanical software at mechanical cost.

Wood: Isn't that a stronger argument for software maintenance, -- that the simplest program to maintain is one that was written in a fairly mechanical, routine way? Some screwballs have produced programs that they themselves could not read at a later time!

Maxwell: You are absolutely right about that. We do get some screwball software. Good software demands good heads, but if you're doing mechanical software, you should be able to reproduce it from files of software. If you cannot manage the economics of your projects from reusable code, then how can you manage it?

Inamoto: I am interested in the reliability of the distributed computer control systems. Mr. Maxwell said that he couldn't adopt the distributed systems unless it's very reliable, more reliable than the non-distributed system. We are designing and manufacturing distributed computer control systems and we would like to sell the system to a customer like Mr. Maxwell. The question is, how can we persuade the customer that it is more reliable than the non-distributed architecture, even though the amount of the hardware components may increase.

Harrison: You seem to be starting out with the premise that a distributed system is intrinsically more suitable than the centralized system.

Narita: My comment is really related to the definition of reliability of the whole system. As I pointed out, DCCS has many components: many components, many peripherals, interconnected by communication channels. Each component has possibilities of failures and therefore, the mean time between failure for DCCS may be much shorter than for the similar centralized control system. My definition of reliability is the information loss caused by the malfunction or failure of components. So, if you employ MTBF for DCCS the reliability of DCCS may be very bad, as compared to that of the centralized one, simply because there are many components involved. I think that we have to find some other appropriate measure to define the systems reliability of spacially or functionally distributed architecture.

Harrison: I think that it is an interesting comment about information loss as a measure of reliability. I'm guessing that Mr. Maxwell would say the reliability is measured by how much soap he didn't make because of the computer, and that goes back to Raul Tavast's point of view on whether we are a user, a system analyst, or a designer. Is it true that a distributed system is intrincically more reliable, as Mr. Inamoto is basically assuming.

Natiello: I could comment on that as the user of distributed systems. What I'd like to say is that failures in distributed systems are less of a problem in an operating plant than failures in a centralized system, simply because a failure in a distributed system disables less of the control aspect of the control system; smaller pieces will "disappear", let's say, in an automatic control or in a data acquisition sense. In a large centralized computer system, a single failure in the right place could disable the entire system, perhaps shut down the entire plant, power station, soap factory, or whatever. So, from this petrochemical user's point of view, distributed systems are preferable to centralized systems because of this partitioning.

Harrison: Let me ask you this: we talk about MTBF -- mean-time-between-failures -- but we can also talk about MTTR -- mean-time-to-repair. From your point of view, does it take longer to find a problem in a distributed system and fix it than it does in a centralized system?

Natiello: My own experience is that it takes less time to find and repair problems in distributed systems than it does in centralized ones. Distributed systems, by their nature, have the ability to isolate problems to a smaller part of the system, making it easier to eventually localize and repair the particular fault; whereas in a centralized system, because of the interactions and the largeness of the data bases, etc., it is more difficult. For example, you get a feeling by some diagnostics on a thousand point multiplexor that there is something wrong with it; but in a set of 250 four-point multiplexors, each one could have its own little diagnostic code and you could quickly get to the right board that has to be repaired.

Harrison: But are you presuming adequate diagnostic aids?

Natiello: Yes

Wood: I think that it is also evident that if you have a distributed system with 30 small modules you can afford to carry one complete spare module and plug it in on a TV repair principle; you can't do that with a central monolithic machine.

Narita: I'd like to say that the MTBF or the MTTR are what we call "old friends" of the reliability people in designing electronic circuits. But today is the age of DCCS and it is not appropriatae to use MTBF or MTTR to measure the reliability of large-scale distributed systems.

Harrison: But if you need to simulate a particular system, do you do it against these parameters?

Narita: I tried to apply the traditional reliability engineering techniques to evaluate very complex systems, but I failed.

Vamos: I fully agree with Professor Narita in that MTBF is a very simplified concept. Graceful degradation is probably a relevant idea for DCCS but surely not mean time between catastrophes or optimum runs.

Tamm: Reliability seems to be the most essential performance index of an implemented DCCS. Yet, my question addressed to Professor Narita is on the optimum design of the system as a mathematical optimization problem. A DCCS is a very complicated thing and I don't see how to optimize it in a mathematical sense with respect to a single performance index. The only practical approach is simulation in DCCS CAD system to get some reasonable alternatives for an expert to decide what would be best in many respects. What is your opinion about optimality of DCCS design? Can you see any principally new optimization methods for this purpose?

Harrison: We should ask Professor Narita to answer that in terms of how long the analysis he presented took and whether or not these analyses gave the optimum configuration for his control system.

Narita: I don't know if I have any direct answer to your question but there are so many theoretical works regarding the design of optimum configuration. Nevertheless, the most common way for the company people is to employ the rules based on past experiences, plus calculations based on very simple models.

Harrison: Now, I think it's time to summarize our conversation which has been rambling in some respects. We have covered a lot of different things: we've touched on software, talked about reliability, about different system configurations and programming productivity. We did touch on a very important question which we didn't follow up on; specifically Mr. Work's comment concerning religious orthodoxy, meaning people who get in a rut in their careers and forever after believe that it is the proper rut for the rest of the world to be in -- or ditch, or canal, or whatever word you'd like to use. They never bother looking over the sides to see if, in fact, there might be smoother water in the canal right next to them. Perhaps that could be a topic for future workshops.

We have been talking about how we keep our people up to speed in a technology which is changing very rapidly, when it takes a lot of personal effort to learn the new things and to make the decisions between the alternatives. I think we've done some of the things that panel discussions are expected to do. I hope that all you, whether you actually participated in the discussion or not, have picked up some ideas at least. I hope that we didn't lose sight of the Coca-Cola bottle as we went through this whole discussion. Thank you all very much.

EMUNET "EMULATOR OF NETWORK SYSTEMS" A GENERAL FRAMEWORK

H. G. Mendelbaum and G. de Sablet

I.U.T. Université Paris, France

ABSTRACT   The project discussed here provides a general tool, the architecture of which allows to emulate various network configurations, and to implement various synchronization algorithms for distributed control, distributed data or functions. This emulation enables the measurement of various characteristics of these algorithms: resiliency, overhead, fairness, response time and so on. This tool resides in a physically local network, built with homogeneous microcomputers. The desired network is simulated on this physical network. The topology of the physical network supporting the simulation is a star. The central node is considered from the other satellite computers as a simple transport medium which emulates the links of the required network to which each one is connected. In the central node a simple interpreter performs the desired emulated network (loop, multidrop, mesh.....) by consulting a description of the links of this network. The simulation is made, using 3 layers of software in the central node emulator:
first level: connections with the various computers of the physical network and data link control protocols. second level: path control where the routing of the various messages is performed between the satellites in the simulation. third level: traffic supervision, keeping track of the different transactions and to develop measurements and statistics on the running of the simulated network.

In the satellites, the distributed systems and applications are implemented. This tool is intended to facilitate the implementation and the testing of algorithms for distributed control as well as algorithms for distributed data: ticketing algorithms, virtual rings, logical clocks or any other. Even more, it is designed to measure the different characteristics of these algorithms so as to permit their comparison. Its modularity authorizes also to perform the simulation of various faults in the satellites.

KEY WORDS: network architecture simulation, distributed systems, distributed control, synchronization, network architecture validation.

PROLOGUE

Main distributed computer systems have been made in the aim of sharing the various (hardware and software) resources of a computer system. The sharing of these resources allows parallel processing and data handling at the location where they are created.

Let us distinguish here 3 classes of distributed systems:

- distributed data: the distributed system is made up of a collection of intelligent terminals located at the point of use in order to optimize their local functions. Data may be distributed on the various terminals or may be duplicated on these terminals in order to optimize access time.

In the latter case a great problem is to ensure coherency between various copies and to order transactions on duplicated data.

- distributed functions: distributed systems may also be used to manage homogeneous or (even more) heterogeneous computers interconnected in a computer network. A useful way to proceed is to dedicate computers to functional classes (for example: compilation on one computer, DBMS on another one and so on....). Another way is to decompose one task into parallel subtasks provided one has solved the problem of decomposing it with minimum communication requirements.

The functional distribution in not necessarily fixed and it is difficult to fix a priori criteria. Control of these functions and synchronization are not trivial.

- <u>distributed control</u>: the decisions of an operating system may be distributed on various interconnected machines: this increases the parallelism of processing.

Another important feature in distributed control systems is the robustness of these systems. If the control is distributed, the failure of any component may not lead to the failure of the whole system: the system becomes fault-tolerant.

The main problem, where no trivial solution has been found in distributed control systems, is the synchronization and the coherency of the control between many processors. This problem is not easy to solve because of the incertainty about the global state. Another problem is the vulnerability of communications.

INTRODUCTION

This short abstract on some problems encountered in distriubted computer systems (data, functions and control) shows the difficulties in finding solutions which can be proved and/or validated.

All problems have not been solved, even if some solutions have been found [1,3,4,6]. Some of these solutions have recieved a priori formal proof in restricted configurations [1,3,6].

Other solutions have been implemented and validated a posteriori on various real applications [4,5].

An interesting way of research can be the study of <u>simulators</u> of network architecture. These simulators may enable to experiment some functions models.

The host computer system of the simulation may be a monoprocessor machine on which the simulation is programmed with the use of cooperating tasks [7,8]. But in this case, the absence of real simultaneity is a great handicap for validating some synchronization algorithms.

Our approach will be the realization of a physical and versatile architecture which will enable the physical <u>emulation</u> of a large number of configurations and the implementation of really distributed synchronization algorithms.

The various concepts defined in this paper and implemented on the computer system described here, will enable the verification of the feasibility of some networks and of distributed algorithms, languages or operating systems. This emulator may authorise measurements on these softwares and on characteristics of these various algorithms.

This paper is not intended to study communication protocols and to measure their performances, the reader interested in these subjects can read some specialized papers [12,13].

CHOICE OF AN ARCHITECTURE FOR THE HOST EMULATOR NETWORK

The basic idea for such a distributed emulator is to separate the simulator of the network links from the simulation of each node.

First Condition

Considering the various computer interconnection structures [11] we have to find a structure which may simulate a great number of network topologies.

Second Condition

The simulation must not make any unintentially wrong projection of the simulator on to the target system being simulated. It means that the general overhead of the architecture simulator must be strongly separated from the simulated system being measured.

The first condition leads to choose among 3 kinds of architecture which allows direct interconnection between the architecture simulator and the nodes simulation:
1°) Complete meshed network (DCD : <u>D</u>irect <u>C</u>edicated path with <u>C</u>omplete <u>I</u>nterconnection).
2°) Star (ICDS : <u>I</u>ndirect <u>C</u>entralized routing with <u>D</u>edicated path <u>S</u>tar).
3°) Bus : global linear bus, linear bus with central switch, single loop, loop with central switch.

The third kind of architecture (bus) does not permit to fulfill easily the second condition because in that case the performances of the various nodes depend strongly on their geographical location in the simulator network; the measures depend also on the unknown repartition of the collisions on the bus.

The first kind of architecture (mesh) is very heavy to manage (for n nodes, nx(n-1) physical links).

We have thus chosen the second architecture (star) so that the architecture simulator is centralized (and easier to write), and the satellites may be used for the system and application simulation.

PRESENTATION OF THE EMULATOR

The architecture of this distributed emulator consists of a microcomputers-network structured in a star network as discussed previously.

Each microcomputer contains a video keyboard and a diskette unit used as a small mass storage which enables to keep some intermediate data and to modify some parts of the simulation. Each microcomputer may be autonomous for processing local data and contains its own operating system.

The central node of the network (centrum of the star) is not seen from each satellite as another node but only as a single transport medium. This enables to emulate various configurations such as loop, multidrop, mesh, linear bus and so on.... The central contains a description of the emulated network. The central can receive messages from each satellite and transmit it to an addressee, according to the rules defined in its simulation description.

The microcomputers are Rockwell 6502 (Apple II) linked through asynchronous communication interfaces (ACIA) in full-duplex mode at a transfer rate of 1200 bps with the central computer (also Apple II).

The central node has the same configuration as the satellite computers plus a typewriter as permanent output.

SIMULATION TOOLS

The simulation tool is built around an interpretation kernel and some tables describing the specific architecture required.

According to the OSI specifications of the ISO [9], the simulator has been organized in multiple layers of protocols implemented in the various satellites and in the central node.

Only one layer is a dedicated software to control our physical emulator network. Some layers describe the architecture simulation, some others describe the application(s).

1.- PURPOSE OF THE SOFTWARE LAYERS

- in the satellites:

<u>Layer 1</u>: Physical connection between the microcomputers (Asynchronous Communication, Interface Adaptor protocols) and line control routines.

<u>Layer 2</u>: Path control and routing of the various messages between the satellite and the central node (emulation of the physical architecture).

<u>Layer 3</u>: Description of the logical network: flow control, transactions log, functional measurement of the network.

<u>Layer 4</u>: Synchronizing algorithms between satellites.

<u>Layer 5</u>: Executive system enabling communication between an application and the network.

<u>Layer 6</u>: Applications to be processed in the distributed system environment.

- In the central node:

The 2 first layers (1c and 2c) are similar to the previous ones, but for a simulation purpose. The central node is only considered as a simple transport medium and is not concerned with the management of the simulated application resources. The layer 3c is concerned with measurement and traffic control routing management.

Layer 4c of the central node enables the management of the various functions of the simulator.

DETAILED DESCRIPTION OF THE SOFTWARE OF THE VARIOUS LAYERS

The Satellites (System & Application Simulation)

Layer 1 (real traffic)

At this level, each satellite is considered as a node of the real host network. The software allows the exchange of groups of bytes (GOB) on the lines betwen each satellite and the central node. These groups of bytes (GOB) are messages produced by the upper layers.

These exchanges depend on the hardware constraints of the host network which support the simulation. The exchanges are programmed using the following primitive inserted in an algorithmic language.

We use logical unidirectional links (receive or emit) for a simpler management of the dialog.

These primitives are assembler written routines which control the line interface circuits.

INIT-NET: This primitive initializes the links (i.e. the buffers, the flags, the interrupt vectors) and resets the interface.

BEGIN-RECEPT: Opens the link, checks the line and waits for the next GOB (Group of Bytes).

GOB-RECEIVE: When a GOB is ready in the interface buffer, this primitive transfers it into the layer area. Depending on the hardware, an interrupt signal may notice that a GOB is available in the buffer (i.e. may be received).

END-RECEPT: Closes the link for reception: the satellite will not receive any other messages and will be isolated from the network until the next opening.

BEGIN-EMIT: Opens the link for emission, notifies the addressee that a GOB will be sent.

GOB-EMIT: Enables to transfer a GOB from the 2nd layer area to the line buffer. Then the ACIA will transfer it to the addressee via the line protocol (this may be done by interrupt handling).

END-EMIT: Sends an end of text (EOT) message to the receiver.

SUSPEND-TRANSMIT: Prevents the sender from sending any GOB until a certain condition has been fulfilled and the receiver resumes the transmission (used only for reception).

It may be used for flow regulation between sender and receiver, it may also be used for hardware purpose (for example if there is only one interrupt level) or for software purpose (for instance queue management).

RESUME TRANSMIT: Authorizes the sender to resume the transmission.

Layer 2 (emulated traffic)

At this level, the satellite is no more considered as a node of the real host network, but as a simple node of the emulated network. The real central node of the host network becomes transparent.

Each satellite exchanges messages with the other satellites through virtual line control protocols. This is done by use of the transmission routines of the layer 1 called by underneath primitives. These routines manage the end-to-end protocols with the real central (as seen above).

This layer provides various functions whose structure has been described in a previous schema.

EMISSION: When the message arrives from layer 3, it does not contain the name of the next emulated node. Layer 2 is in charge to furnish it from the identification of the next logical node given by layer 3. In each satellite, layer 2 contains a list of the next emulated nodes.

It has to choose an emulated path to this logical node and to insert it at the right place in the message and, either put this message in a queue for layer 1, or send this message directly to layer 1.

Here is also performed a flow regulation with layer 1.

RECEPTION: When layer 1 transmits a GOB to layer 2, layer 2 does not know if the message is addressed to this Processing Module (PM) or if it has only to route it toward another node in the emulated network.

At this level the satellite's software must extract the identification of the logical addressee and compare it with the local one.

In the first case (the messge has been addressed to the actual PM) the message is prepared to be transmitted to layer 3 of the same satellite in order to be processed.

In the second case, the message is destinated to another PM and must be sent through the emission procedures.

For the communication with layer 1, three simple procedures are necessary.

EMULATED-EMIT: Sends a message to layer 1 and allows to continue processing in layer 2.

EMULATED-RECEPT: Ask layer 1 to move the received message in the layer 2 buffer.

EMULATED-LOOKUP: Allows layer 2 to test if layer 1 is able to transmit a received message to it.

Layer 3 (logical traffic)

This layer has been designed for the case in which one has to simulate a logical architecture implemented on a physical network which can be different from the logical architecture.

For instance a logical ring [10] may be implemented on a mesh network. So, in our EMUNET, we should have the host simulator network (layer 1), the simulated mesh network (layer 2) and the logical ring (layer 3).

Another function of this level is the adaption of the messages to the communication medium: the user messages are sent split into smaller packets (subdata) for better flow regulation and error detection. In the other way, layer 3 reassembles packets in the right order to form a correct user message.

The other software of this layer 3 is similar to the layer 2, since it simulates another architecture network supported by the underneath layer.

In a satellite, messages coming from layer 2 (supporting network) may be retransmitted to layer 2 if they are not destinated to this satellite application, or transmitted to layer 4 if the message is to be processed on this node.

If messages are to be routed in the logical simulated network, layer 3 must include the corresponding path in the message and transmit it to layer 3 (emulated physical network). This is done by use of the primitives:

```
LOGICAL-EMIT
LOGICAL-RECEPT
LOGICAL-LOOKUP
```

Which are similar to the set of primitives of layer 2 (emulated primitives). In each satellite, layer 3 contains a list of the next logical nodes.

Layer 4 (synchronizations)

Layer 4 of the satellite communication protocols is only concerned with the concurrency of access to the various resources in the network.

The user of the network emulator can then implement here the various synchronization algorithm he wants to be tested on the network. Layer 4 ensures then that the messages transmitted through layer 3 follows the synchronization rules authorizing the correct sharing of network resources.

Layer 4 may also produce or retransmit some synchronization messages like tickets, clocks or anything else, using lower layers.

This is done by a set of primitives similar to the lower layer:

```
SYNC-EMIT
SYNC-RECEIVE
SYNC-LOOKUP
```

Layer 4 receives packets of messages from layer 3 (logical network) and separates data messages from synchronization messages, in order to transmit data messages to layer 5 or to apply the synchronization rules.

Layer 5 (executive system)

Here is described the executive system. At this level the communication between two nodes are made as if there were no transport network and the communication protocols are end-to-end protocols using a single transport medium simulated by layer 3.

The executive system furnishes facilities for data communication between applications invoked by standard READ, WRITE and LOOKUP requests.

The upper layer application has to indicate in its request the identification of the other end of the communication and a data buffer name for correct transfer.

The various messages sent to layer 4 must indicate the type of sharing which the server of the hardware or software resources must apply for processing of this message:

- exclusive access
- protected access
- shared access

so that layer 4 should use the synchronization algorithm or not. This can be ordered by the executive at layer 5.

Layer 6

At this layer the application running on the simulated network is programmed. It is only concerned with manipulating data and data structures without any regard to the geographical distribution of the network.

The applications may be programmed using any programming language provided it uses the correct interface with the layer 5 defined executive system.

THE CENTRAL NODE (Architecture Simulation)

Layer 1c (host network lines management)

The central node controls the various physical communication lines with the satellite nodes. It receives and emits GOB from or to these nodes, and manages the corresponding physical buffers and status words.

The primitives used at this level are similar to those used in the layer 1 of the satellites (BEGIN-RECEPT, GOB-EMIT and so on.....). The programming procedures depend strongly on the host hardware.

Layer 2c (emulated network routing validation)

At this layer the central node has tables describing the authorized links between satellite nodes in the emulated network. These tables are prepared at the simulator initialization time. When a message arrives from a satellite through layer 1, the central layer 2c looks for the identification of the requested next node, verifies the validity of the emulated path according to its tables. Then it puts the message in a buffering queue for the indicated link. For each emulated path, layer 2c has a buffering queue.

Then the control is given to layer 3c for routing measurements and traffic control on the paths. At level 3c the message is moved to the emulated addressee buffering queue. When this is done, control is given back to level 2c which will transfer a GOB from this buffering queue into the physical communication buffer of the addressee node to prepare the work of level 1c. Then level 1c emits the GOB to the physical node.

Layer 3c (emulated traffic management)

For each addressee, this layer has a buffering queue where messages are stored, coming from the corresponding emulated path buffering queue (layer 2c) and going to the addressee physical communication buffer (layer 1c).

Layer 3c has also data structures enabling to collect the various measures. Further more, for each emulated path, this layer has a list of measurement and traffic control subroutines. These modular routines are prepared at the time of the simulator generation, according to the aims defined by the engineers.

When a message comes from a sender (through layer 2c, in the path buffering queue), layer 3c looks at its routines list, and executes the corresponding measurements and traffic control. Then the message is transferted to the corresponding addressee buffering queue, and control is given back to level 2c which will send a GOB (group of bytes) to this addressee (through level 1c).

Various actions performed by layer 2c and 3c are controlled by use of control automatas or synchronizers [14,15,16].

Layer 4c (simulator supervision)

This level manages the various simulation operator's functions:

- at initialization time:
    * defining the simulated architecture description (layer 2c tables)
    * defining measurement and traffic control subroutines and tables (layer 3c)
    * defining physical node buffers, path buffering queues, addressee buffering queues etc.
- at running time:
    * interrogation of the measurement tables, buffers, queues
    * modifying traffic control parameters
    * printing of results etc.

All these functions are in interactive mode, since tables, parameters and subroutines are built in modular form.

CONCLUSION

This emulator is intended to facilitate the implementation and the testing of algorithms for distributed control as well as algorithms for distributed data: ticketing algorithms, virtual rings, logical clocks or any other. Even more, it is designed to measure the different characteristics of these algorithms so as to permit their comparison. Its molularity authorizes also to perform the simulation of different faults in the satellites.

It also is intended to facilitate the implementation of languages describing the control of distributed applications.

Another advantage of this centralized architecture is to enable the implementation of broadcasting protocols and to control the transfer delays between different computers.

The problems of a priori proofs or of a posteriori validations of distributed systems (or applications) are far to be solved.

The approach of this paper is promising: building various emulated scenarios of such systems, comparing them, measuring them.

It can be seen as a priori validation of a system before its real implementation.

BIBLIOGRAPHY

1. L. LAMPORT
   Time, clocks and the ordering of events in a distributed system. CACM, Vol.21, No.7, July '78. p.28-38.

2. J.N. GRAY et al:
   Granularity of locks in a shared data base. Proc. Int. Conf. on VLDB. FRAMINGHAM Mass. Sept.'75. p.428-451.

3. D.P. REED, R.K. KANODIA
   Synchronization with even counts and sequences. Proc. 6th ACM Symp. on Operating Syst. Principles. Nov. '77.

4. G. LE LANN
   Algorithm for distributed data sharing systems which use tickets. Proc. 3rd Berkeley Workshop on Distributed Data Management and Comp. Systems. San Francisco. Aug.'78. p.259-272.

5. J.M. COHEN, H.E. MOSES
   New test of the synchronization procedures in the noninertial systems. Physical Review letters. Dec.'77. Vol.39.

6. J. MOSSIERE et al
   Sur l'exclusion mutuelle dans les réseaux informatiques. Internal Publication IRISA 1977. No.75. RENNES University, France.

7. S. SCHOEMAKER
   Computer networks and simulation. North-Holland, 1978.

8. H. VAN ISSENDORFF & W. GRUNEWALD
   An adaptable Network for Functional Distributed systems. Proc. of the 7th Symposium on Computer Architecture. LABAULE (France) May 1980.

9. I.S.O.
   The reference model for Open System Interconnection. ISO, TC 97/SC16 No.227. June, 1979.

10. A. HOPPER
    Data ring at computer laboratory, University of Cambridge Computer Science and Technology Local Area Networking. Washington NBS Special Publication. August 1977.

11. C.A. ANDERSON & E.D. JENSEN
    Computer Interconnection Structures: Taxonomy, Characterisitcs and Examples. Computing surveys. Vol.7, No.4, Dec. 1975.

12. D. DROMARD & D. SERET
    PERFORMANCES REELLES d'une liaison telematique. La conception des systemes telematiques NICE (France) Juin, 1981.

13. D.R. POWELL
    Performance Evaluation and Comparison of Dependable Channel Access Technologies for Locally-Distributed Computer Systems. Proc. of the 2nd International Conference on DCS Paris, France. April 8-10, 1981.

14. H.G. MENDELBAUM & F. MADAULE
    Automata as structured tools for easy real-time programming. IFAC-IFIP Workshop on Real-Time Programming. Boston, Aug. 1975.

15. Y. LE MOAN & H.G. MENDELBAUM
    "Le Synchroniseur" un outil de mise en oeuvre de systèmes parallèles décrits globalement. Symposium Europeen sur le Calcul Distribue Toulouse 1979.

16. R. SAMUEL & H.G. MENDELBAUM, F. MADAULE
    A Fault Tolerant Distributed Real-Time Machine Microcomputer Architecture EUROMICRO. 1977.

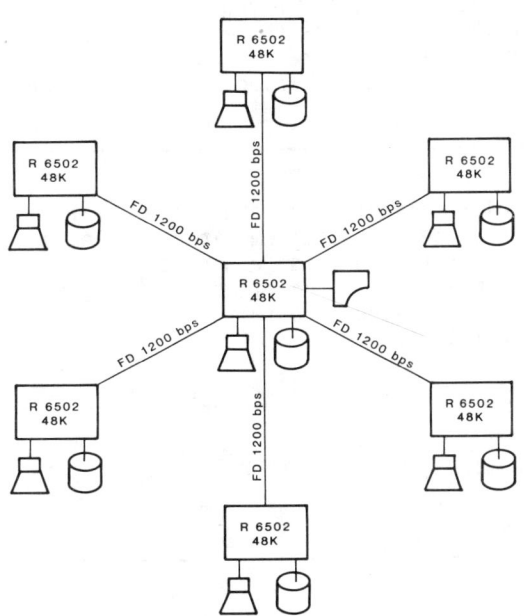

Figure 1. Structure of the Host Network

Messages are produced by the application running at layer 6 in the satellites. Let us see the transformations of a message in the system through the various layers in the satellites according to the following way:

LAYER	EMISSION	RECEPTION
6 APPLICATION	Produce the data	Process the data transmitted in the message.
5 EXECUTIVE SYSTEM	- Insert the identification of the addressee - Insert the type of transaction (need synchronization or not, and which one)	extract the user data, transmit it in the application buffer
4 SYNCHRONIZATION	Insert the identification of the sender. Need a synchronization? If so, call the synchronization routine. If not synchronization, wait for synchronization, send some message or part of it to layer 3.	Is it a data message, or a synchronization one? If data, build a message and, if complete, transmit it to upper layer. Else, if necessary, process the synchronization (may awake proces waiting for it). Send eventually a new sync-message to layer 3.
3 LOGICAL TRAFFIC	read the identification of the addressee, look for the next node of the logical simulated network to which the message has to be sent and insert it in the message, transmit the message to layer 2.	read the identification of the addressee, if same as the actual one, extract it and transmit to layer 4, else look for the next logical node, insert it and retransmit it to layer 2.
2 EMULATED TRAFFIC	read the identification of the next node of the physical emulated network to which the message has to be sent, insert it at the right place in the message, transmit it to level 1.	read the identification of the logical node, if same as the actual one, extract it and transmit to upper else look for the next emulated node, insert it and retransmit to layer 1.
1 REAL TRAFFIC	send the message to the central node	receive the message from the central node and transmit to upper layer.

Figure 2. Using the layers in the satellites.

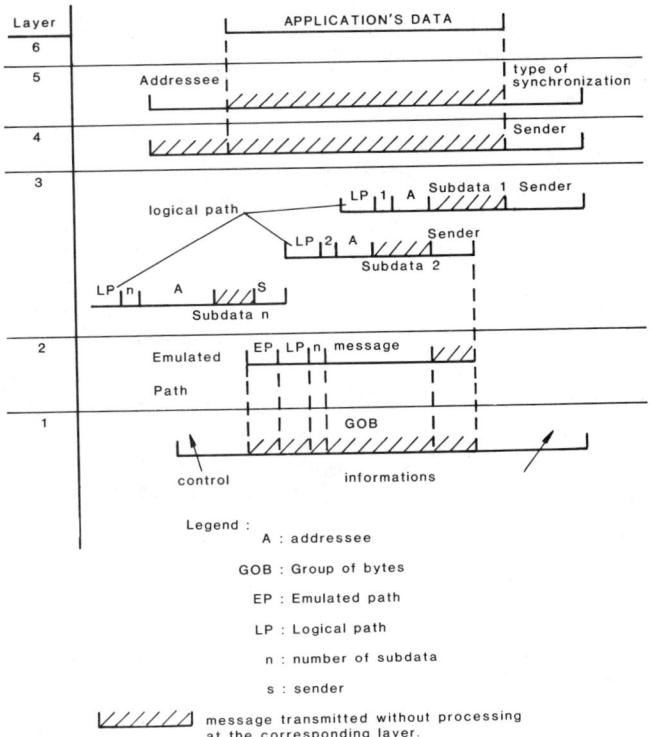

Figure 3. Building the messages in each layer of the satellites.

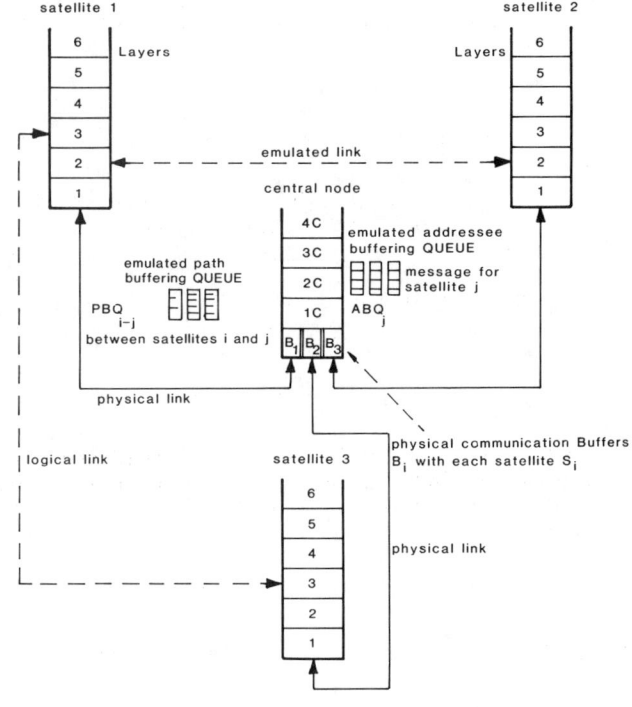

Figure 4. Flows between Layers.

N.B. In the central, a message coming from satellite i, arrives to B_i (layer 1c), transfers to PBQi-j (level 2c), then to ABQj (level 3c), then back to B_j (level 1c) and is sent to satellite j.

THE COMMUNICATION SOFTWARE ON A NODE OF THE RDC NETWORK

Wu Zhimei, Zhang Wenkuan, Zhang Yingzhong and Cheng Yunyi

Institute of Computing Technology, Academia Sinica, Peking, China

Abstract. RDC is an experimental packet switched network, aiming at the real-time distributed control and resource-sharing. It uses the DJS-100 computers, which are produced in China, as its node computers. With the support of original operating system (OS) on these nodes, we developed the communication software. It is hierarchical, and consists of link level, packet level, transport level and other levels. In this paper, several aspects of it are described as follows: external environment; major facilities; main modules and the relations between them.

Keywords. Computer communication network; communication software; module; finite automata; parallel processing.

INTRODUCTION

RDC is an experimental computer network, aiming at real-time distributed control and resource-sharing. Using DJS-100 computers as nodes, we designed and developed this network. The DJS-100 is a series of products and ranks first among the Chinese made computers in terms of the numbers of products. The necessary software has been built for it. DJS-100 is not only used for scientific computing, but also widely used for data processing and realtime control, such as information retrieval for various purposes, process control and so on. At present, DJS-100 is handling an ever-increasing amount of data, and the geographical area concerned is spreading ever wider. Therefore, there arises the need for a computer communication network. It was against this background that the work on RDC network began.

RDC is a packet switched network based on the CCITT recommendation X.25 and is implemented in the form of a hierarchical structure. Its communication system consists of five levels, i.e. physical level, link level, packet level, transport level and session-presentation level. The communication software (CS) for DJS-100 is built in accordance with the levels mentioned, and added to the original real-time OS. These computers then are connected to form a computer communication network. In this way, a user of the RDC network can communicate with another over a long distance. It can ask for service from or provide service to another, and access any resources of the network. As DJS-100 has multi-task realtime OSs (RDOS, RTOS) capable of fulfilling real-time tasks effectively, realtime distributed control can be achieved with the support of the communication network. In future, we are going to connect, step by step, large scale hosts, special devices and various kinds of terminals to RDC, so as to enlarge the facility of distributed control and distributed processing. The logical configuration of RDC is shown in Fig. 1.

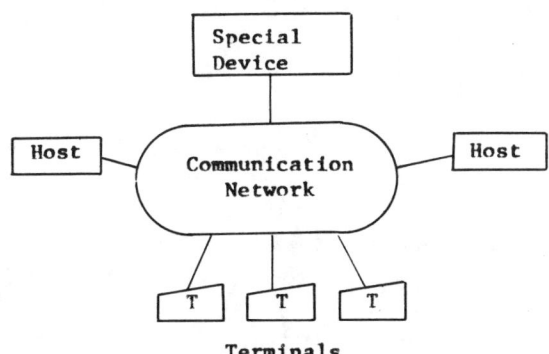

Fig. 1. Logical configuration of RDC network

This paper describes the structure of the communication software and the functions of different modules on a node.

ENVIRONMENT

At this stage RDC is formed by connecting some DJS-100 computers. In our opinion, front-end processors and concentrators also belong to the communication network, therefore communication in the network is undertaken by minicomputers. In Fig. 2 the external environment of CS is shown. A host (or a special device) may interface to RDC by means of a channel, so that the communication load of the hosts and special devices may be decreased as much as possible and they may be allowed to focus on data processing and process control.

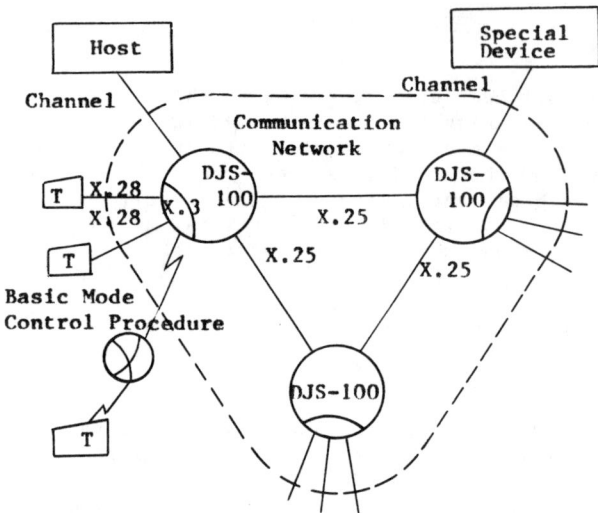

Fig. 2. External environment of CS

Interfaces between the communication network and terminals are of various forms, in order that different kinds of terminals can be connected to the network. The characters used by terminals may be 5, 7 or 8 bits/ character. The lines may be Private, leased or switched line. Transmission rate varies from 50 to 9.6 kb/s, thus user is free to chose the appropriate terminal to access the resources of the network.

In the implementation, the interface between node computers follows the CCITT recommendation X.25. When a new node is added to the network, the software scarcely needs any modification. As an internetwork connection protocol, the CCITT recomendation X.75, is very similar to the X.25, therefore when we want to connect the RDC network to another one, the software needs very little modification.

FUNCTIONAL STRUCTURE

Every node has a CS responsible for communication, which is hierarchical and includes five levels. Every level consists of one or more modules and each module has its own functions. In Fig. 3 the main modules and their relations are shown.

Every module has one or more queues. CS pro-

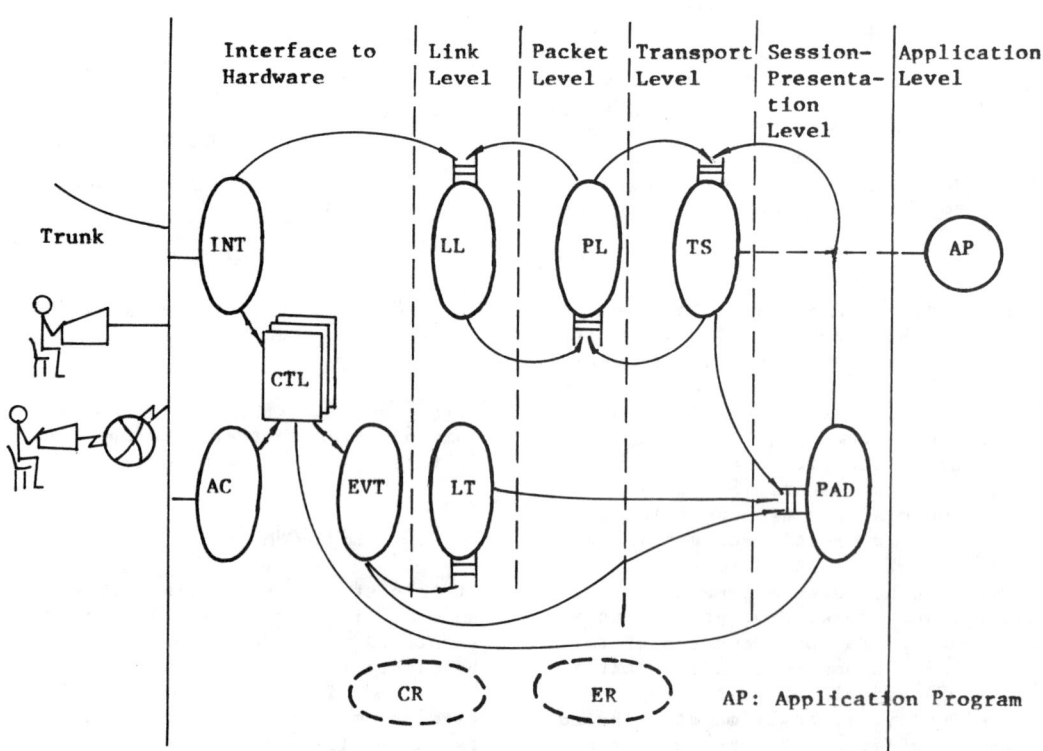

Fig. 3. Modules of CS and their relations

vides two queuing primitives, GETQ and PUTQ. The function of PUTQ is to put a cell into the desired queue. The GETQ is to get a cell from a specified queue.

In general, a cell contains the following parameters: the origin of the cell, operation code of service required, the address of data block and so on. With the introduction of cells and queuing primitives, modules can coordinate each other. When a module requires another module's service, it puts a cell into the proper queue by the PUTQ primitive. When a module uses the GETQ primitive and gets a cell from its own queue, it performs the required service.

For every outer line (i.e. a line that connects a terminal to a DJS-100 computer or a line between two computers), there is a characteristic table of link (CTL) which records the sending and receiving state of the link, the pointers of queues for sending data and receiving data, and the characteristics of the terminal connected to the link. CTL is one of the most important data structures, which many modules access.

The interrupt module (INT) is responsible for the reception of information from an outer line and for the transmission to it. It comprises of a link driving program and an interrupt handler routine. INT supports four kinds of lines: trunks, long distant lines across the telephone switching network, private (or leased) lines connected to 7-unit devices and lines to 5-unit devices. Different lines are handled by INT in different ways. A frame received from a trunk will be put to the LL queue. And after sending a frame, it will go on sending the next one according to the CTL's information. A character received from a line which connects a terminal and a computer will be put into a buffer specified by CTL. If it is not an ASCII code, the corresponding ASCII character will replace it. During this process, if the buffer is full or a "CR" character is received, a mark will be made in the CTL so that the EVT module can analyse it. Conversely, after a character is sent to a terminal, INT will take out the next character according to the CTL, substitute it with an appropriate one if necessary, then send it to the terminal. After the whole block of data has been sent, a mark will be made in the CTL.

The event analysis module (EVT) is a timing task. It examines CTL and analyses various events which occur on links. When it detects an event (e.g. a buffer used for receiving data on a link is full, a data block has been sent to a link, some control signals are received from a link and so on), EVT produces a cell to describe what has happened and puts the cell to an appropriate PAD's queue or LT's queue. After sending a data block, EVT changes the state of the link or asks INT module to send another data block.

The answer of calling module (AC) is also a timing task. It examines states of the lines. As soon as it detects a bell signal on some line, it switches on the data set at the side of computer. If a carrier signal arrives in a specified time interval, for example 2 minites, the initialization of the line is performed, the initial data will be written into the corresponding CTL, and the line enters its communication stage. Otherwise the line will be closed and will wait for another user.

The computer-computer link management module (LL) carries out the link level protocol of X.25. It uses window mechanism to control data flow, uses CRC to detect errors in transmission and uses retransmission to correct transmission errors. All of these aim at ensuring errorfree transmission between two adjacent nodes. The state transition of a link is described by a finite automata. The software of the LL module implements this automata.

The terminal link management module (LT) carries out basic mode control procedure. It uses waiting mechanism to control data flow, uses vertical and horizontal redundancy checking to detect the failure of transmission and uses retransmission to correct transmission errors.

The terminal which connects to a computer by private or leased line does not need to be treated by the LT module. It only needs a MODEM at its side. But for a terminal which connects to a computer across switched circuits of a telephone network, a transmission controller to carry out basic mode control procedure is needed.

The packet level module (PL) implements the X.25 packet level protocol. Based on multiplexing data link, it provides virtual circuit between two DTEs. In establishing a virtual circuit (VC), the PL module uses a routing algorithm to select a path across the network. Then PL module is responsible for transmitting data and interrupt packets and performing data flow control.

The transport station module (TS) implements the transport level protocol and provides interprocess communication facility within the network. As there is no formal international standard for this level protocol now, the software at this level was designed with consultation of the IFIP WG6.1 proposal. The main functions of TS are as follows: providing an uniform name space which consists of ports; managing the correspondence between a process and a port; on the basis of VC provided by the packet level, establishing, maintaining or closing the liaison between two ports according to user's requirements; transmitting two types of data, i.e. letter and telegraph when liaison are established;

and performing the function of switching an established liaison to another port.

The packet assembler/disassembler (PAD) module performs the main functions specified by the CCITT recommendations X.3 and X.28. In accordance with user's commands, PAD manages the session between a user's terminal and its associated service program, that is setting up the association, disconnecting the association and transmitting data or interrupt. It also modifies, according to user's commands, the parameter which specify the terminal characteristics, so that CS can output information at the terminal accordingly. These parameters indicate that, for example, after how many characters a "CR" character should be inserted, whether or not a certain number of padding characters should be inserted after a "CR", and so on. PAD is also in charge of assembling character strings from a terminal into packets so as to transmit them across the network, and disassembling packets from the network into character strings so as to send them to terminals.

Common routine (CR) and error-handling routine (ER) are used by all the modules. They do not belong to any level. Under normal conditions, CR provides various services, such as the queue management and the buffer management. When CS detects an error, ER provides some means of statistics and protection, and reports the system failure to the operator.

In general, these modules provide two kinds of communication services (shown in Fig. 4). One of them is the communication between application programs in the network; another is the communication between user's terminal and their associated service programs which may be in the local node or in the remote node.

INTERFACES

Having defined the functions and the structure of CS, we define the functions of various interfaces in detail. There are three kinds of interface: (1) the interface to the user, (2) the interface to the hardware, (3) the interface between the modules of CS.

CS has two kinds of users, one of them is the terminal user, the other is the application program. We provide simple interfaces for the users (shown in Fig. 5), so that they can use CS easily.

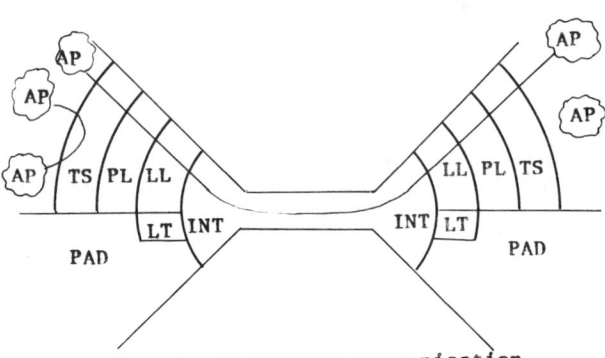

a. process-to-process communication

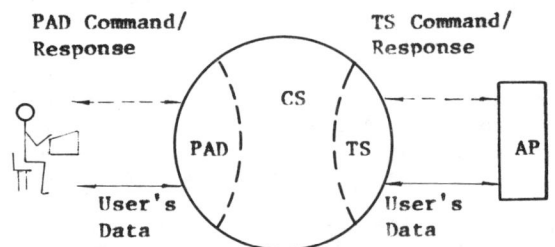

Fig. 5. Interfaces to users

A character string is entered by a user at the terminal. If the first character is "esc", the string is recognized as a PAD command, otherwise, as user's data. The PAD commands are listed (shown in TABLE 1).

By means of TS commands, the application program can use the service provided by CS. The TS commands are listed (shown in TABLE 2).

Usually, the service program (one of the application programs) first sends a LISTEN on its port. When a user sends a CONnect at the terminal, CS sets up an association between the service program and the terminal, i.e. a two way path is established. Then the conversation between the user and its service program can start.

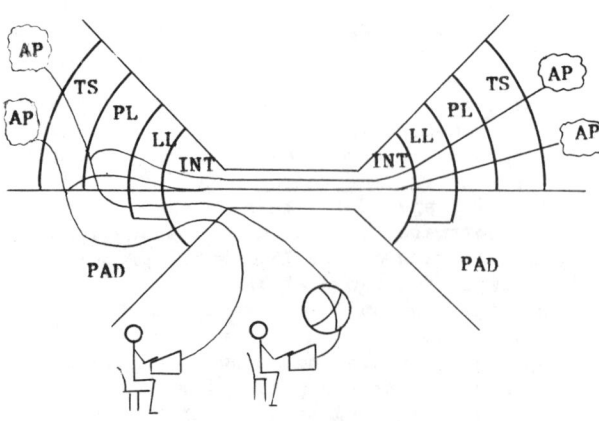

b. terminal-to-AP Communication

Fig. 4. Information flow

TABLE 1 PAD Commands

Name	Meaning
PROF,n	Specify default parameters
PAR?n1...	Read specified PAD parameters
SET:n1:v1...	Write specified PAD parameters
SET?n1:v1...	Write and read specified PAD parameters
CON,m	Create an association
RESET	Reset the association
INT,x	Send an interrupt
CLEAR	Terminate an association
STAT	Enquire the state of the link

In designing the interface between CS and hardware, we tried to follow the international standard as close as possible. We hoped to get a good performance-cost ratio by dividing the functions properly. But in fact, we were limited by the equipment available to us. For the high speed synchronous line, CRC is accomplished by hardware, and an interrupt signal is issued after receiving or sending a frame. For the low speed line connecting the terminal, an interrupt signal is issued by hardware after receiving or sending a character. For the line across the telephone switching network, the parity check of character is accomplished by hardware whereas the horizontal parity check of block is accomplished by software.

Within CS, the interface between modules should be as simple as possible. Each module should be independent so that when some function is improved or changed, the other parts of the system will be affected as little as possible.

TABLE 2 TS Commands

Name	Meaning
OP-PT	Open a port
CL-PT	Close a port
OP-LI	Create a liaison
LISTEN	Listen on a port
CL-LI	Terminate a liaison
RESET	Reset a liaison
SEND-LY	Send a letter
SEND-TG	Send a telegraph
RECEIVE	Receive a letter
SWITCH	Switch a liaison to another port

CONSIDERATIONS FOR THE IMPLEMENTATION

There are three possible ways to implement the RDC communication software.

1. The CS may be implemented as a set of user tasks of the operating system. In this way, it is unnecessary to modify the original OS. The implementation will be fairly easy, but its efficiency will be low.

2. The CS may be implemented as part of OS. In this way, the efficiency will be high, but the OS needs to be modified, and the implementation will be more difficult.

3. We may make a compromise between the first two ways.

Considering the efficiency in execution and the manpower in programming, we have chosen the third scheme. That is to say, the main modules of the CS are implemented as a set of user tasks of the OS, and these tasks are managed by multitasking scheduler of the OS, while a few changes are to be made. The modification concerns the following aspects:

1. We supplemented the OS with the buffer management system and the queuing system.

2. The interrupt handler routine for the asynchronous multiplexor was changed. In order to meet the requirements of network communication actions, interrupt handler routine for the trunk was added.

There exist the problems of parallel processing at any level of CS. To solve these problems, we made use of some appropriate data structures.

CTL describes sending and receiving states of a link. If an interrupt signal is issued, then interrupt handler routines will access CTL in accordance with the subchannel numbers. At the link level, states and state transitions of a link are described by a finite automata. Current state of the automata is recorded in CTL. In this way, the INT, EVT, LL and LT modules manage various communication lines.

A logical channel table (LOCT) is utilized to describe the current state and information related to the respective logical channel. By means of LOCT, PL manages all the logical channels.

At the transport level, an entry of liaison table (LIAT) is established for each active liaison. The current state and all the information related to this liaison are recorded in LIAT. Thus, TS can deal with various events on many liaisons without any confusion.

A set of parameters recorded in CTL describes the characteristics of a terminal. According to these parameters, PAD is able to support the conversation between a terminal and its service program.

CS is written in assembly language. We try

to save memory and runtime of CPU as much as possible.

In view of the fact that the memory of minicomputer is limited, the most important thing is to save memory. But INT is different. For it, the most important thing is to save runtime, so as to support many lines without losing any signal.

In short, CS runs as a user of the OS. Every module (shown in Fig. 3) is implemented as a task. According to its state table, every task deals with all the parallel events to accomplish expected functions. The GETQ and PUTQ primitives are used for the communication between tasks of the CS.

The advantages of this scheme are:
a. simple and easy to implement,
b. explicit in system structure,
c. efficient in the use of memory and CPU,
d. convenient for users to develop new service systems and application softwares.

TOPICS FOR FURTHER STUDY

In a specific network configuration, each DJS-100 computer has its specific role. For example, some work as front-end processors of hosts; some are engaged in data transmission, working as DCE nodes of the network; some are used to collect data from terminals, working as concentrators. And some may even function as hosts. CS is built in the form of a hierarchical structure, so it is easy to tailor and modify these modules to meet the different requirements.

Different statistical programs, error recovery procedures and methods of measurement are to be perfected in future. And, the practicability are to be improved.

At present, CS has not provided the facility, by which application programs can call user's terminals. Whether this function should be incorporated or not depends upon the requirements of users.

CONCLUSIONS

The main parts of RDC have been debugged, we shall put it in operation soon. We shall select several places for practical use, and improve CS. The software we have developed will be useful for on-line system, distributed control and data collecting and processing in China. In the near future, we are going to connect large scale hosts, special devices to form a two-level structured computer network (resource subnet and communication subnet), so as to get certain experiences for building large scale computer network.

ACKNOWLEDGEMENTS

The communication software of RDC was developed under the guidance of Prof. Cao Dongqi And Mr. He Rengdong, Mr. Song Huangang, Ms. Sun Simin, Ms. Xu Guirung and the hardware engineers have participated in its designing, programming and debugging. We are grateful to our advisor and colleagues.

REFERENCES

(1) CCITT Recommendation X.25 Interface between data terminal equipment (DTE) and data circuit-terminating equipment (DCE) for terminals operating in the packet mode on public data networks (Geneva, 1976, amended at Geneva, 1977).
(2) CCITT Recommendation X.3. Packet assembly/disassembly facility in a public data network (Geneva, 1980).
(3) CCITT Recommendation X.28. DTE/DCE interface for a start-stop mode data terminal equipment accessing the packet assembly/disassembly facility in a public data network situated in the same country (Geneva, 1980).
(4) ISO-3309, Data communication-high-level data link control procedures-frame structure. 1976.
(5) ISO-4435, Data communication-high-level data link control procedures-element of procedures. 1979.
(6) ISO-1745, Basic mode control procedures for data communication sytem.
(7) IFIP-WG 6.1 Proposal for an internetwork end-to-end transport protocol. INWG Note 96.1, 1978.
(8) Davies, D.W., D.L.A.Baber, W.L.Price and C.M.Solomonides. (1979). Computer Networks and Their Protocols. Wiley, New York.
(9) Cao, Dongqi. (1979). Introduction to computer network software. Beijing.
(10) UNINETT Reports. No. 1,2,3,4,5. Oslo.

AUTHOR INDEX

Bean, D.P.A. 141
Britton, D.G. 45

Davidson, J. 83
Dimmler, D.G. 59
Duffy, J.M. 45

Goscinski, A. 167

Halloway, F.W. 45
Harrison, T.J. 13, 181
Houle, J.L. 83

Inamochi, C. 33
Inamoto, A. 33

Jin Lan, 75

Kaaramees, K. 93
Kasahara, H. 103
Kato, S. 33
Kikuts, Ya.A. 153
Krammen, I.R. 45

Maxwell, M. 189
Mendelbaum, H.G. 201
Motus, L. 93
Muto, T. 33

Narita, S. 103, 183

Ollus, M. 117
Ozarski, R.G. 45

Peberdy, N.J. 141

Rodd, M.G. 141
Rotanov, S.V. 153

de Sablet, G. 201
Sevevyn, J.R. 45
Shuraits, Yu.M. 1
Sheng, Mei-ming. 75
Suski, G.J. 45

Takezawa, K. 129
Tamm, B. xix
Tavast, R.-R. 193
Trakhtengerts, E.A. 1

Vamos, T. xvii
Van Arsdall, P.J. 45

Wahlstrom, B. 117
Walasek, T. 167
Wang, Ding-xing. 75
Weehuizen, H.F. 141
Wenkuan, Zhang. 211
Wood, G.G. 27, 191

Yakubaitis, E.A. 153
Yingzhong, Zhang. 211
Yunyi, Cheng. 211

Zheng, Wei-min. 75
Zhimei, Wu. 211
Zielinski, K. 167